Advanced Treatment Technologies for the Removal of Microplastics in Wastewater

Microplastics have become a significant environmental concern due to their persistence, potential to accumulate in aquatic ecosystems, and ability to infiltrate the food chain. Given the limitations of conventional wastewater treatment processes, advanced treatment technologies are being developed to enhance microplastic removal and mitigate their environmental impact. *Advanced Treatment Technologies for the Removal of Microplastics in Wastewater* presents the pervasive issue of microplastic pollution in wastewater, examining its origins in everyday activities and industrial processes and its spread through aquatic ecosystems. This book provides a deep dive into advanced treatment technologies including physico-chemical and biological methods, highlighting both the progress and challenges in effectively removing microplastics from wastewater. It also addresses the potential health risks posed by microplastic exposure, discussing how these particles act as carriers for toxic substances and their suspected links to health issues in humans and wildlife. With cutting-edge research, practical insights, and a comprehensive overview, this book equips readers with the knowledge needed to understand and address the critical issue of microplastics in our water and wastewater systems.

- Provides updated occurrence and characteristics of microplastics in various industrial wastewaters
- Presents advanced wastewater treatment technologies for microplastics removal
- Gives a detailed account of the toxic effects of microplastics on animals, plants, and humans
- Covers innovative approaches for the management of emerging toxic compounds in industrial wastewaters.

Advanced Treatment Technologies for the Removal of Microplastics in Wastewater

Edited by
Izharul Haq and Maulin P. Shah

CRC Press
Taylor & Francis Group
Boca Raton London New York

CRC Press is an imprint of the
Taylor & Francis Group, an **informa** business

Designed cover image: Shutterstock

First edition published 2025
by CRC Press
2385 NW Executive Center Drive, Suite 320, Boca Raton FL 33431

and by CRC Press
4 Park Square, Milton Park, Abingdon, Oxon, OX14 4RN

CRC Press is an imprint of Taylor & Francis Group, LLC

ISBN: 978-1-032-45947-9 hbk)
ISBN: 978-1-032-45949-3 (pbk)
ISBN: 978-1-003-37938-6 (ebk)

DOI: 10.1201/9781003379386

Typeset in Times
by Newgen Publishing UK

Contents

Preface

The book, ***Advanced Treatment Technologies for the Removal of Microplastics in Wastewater***, aims to shed light on the scope and impact of microplastics in wastewater while presenting the latest advancements in treatment technologies designed to address this issue. We delve into physico-chemical and biological methods, offering readers a thorough understanding of the techniques currently available and under development for microplastic removal. Emphasis is placed on innovative approaches, as well as emerging biological strategies that harness the degradative potential of microorganisms.

In addition to treatment solutions, this book explores the environmental and health impacts of microplastics, addressing the toxicological implications of these particles, which often carry harmful substances. Our goal is to equip researchers, environmental scientists, policymakers, and students with the insights needed to advance research, improve wastewater treatment, and ultimately minimize microplastic pollution.

This book has been enriched by the expertise and insights of global experts who contributed their extensive knowledge on hazardous pollutant toxicity, characteristics, effects, and the latest methods for detection and treatment. Their dedication has provided a comprehensive resource, essential for scientists, students, and anyone seeking to understand and mitigate the impacts of microplastics and other pollutants in wastewater.

We extend our deepest gratitude to all chapter authors for their invaluable contributions, compiling critical information and innovative perspectives to create an accessible and impactful reference. Their efforts will undoubtedly benefit the scientific community and broaden the understanding of environmental sustainability practices.

Our sincere thanks also go to the expert reviewers whose meticulous feedback and suggestions were instrumental in enhancing the quality and cohesion of each chapter. Their guidance helped shape the scientific discussions, and their insights were vital to refining the book's organization. Finally, we would like to express our appreciation to the entire CRC Press team, including the Senior Book Acquisition Editor, Editorial Project Manager, and production team, for their unwavering support in bringing this work to publication. Their professionalism and dedication made this collaborative effort possible.

Editors
Izharul Haq
Maulin P. Shah

About the Editors

Izharul Haq is an Assistant Professor at the Dr. B. Lal Institute of Biotechnology in Jaipur, India, where he specializes in environmental biotechnology, with a focus on waste management using microbial processes and evaluating the toxicity of pollutants. He completed his PhD in Microbiology at the CSIR-Indian Institute of Toxicology Research, Lucknow, India, and later conducted postdoctoral research at the Indian Institute of Technology, Guwahati. Dr. Haq has authored 65 publications, holds an h-index of 20, and has over 1,900 citations, reflecting his impactful contributions to the field of environmental microbiology and pollutant remediation.

Maulin P. Shah, currently a Deputy General Manager – Industrial Wastewater Research Lab, Division of Applied and Environmental Microbiology Lab at Enviro Technology Ltd., Ankleshwar, Gujarat, India, received his Ph.D. (2002–2005) in Environmental Microbiology from Sardar Patel University, Vallabh Vidyanagar, Gujarat. He served as an Assistant Professor at Godhra, Gujarat University in 2001. He is a Microbial Biotechnologist with diverse research interests. A group of research scholars is working under his guidance on the areas ranging from Applied Microbiology, Environmental Biotechnology, Bioremediation, and Industrial Liquid Waste Management to solid state fermentation. His primary interest is the environment, the quality of our living resources and the ways that bacteria can help to manage and degrade toxic waste and restore environmental health. Consequently, he is very interested in genetic adaptation processes in bacteria, the mechanisms by which they deal with toxic substances, how they react to pollution in general and how we can apply microbial processes in a useful way (like bacterial bioreporters). One of our major interests is to study how bacteria evolve and adapt to use organic pollutants as novel growth substrates. Bacteria with new degradation capabilities are often selected in polluted environments and have accumulated small (mutations) and large genetic changes (transpositions, recombination, and horizontally transferred elements). His work has been focused on assessing the impact of industrial pollution on microbial diversity of wastewater following cultivation dependent and cultivation independent analysis. His major work involves isolation, screening, identification, and genetic engineering of high impact of microbes for the degradation of hazardous materials. He has more than 200 research publications in highly reputed national and international journals. He directs the Research program at Enviro Technology Ltd., Ankleshwar. He has guided more than 100 postgraduate students in various disciplines of Life Science. He is an active Editorial Board Member in more than 150 highly reputed journals in the field of Environmental & Biological Sciences. He was Founder Editor-in-Chief of *International Journal of Environmental Bioremediation and Biodegradation* (2012–2014), as well as *Journal of Applied and Environmental*

Microbiology (2012–2014). He is actively engaged as an Editorial Board Member in 32 journals of high repute (Elsevier, Springer, Taylor & Francis, RSC, Wiley, KeAi, De Gruyter). He is also a reviewer in various journals of national and international repute. He has edited more than 200 books in wastewater microbiology, environmental microbiology, bioremediation, and hazardous waste treatment.

Contributors

Saman Aftab
Institute of Industrial Biotechnology,
Government College University,
Lahore, Pakistan

Anouar Bardi
Higher Institute of Management of Gabés,
Gabés, Rue Jilani Habib,
Gabès, Tunisia.

Annesha Basu
Department of Chemical Engineering,
Jadavpur University,
Kolkata, India

Chiranji Bhattacharjee
Department of Chemical Engineering,
Jadavpur University, Kolkata, India

Sarbartha Chakraborty
Department of Microbiology,
Savitribai Phule Pune University,
Pune, Maharashtra, India;

Aishwarya Das
Department of Chemical Engineering,
Jadavpur University, Kolkata, India

Ranjana Das
Department of Chemical Engineering,
Jadavpur University,
Kolkata, India

Arundhuti Devi
Environmental Chemistry Laboratory,
Resource Management and Environment Section,
Life Science Division,
Institute of Advanced Study in Science and Technology (IASST),
Guwahati, Assam, India.

Bhaswati Devi
Environmental Chemistry Laboratory,
Resource Management and Environment Section,
Life Science Division,
Institute of Advanced Study in Science and Technology,
Guwahati, Assam, India.

Devki
Department of Microbiology,
JECRC University,
Jaipur, Rajasthan, India

Patricio J. Espinoza-Montero
Escuela de CienciasQuímicas,
Pontificia Universidad Católica de Ecuador,
Quito, Ecuador

Sougata Ghosh
Department of Microbiology, School of Science,
RK. University, Rajkot, Gujarat, India; and Department of Physics,
Faculty of Science,
Kasetsart University,
Bangkok, Thailand

Varsha Gupta
Department of Microbiology, JECRC University,
Jaipur, Rajasthan, India.

Izharul Haq
Department of Biotechnology,
Dr. B. Lal Institute of Biotechnology,
Jaipur, Rajasthan, India

Muhammad Uzair Javed
Department of Biotechnology,
Quaid-i-Azam University,
Islamabad, Pakistan

Meena Khwairakpam
School of Agro and Rural Technology,
Indian Institute of Technology Guwahati,
Guwahati, Assam, India

Hamid Mukhtar
Institute of Industrial Biotechnology,
Government College University,
Lahore, Pakistan
Hamid Mukhtar Institute of Industrial Biotechnology,

Deepesh Kumar Neelam
Department of Microbiology,
JECRC University,
Jaipur, Rajasthan, India

Pranav Pandya
Department of Microbiology, School of Science,
RK. University, Rajkot,
Gujarat, India

Mandeep Prakash Parhi
School of Bioengineering and Biosciences,
Lovely Professional University,
Jalandhar, Punjab, India.

A. A. Narmada Thathsarani Perera
Amrita School of Nanosciences and Molecular Medicine,

Amrita Vishwa Vidyapeetham,
Kochi, Kerala, India.
Sujata Phurailatpam
School of Agro and Rural Technology,
Indian Institute of Technology Guwahati,
Guwahati, Assam, India
Netra Prova Baruah
Environmental Chemistry Laboratory, Resource Management and Environment Section,
Life Science Division, Institute of Advanced Study in Science and Technology,
Guwahati, Assam, India.
Wendy Quilumbaquin
Escuela de CienciasQuímicas,
Pontificia Universidad Católica de Ecuador,
Quito, Ecuador

Hayfa Rajhi
Hospital Gabès University,
Gabès, Tunisia.

Sonia Sethi
Dr. B. Lal Institute of Biotechnology,
Malviya Industrial Area,
Malviya Nagar Jaipur, Rajasthan, India

Muskan Sharma
Department of Microbiology,
JECRC University,
Jaipur, Rajasthan, India.

Katherine Vaca-Escobar
Escuela de CienciasQuímicas,
Pontificia Universidad Católica de Ecuador,
Quito, Ecuador

Ruchi Vyas
E-YUVA Centre, Centre for Covering Technologies,
University of Rajasthan,
Jaipur, Rajasthan, India

1 Removal of Microplastics from Industrial Wastewater Using Microalgae

Pranav Pandya, Sarbartha Chakraborty, and Sougata Ghosh

1.1 INTRODUCTION

Microplastic pollution is becoming a global concern due to its threat to the environment and its refractory nature. Being generated from the urban and developing areas, microplastics end up in water-bodies such as, lakes, rivers, and oceans, which is having far-reaching consequences (Hurley et al., 2018). There are numerous sources and pathways through which microplastics enter our ecosystems and cause health issues to living organisms and also to the environment in general. Microplastic types can be morphologically classified into six categories; fiber, spherical, foam, sheet, fragment, and film (Chaudhry and Sachdeva, 2021). Microplastics can originate from primary and secondary sources (Cole et al., 2011). The primary hotspots of microplastics can be considered from the hubs where plastic materials are produced and processed. The disposal grounds of plastics packages, products (bottles, bags, and single-use plastics) and microbeads or plastic particles used in cosmetics, toothpaste, body washes etc. are also considered as the primary sources of microplastic pollution as they disintegrate and emit microplastic particles which amalgamate into soil, water bodies, and air. The secondary sources of microplastic pollution, which contribute significantly to the culminating problem, include the breakdown of large plastic-based vehicle tires, synthetic fibers, plastic bottles, and bags due to mechanical stress, sunlight (UV rays), and environmental actions (Eriksen et al., 2014). They disintegrate to form microplastics. Other sources of microplastic pollution are caused by the wear and disintegration of plastic-made equipment in agriculture and household applications. Improper disposal of the plastic wastes in landfills are also a common cause of microplastic pollution as it leaches into the surrounding water and land.

1.2 SOURCES

The primary and secondary sources can be further divided into various categories. The disintegration of plastic items like bottles, bags, and packaging materials can be

DOI: 10.1201/9781003379386-1

termed as the leading cause of microplastic pollution. The release of microplastics from plastic items is caused due to overexposure to weathering, UV radiation, and physical abrasion (Alimi et al., 2018). Synthetic fibers used in clothing like polyester, fleece, acrylic, and nylon, release microparticles of plastics during washing and day-to-day uses. Around 1 million tons of these microparticles are estimated to be released into the environment every year (Sharma and Kaushik, 2021; Haq et al., 2016ab; 17, 18, 20; Haq and Kalamdhad, 2021; Haq et al., 2022abc; Haq and Kalamdhad, 2022, 2023; Haq et al., 2024). According to the study of Habib et al. (1998), these particles reach greater than 100 particles per liter of effluent, more than half of which escapes the filtration systems, making their way into water sources and accumulating, which leads to the pollution of the aquatic ecosystems where they pose a threat to the aquatic flora and fauna and also humans when they come into contact with it.

Styrene butadiene tires of vehicles are potent sources that can release countless microplastic particles due to the movement and friction of the road. About 20 g of tire dust is produced every 100 km of the vehicle runs on the road. Norway produces 1 kg of average tire and road dust each year for every citizen in the country (Sharma and Kaushik, 2021). These particles, comprised of synthetic rubber and additives, are carried by air, dust, and rainwater runoff, and accumulate in the urban areas, rivers, and lakes leading to air and water microplastic pollution.

Microbeads used in personal care products like face-wash, scrub and body cleansers, as well as toothpaste, can easily get mixed with the water system with daily use. Being microparticles, the conventional wastewater treatment plants are inefficient at filtering these microplastics. About 8 million of these microbeads pollute the US waterways, making it banned in the USA and Canada (Sharma and Kaushik, 2021). These microbeads accumulate and contribute to the burden of microplastic pollution. Various industries, like plastic manufacturers, processing, and recycling industries, generate microplastic particles in large amounts on a daily basis and they get mixed with the land and water sources. Effluents loaded with microplastics from industries are released into the sewage systems and eventually contaminate the environment (McDevitt et al., 2017). Mechanical actions during the process, such as grinding, cutting, and molding plastic materials also release these microparticles into the surrounding environment, leading to air pollution with microplastics.

Additionally, the use of plastic mulch films in agriculture to enhance crop yields and the use of plastic-made equipment can lead to soil pollution (Bläsing et al., 2018). Over time, these equipments degrade and erode due to friction and use, thus releasing microplastics into the soil slowly. The application of microplastic-containing fertilizers or biosolids derived from sewage sludge or compost can introduce the pollutant (as they are already contaminated with microplastics) into agricultural fields, potentially affecting food chains. Inappropriate or illegal methods of waste management and disposal practices can be the leading cause of microplastic pollution in soil. When plastic wastes from household, hospital, and industry use are disposed of in landfills, it can disintegrate over time, releasing microplastic particles into the surrounding soil and in more severe conditions it can leach into the groundwater reservoirs, leading to severe levels of pollution. Fuller et al. (2016) extracted and showcased the occurrence of microplastics in the soil samples from the industrial area of Sydney, Australia that showed the concentrations of microplastics in the range of 300–67,500 mg kg^{-1}.

Fishermen lose their fishing gears and sometimes discard them in water bodies. Most of the gears are made up of plastic material. Hence, with time, microplastics leach out from these gears into the water. The most abundant microplastic forms available are polypropylene (PP), polystyrene (PS), polyethylene (PE), polyvinylchloride (PVC), and polyethylene terephthalate (PET) (Guzzetti et al., 2018). Hidalgo-Ruz et al. (2012) stated that the plastic load in the ocean is around six orders of magnitude that leads to the accumulation of the microplastics inside the marine organism and contribute majorly to marine microplastic pollution. With the aim of reducing the volume of huge amounts of plastic waste, often it is set on fire. The burning of plastic wastes not only releases toxic fumes but also fine microplastic particles in the atmosphere as aerosols, which can lead to air pollution and breathing problems. On average, a human with regular physical activity inhales 272 microplastic aerosols every day (Vianello et al., 2019). So, this method of getting rid of plastic waste is one of the prominent sources of microplastic pollution.

1.3 ADVERSE IMPACT

Microplastics are found in various environments, which include oceans, rivers, lakes, underground water reserve, air, and even soil. Abandoned polymers of plastic categorized as high- and low-density polyethene, nylons, and polyvinyl chloride, are a major threat to the environment as they disintegrate to release microplastics (Barnes et al., 2009). They can be sustained in the environment, without further disintegration leading to the accumulation and release of toxic substances. This has a broad-spectrum effect on the ecosystem by affecting the nutrient cycle, altering the food webs, and reducing biodiversity as listed in Table 1.1.

TABLE 1.1
Types of microplastics and their impact

Type of Microplastics	Impact	References
Plasticizers	absorb pollutants after ingestion by bioaccumulation and biomagnification, releasing toxins and chemicals which leads to cancer	Teuten et al. (2009)
Polyvinyl chloride, polyethylene, polypropylene and polystyrene,	increases the sorption capacity for carcinogens like- DDTs, polycyclic aromatic hydrocarbons (PAHs), hexachlorocyclohexanes, and chlorinated benzenes	Lee et al. (2014)
Endocrine disrupting chemicals (EDCs) like -nonylphenol and bisphenol-A	interference with the reproductive system, affects the immune and nervous systems and can increase the chances of cancer	Hirai et al. (2011); Fries et al. (2012)
Polyethylene, silicone, polycarbonate and polystyrene	leach bisphenol-A and nonylphenol leading to carcinoma	Fasano et al. (2012)

(continued)

TABLE 1.1 (Continued)
Types of microplastics and their impact

Type of Microplastics	Impact	References
Phthalates or polychlorinated biphenyls (PCBs)	effects on the digestive system as it enters through the food source	Karbalaei et al. (2018)
Polyethylene (PE), polypropylene (PP)	present in skin care products so absorbs into the skin resulting in skin rashes and injuries	Duis et al. (2016)
Bis(2-ethylhexyl) phthalate (DEHP)	pregnant women and babies are affected, these are also known as endocrine disruptors. It causes endometriosis and prostate cancer	Rowdhwal and Chen, (2018)
Bisphenol A (BPA)	interferes with our hormonal system and imbalances the hormones, linking it to the cause of breast cancer, infertility, and miscarriage	Lithner et al. (2011)

Both aquatic and terrestrial organisms face danger from microplastic pollution. However, aquatic life is particularly more vulnerable as microplastic contamination is recorded in fresh as well as sea water (Imhof et al., 2013). Marine animals, such as amphipods, copepods, lugworms, barnacles, mussels, decapod crustaceans, gulls, fish, and turtles can mistake microplastics for food, leading to ingestion and potential health issues (Cauwenberghe et al., 2015; Chaudhry and Sachdeva, 2021). Microplastics can also bioaccumulate through the food chain; that is, the predators who are at the peak of the food chain system generally accumulate higher concentrations of microplastics than the members below in the chain. This can have severe metabolic impairment and reproductive failure in marine organisms like fish, seabirds, and aquatic mammals. Microplastics can be ingested by marine animals and potentially other wildlife as food material, leading to blockages in their digestive systems, reduced nutrient absorption, and malnutrition (Thompson et al., 2009). Microplastic's presence is detected in numerous human food sources, such as drinking water and seafood. Ingested microplastics can also cause physical damage to organs and tissues. The ability of microplastics to accumulate in human tissues potentially leads to localized inflammation, oxidative stress, and the transport of chemical contaminants into the body.

Microplastics in the form of aerosols is a growing concern as we breathe it in. Fasano et al. (2012) found that polyethylene, silicone, polycarbonate and polystyrene can leach out bisphenol-A and nonylphenol on entering the respiratory system. Microplastic fibers, originating from textiles and other sources, can become airborne and be inhaled, potentially reaching the respiratory system. While the health effects of inhaling microplastics are still being investigated, there are concerns about respiratory inflammation, lung damage, and the possible introduction of chemical pollutants into the lungs. Microplastics accumulation increases the stress-related genes expression recorded in gilthead seabream *Sparus aurata* L., nematode *Caenorhabditis elegans*, and Chinese mitten crab *Carcinusaestuarii* (Piarulli et al., 2019). Microplastics can accumulate in the tissues of animals. They can pass through the food chain, with predators accumulating higher concentrations of microplastics than their prey. This bioaccumulation can lead to higher exposure levels and potential adverse effects on animals higher up the food chain. Microplastics have severe, if not lethal, impacts on aquatic organisms in several reports studied at different levels like individual, cellular, and molecular (Browne et al. 2015). The chemical composition of microplastics, and their ability to absorb and concentrate toxic pollutants from the environment, can pose risks. These pollutants can include persistent organic pollutants (POPs) and also micro-particles of heavy metals, which can be associated with numerous health hazards in living beings.

Ingested microplastics can interfere with the feeding and reproductive behaviors of animals. They can cause reduced appetite, altered feeding efficiency, and impaired reproduction, leading to population declines and ecological imbalances. Alterations in chromosomes can have a consequence on the intake ability of microparticles, resulting in infertility, obesity, and cancer (Campanale et al., 2020). The impact of microplastics on human health is not yet fully understood, because the exposure is diverse. The level of microplastic exposure through food and water is relatively low. Ongoing research on evaluation of the potential health hazards shows the presence of microplastic in blood, lungs, placentas, and stools (Jung et al., 2022). The gastrointestinal system is vulnerable to the effects of microplastic ingestion. Microplastics get assembled in the digestive tract, leading to localized inflammation and tissue damage. The presence of microplastics in the gastrointestinal tract may interfere with nutrient absorption, disrupt the gut microbiota, and impair overall digestive function. Toothpaste associated microplastics and microbead particles can be swallowed in regular day-to-day activity and get absorbed through the gastrointestinal tract (Ustabasi et al., 2020). In severe cases, large accumulations of microplastics can cause blockages or obstructions in the intestines, eventually requiring medical intervention. Bioaccumulation of microplastics for a long period of time may alter our human chromosomes and can lead to infertility, chromosomal disorders, obesity, and cancer in extreme cases (Wang et al., 2016).

Although research on the impacts of inhaling microplastics is still in its early stages, there are concerns about potential respiratory effects. Microplastic fibers and particles in the air can be inhaled that reach the respiratory system. They may trigger inflammation, irritation, and respiratory distress. Prolonged exposure to microplastics may lead to the development or enhancement of exacerbating respiratory conditions such as asthma and chronic obstructive pulmonary disease (COPD).

The liver plays a crucial role in detoxification and metabolism. Studies have shown that microplastics can accumulate in the liver, potentially leading to oxidative stress mediated by the reactive oxygen species (ROS), resulting in an increased expression of oxidative stress associated genes and inflammation (Liu et al., 2019). The presence of microplastics in the liver may interfere with its normal functions, impairing the detoxification process and affecting overall liver health. Also, the endocrine-disrupting chemicals (EDCs) in the microplastics can cause cancer and impair the functions of nervous, and digestive systems (Hirai et al., 2011; Fries et al., 2012). Additionally, microplastics can serve as carriers for toxic substances, which can be released in the liver, causing further damage. Reports suggest that microplastics can even bioaccumulate in the kidneys and potentially lead to inflammation and tissue damage. The presence of microplastics in the kidneys may interfere with their filtering capabilities, impairing kidney function that can lead to the development of renal diseases. Karbalaei et al. (2018) found that phthalates or polychlorinated biphenyls (PCBs) entering through the food source can drastically affect kidneys apart from the digestive system. Microplastic exposure may have immunomodulatory effects as well. Microplastics can trigger immune responses, leading to inflammation and the release of pro-inflammatory cytokines. Hirai et al. (2011) stated that endocrine disrupting chemicals (EDCs) like nonylphenol and bisphenol-A can drastically affect the nervous system. Prolonged immune activation and inflammation may contribute to chronic inflammatory conditions and immune dysregulation. Microplastics can interfere with the hormone signaling pathways, potentially leading to hormonal imbalances and impairment of the endocrine system. These disruptions may contribute to reproductive disorders, developmental abnormalities, and other endocrine-related health issues.

Plastics have a huge socio-economic influence. The output of plastic manufacturing in the world has reached 348 million tons per annum (Mtpa) in 2017 and 8.8 million metric tons (MT) of plastic are entering the world's oceans annually (Chen et al., 2020; Jambeck et al., 2015). Microplastic pollution can have economic implications as well. It can damage fishing and aquaculture industries, impacting the livelihoods of communities that depend on these resources. It is believed that microplastics can be ingested by soil-dwelling organisms, potentially affecting their physiology and altering soil fertility. The transfer of microplastics through the food chain can also impact the terrestrial predators and scavengers Additionally, the cost of cleaning up microplastics from the water bodies and beaches is substantial. The consequences of microplastic pollution are far-reaching and can disrupt ecosystem dynamics. It can affect nutrient cycling, alter food webs, and reduce biodiversity. The use of medical masks became compulsory after the emergence of the COVID-19 pandemic and reached up to 89 million per month. These were made up of polymers, such as polystyrene, polyurethane, polypropylene, poly-acrylonitrile, and polycarbonate. The disposal and disintegration of the masks converts them into small sized particles ranging in the size of 5 mm that emerged as a new source of microplastics pollution (Fadare et al., 2020). The ingestion of the microplastics by key species can lead to population declines and imbalances, impacting the stability and resilience of ecosystems. Furthermore, the microplastics can be transported through water and air currents that facilitate the spreading of the invasive species and pathogens, further exacerbating ecological disruptions.

1.4 MICROPLASTIC DEGRADATION BY MICROALGAE

Various microalgae from diverse ecological niches are explored and exploited for the effective degradation of microplastics that are discussed in this section in detail and listed in Table 1.2.

1.4.1 Polystyrene

Long et al. (2015) reported on the removal of polystyrene (PS) from the marine environment by employing *Chaetocerosneogracile* and *Rhodomonas salina* as well as their consortium. In the first stage, the microalgal species were cultivated in an f/2 medium prepared using filtered seawater having a density of 1.025 kg/L. The culture was incubated at 16°C under light and dark cycles of 12:12 h. After entering the stationary phase, the cultures were kept in dark conditions for a few days in order to provoke the stress which resulted in the induction of transparent exopolymer particles (TEP) excretion. The reaction mixtures of three different types were prepared, of which, first and second monoculture mixtures were comprised of 4,00,000 cells/mL of *C. neogracile* and 1,00,000 cells/mL of *R. salina*, respectively. The third mixture was comprised of the mixed cultures of *C. neogracile* (1,60,000 cells/mL) and *R. salina* (3,00,000 cells/mL). In order to form the aggregates, each of the three mixtures was added into two cylindrical tanks in which one was a flow-through tank and the other was a non-modified roller tank. Both the tanks were kept under dark conditions at 16°C. The speed of 1.6 rpm was maintained in order to keep aggregates in the suspension as well as to allow collisions via differential sinking rates among particles. With the formation of aggregates, the flow-through roller tank was attached to the flow and the solution containing 2 μm PS microbeads having a density of 1.050 kg/L exhibiting yellow-green fluorescence was introduced at the concentration of 10^4 beads /mL and a flow rate of 680 mL/h. The aggregates were exposed to increasing concentrations of microbeads up to 5500 microbeads/mL after 6 h. In the following 6 h of the dilution phase, the microbead suspension was replaced by filtered seawater in order to lower the tank concentration to 2200 microbeads/mL. The size of aggregates for *C. neogracile* was in the range of 0.9 to 5.7 mm with a mean size value of 3.2 ± 0.13 mm, for *R. salina* was in the range of 1.2–3.5 mm with an average value of 1.8 ± 0.07 mm. Similarly, for the consortium, it was ranging from 1.2 to 8.8 mm with an average size of 3.5 ± 0.15 mm. At 2.5 mm size of aggregates, the density for *C. neogracile*, *R. salina*, and the consortium was 35×10^{-4} kg/L, 4.6×10^{-4} kg/L, and 5.0×10^{-4} kg/L, respectively. The Brinkman permeability factor was around 6.27 (*C. neogracile*), 6.46 (consortium), and 8.87 (*R. salina*) for different aggregates. The concentration of microbeads in the *C. neogracile*, *R. salina*, and the consortium aggregates was $1.9 \times 10^4 \pm 5 \times 10^3$, $3.6 \times 10^5 \pm 8 \times 10^4$, and $4.3 \times 10^4 \pm 5 \times 10^3$ beads/mL, respectively. After PS exposure, the sinking rate of *C. neogracile* aggregates (3.2 mm) decreased from 473 m/day to 165 m/day, whereas, an increase in the sinking rate from 76 m/day to 125 m/day was noted in case of the aggregates formed from *R. salina* (1.8 mm). However, in case of the consortium aggregates (3.5 mm), the sinking rate decreased from 144 m/day to 122 m/day. The increase in the fractional dimension was noted after the PS exposure for aggregates derived from *C. neogracile* (from 1.65 to 1.78) and for the consortium (1.78 to 1.94), whereas, it showed a reduction in the case of aggregates developed from *R. salina* (1.92 to 1.82). The permeability factor after incorporation of the microbeads was 9.86, 6.22, 10.86, for *C. neogracile*, *R. salina*, and consortium, respectively. Similarly, the excess density after microbeads incorporation was 5.7×10^{-4} kg/L, 1.2×10^{-3} kg/L, and 3.0×10^{-4} kg/L for *C. neogracile*, *R. salina*, and consortium, respectively.

Li et al. (2020) studied the effect of PS on the development, photosynthesis ability as well as aggregation of microalgae *Chlamydomonas reinhardtii* as shown in Figure 1.1. Initially, the stock solution of PS with 1 g/L concentration was prepared followed by sonication for 30 minutes in order to form a uniformly dispersed solution. The algal species was cultivated in the SE medium and was further subjected to incubation under light and dark cycles of 12 h each at 25 ± 1 °C for a period of 6 days. The decrease in optical density of algal species was observed with respect to the increase in microplastic concentration as well as the time of treatment. The highest inhibition ratio of 49.2% was observed at 100 mg/L PS concentration followed by 43.9, 33.9, and 26.6% at 50, 25, and 5 mg/L PS concentrations, respectively. Upon evaluation of the chlorophyll a fluorescence after 10 days of exposure to PS, a rapid decrease in the yield was noted that further suggested the effect on photosynthetic apparatus, as well as the closure of PSII. The exposure of algal species to the contaminants resulted in the increase of production of reactive oxygen species (ROS), enhancement in the malondialdehyde (MDA) protein production, as well as reduction in the rate of the electron transport chain. The number of soluble proteins was also increased upon PS exposure. In the absence of PS exposure, the highest settlement rate (ST) of around 82% was noted within 24 h. This suggested that the decrease in the secretion of extracellular polymeric substance (EPS) after PS exposure directly affected the ST rate.

Tunali et al. (2020) evaluated the effect of PS on the growth of microalgae, *Chlorella vulgaris*. Firstly, the algal species was cultivated in the 3N bold basal medium (BBM) under constant illumination of light at room temperature and pH between 6.8 to 7.2. The size of PS was around 0.5 μm in the form of aqueous suspension having spherical morphology. Around 0.52×10^6 algal cells were inoculated and different concentrations of PS (1, 5, 10, 50, 100, and 1000 mg/L) were added into the medium after setting in of the log phase (4.68×10^6 algal cells). After a certain level of PS (≥ 50mg/L) exposure and time (≥ 2 days) the microalgal growth was affected. The inhibition ratios of species upon treatment with 50, 100, and 1000 mg/L PS was 15.71, 16.43, and 28.85%, respectively. Hence, it was confirmed that the inhibition ratios were directly proportional to the PS concentration. The concentration of chlorophyll a in algal species was also affected by the increasing concentration of PS. The reduction of 9.2, 10.8, and 21.3% were noted in the chlorophyll a concentration upon exposure to 50, 100, and 1000 mg/L of PS, respectively. SEM images identified the aggregation of microalgal species upon treatment with PS and the amount of aggregation showed positive effect with the increasing PS concentration as evident from Figure 1.2.

Sundbaek et al. (2018) investigated the adsorptive removal of fluorescent PS particles by the edible algae *Fucus vesiculosus*. The seaweed collected were placed at 10 °C under dark conditions as well as in an aquarium comprising of aeration and artificial lighting for 12-/12- h of dark/light cycle. The green fluorescent PS microplastic particles having 20 μm diameter, 1% soil suspension containing 0.1% Tween 20, and 2-mM sodium azide (1.06 g/cm^3), which acted as an antimicrobial agent were used in the experiment. In order to remove the suspended particulate matter the filtration of the reaction mixture was performed. The concentration of PS microplastic was around 2.65 mg/L (597 particles/mL). Each reagent bottle contained four pieces of

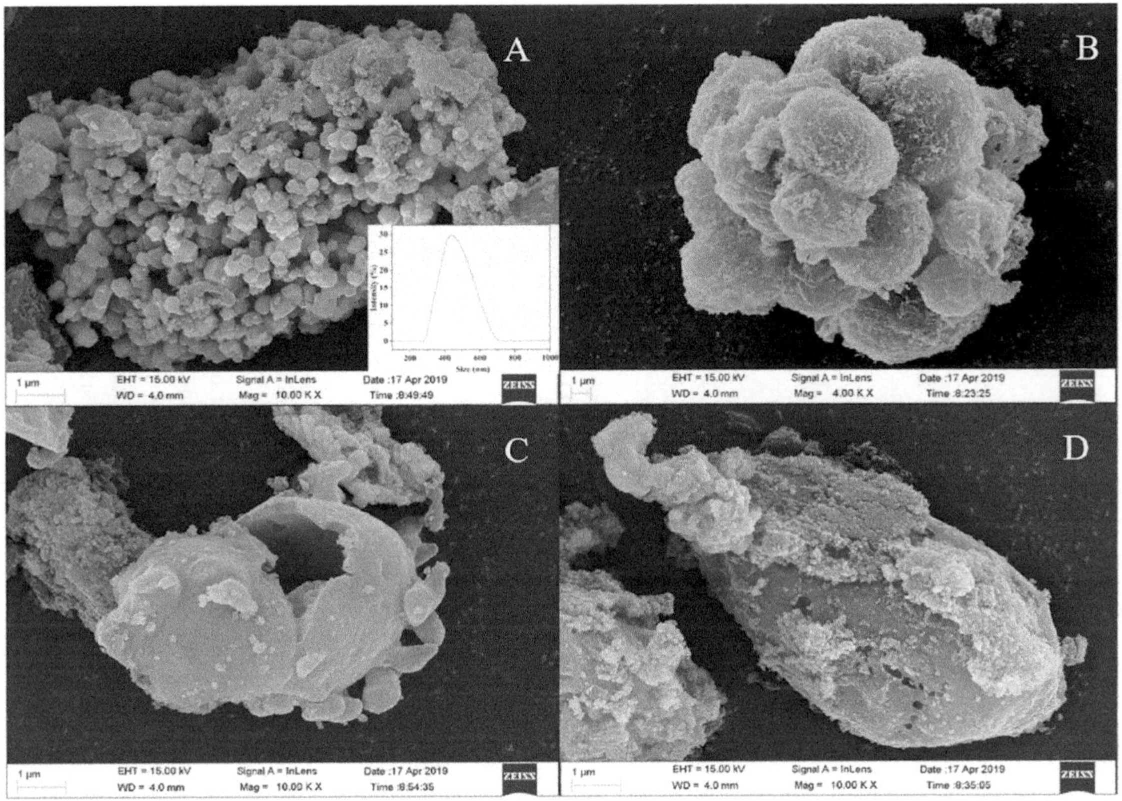

FIGURE 1.1 The scanning electron microscope (SEM) images of PS microplastics with different sizes and shapes A), *C. reinhardtii* homo-aggregated with others; B) PS microplastics attached to *C. reinhardtii* forming hetero-aggregation (C, D). Reprinted with permission from Li, S., Wang, P., Zhang, C., Zhou, X., Yin, Z., Hu, T., Hu, D., Liu, C., Zhu, L., 2020. Influence of polystyrene microplastics on the growth, photosynthetic efficiency and aggregation of freshwater microalgae *Chlamydomonas reinhardtii*. Sci. Total Environ.714, 136767. Copyright © 2020 Elsevier B.V.

FIGURE 1.2 SEM Images of the microplastic and *C. vulgaris* tested under different concentrations of microplastic. a) Microplastic used (Mag: 20.00 K); b) *C. vulgaris* under 1 mg/L microplastic concentration (Mag: 16.00 K); c) *C. vulgaris* under 50 mg/L microplastic concentration (Mag: 12.00 K); d) *C. vulgaris* under 1000 mg/L microplastic concentration (Mag: 8.00 K). Reprinted with permission from Tunali, M., Uzoefuna, E. N., Tunali, M. M., Yenigun, O., 2020. Effect of microplastics and microplastic-metal combinations on growth and chlorophyll a concentration of *Chlorella vulgaris*. Sci. Total Environ.743, 140479. Copyright © 2020 Elsevier B.V.

algal strain followed by sealing and placing at 110 °C for 2 h at 1 rpm. The washing of the algal samples was performed by taking 50 mL filtered seawater followed by agitation for 1 h. After the washing step, the number of PS particles reduced from 2.63 PS particles/mm to 1.07 PS particles/mm. The concentration of 0.9–6.1 PS particles/mm^2 also reduced to 0.13–0.32 PS particles/mm^2 after exposure to PS microplastic as well as washing which showed average reduction efficiency of 94.5%. Even after washing of the seaweed, on an average 515 PS particles/g wet weight and 2787 PS/g dry weight was left. About 17–112% (mean 74%) and 2.4–5.9% (mean 4%) adherence of the particles in the exposed suspension before and after washing were noted.

Hadiyanto et al. (2021) evaluated the effects of interactions between *Spirulina platensis* and Styrofoam as depicted in Figure 1.3. The microalgal culture was initially subjected to incubation under illumination for 24 h of light conditions at 23 ± 2 °C. After the incubation, Styrofoam, in four different concentrations (150, 250, and 400 mg/ 500 mL), was added to the culture medium separately for a period of 30 days. The cells were further harvested via filtration followed by drying at 35–40 °C.

FIGURE 1.3 Microplastic Styrofoam with a diameter of 2 mm (left) and implementation of Styrofoam in microalgae culture (right). Reprinted from Hadiyanto, H., Haris, A., Muhammad, F., Afiati, N., Khoironi, A., 2021. Interaction between styrofoam and microalgae *Spirulina platensis* in brackish water system. Toxics.9(3), 43.

The growth rate of microalgae was around 0.035925, 0.03525, 0.02675, and 0.020425 day^{-1} for cultures containing Styrofoam concentrations of 0, 150, 250, and 400 mg/500 mL, respectively. FTIR analysis indicated the degradation of Styrofoam that was confirmed by appearance of the absorbance peaks in the range of 3378–3345 cm^{-1} that was attributed to the O-H group typically absent in the raw Styrofoam. This change in the functional group of Styrofoam suggested the formation of an alcohol group (-COOH). SEM images identified the formation of EPS having spherical morphology on the microalgal cells exposed to Styrofoam as seen in Figure 1.4. It was also noted that the surface of Styrofoam-exposed algal cells was damaged while that of control remained intact. EDX analysis revealed an increase in the carbon content by 24.56% and 4.24% in the cells treated with 150 and 250 mg/500 mL Styrofoam whereas, in the case of cells exposed to 400 mg/500 mL Styrofoam, the carbon content was reduced by 2.14%.

Long et al. (2017) evaluated the formation of hetero-aggregates between PS and microalgae species *Tisochrysis lutea*, *Heterocaspa triquetra*, and *C. neogracile*. The species such as *T. lutea* and *H. triquetra* were cultivated in F/2 medium prepared using filtered natural seawater, whereas *C. neogracile* was cultivated in a similar medium enriched with 1.07×10^{-4} M silica. The culture medium was further subjected to illumination in light and dark cycles of 12 h each at 16 °C. After inoculation of the microalgal species, the PS was added to the culture medium at 3.96 µg/L and also in 10 times higher concentration (39.6 µg/L). The percentage of the PS after 24 h was 91 ± 4%, which further reduced to 47 ± 7% after 28 days and again increased to 66 ± 5% after 35 days. In case of *C. neogracile*, around 81 ± 3% PS was suspended initially on day 1 (no hetero-aggregates) which increased to 96 ± 1% after 13 days of treatment. At the stationary phase, the number of hetero-aggregates was 19 ± 6%. Treatment with *T. lutea* exhibited the formation of 3 ± 1% hetero-aggregates while 97 ± 1% of PS was adsorbed on the glass surface. In presence of *H. triquetra*, initially 78 ± 2% PS was free in the suspension and 22 ± 2% was adsorbed on the glass surface. After 24 days of treatment, the number of free PS reduced to 15 ± 2% (at stationary phase). The growth rates of *T. lutea*, *C. neogracile*, and *H. triquetra*, after exposure to PS were 0.440 ± 0.005 per day, 0.227 ± 0.011 per day, and 0.198 ± 0.005 per day, respectively.

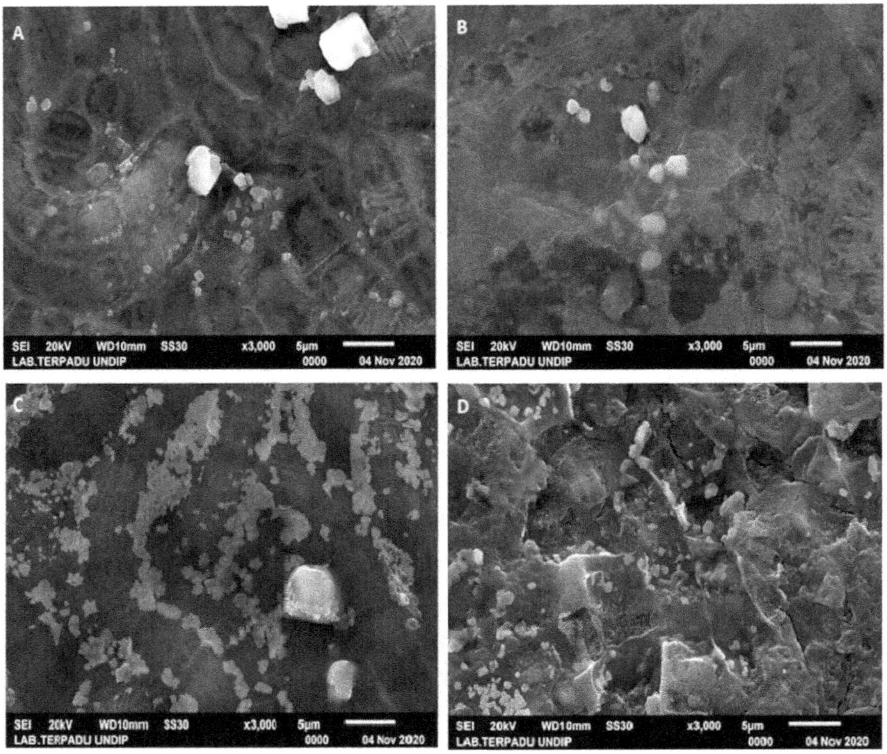

FIGURE 1.4 SEM analysis results of brackish water culture *S. platensis* for 30 days. A) *S. platensis* without Styrofoam treatment; B) *S. platensis* treated with Styrofoam 150 mg/500 mL; C) *S. platensis* treated with Styrofoam 250 mg/500 mL; D) *S. platensis* treated with Styrofoam 400 mg/500 mL. Reprinted from Hadiyanto, H., Haris, A., Muhammad, F., Afiati, N., Khoironi, A., 2021. Interaction between styrofoam and microalgae *Spirulina platensis* in brackish water system. Toxics.9(3), 43.

1.4.2 Polyvinyl Chloride

Manzi et al. (2022) investigated the effect of microalgae *Chlorella* sp. GEEL-08 strain on the removal of polyvinyl chloride (PVC). The PVC particles of size 0.2 μm were initially added to the BBM medium inoculated with the microalga followed by incubation under shaking conditions (150 rpm) at 25 °C and illumination of white light for around 20 days. The highest microalgal growth rate of 0.27 d^{-1} was observed upon treatment with 50 mg/L PVC, whereas the reduction in the growth with highest inhibition of 12.39% was noted upon treatment with 200 mg/L PVC. The pH of the medium increased from 6.3 to 10–10.2 after PVC treatment, which also showed highest microalgal growth suggesting highest photosynthetic activity. At 50 mg/L PVC concentration, the highest removal capacities of 99.59 and 99.73% were noted for total nitrogen and total phosphorous content, respectively by the microalgae. The settling efficiency increased up to 16.9% after the addition of 50 mg/L PVC. At 150 and 200 mg/L PVC concentrations, the settling efficiencies increased to 13.8 and 14.1%, respectively after 60 min of treatment. FTIR analysis identified the stretching vibrations of protein associated OH and NH groups (3550-3200 cm^{-1}), aldehyde C-HO (2830-2695 cm^{-1}), C=C (1750-1735 cm^{-1}), ester (1124-1087 cm^{-1}), carboxylic group (1420-1330 cm^{-1}), and C=C (730-665 cm^{-1}). The peak at 3434.9 cm^{-1} was attributed to the O-H bonding in the PVC. The peaks associated with the CH_2 asymmetric (2968.4 and 2910.7 cm^{-1}), C-H aliphatic bending (1428.6 cm^{-1}) and near Cl (1254.2 cm^{-1}), C-C stretching in PVC back bone (1096.5 cm^{-1}), and C-Cl bonding (607.2 and 427.2 cm^{-1}) were also observed. The microalgal biomass treated with 200 mg/L PVC showed strong absorbance peaks in the range from 3418.6 to 1028.4 cm^{-1}. The C-Cl vibrational stretching at 607.2 cm^{-1} was observed in the presence of both 50 and 200 mg/L of PVC.

Fu et al. (2019) evaluated the interaction of aged PVC with microalgal species *C. vulgaris*. The cultivation of algal species was performed in the Ningbo no. 3 medium prepared using the seawater followed by incubating the medium under the fluorescent illumination for 12 h light/dark cycles at 27 ± 1 °C. The concentration of the cells required in the inoculum was 1×10^6 cells/mL. The aging of the PVC was carried out by submerging the virgin microplastic in the synthetic seawater followed by placing it in the aging chamber for two months under continuous UVC irradiation. After the completion of the aging process, the PVC was washed and dried at 60 °C. The surface of virgin PVC was smooth and had uniformity along with compact textures which further changed to wrinkled surface possessing rough cracks as well as broken textures after the aging process. FTIR analysis revealed the presence of absorbance peaks at 3417, 2912, 1139, 1633, and 603 cm^{-1} in both virgin as well as the aged PVC. The stronger peaks of -OH groups (3400 cm^{-1}) and weaker CH_2 peaks (2926 cm^{-1}) were noted in aged PVC compared to that of virgin PVC. The free radicals were formed in the aged PVC due to the cleavage of C-H bond in CH_2 which was suggested from the FTIR analysis. The peak of C=C aromatic showed enhancement (1642 cm^{-1}) whereas that of -COOC showed reduction (1165 cm^{-1}) in the aged PVC. The enhancement in -OH and C=C as well as reduction in C-H suggested the hike in aromaticity as well as affinity of aged PVC. The highest inhibition ratios of 28.25% (fourth day) and 35.26% (third day) were noted in microalgal species upon treatment with 10 mg/L concentration of virgin and aged PVC, respectively. The maximum microalgal biomass production of 34.83 mg/L/d was noted upon treatment with 1000 mg/L PVC for the period of 10 days.

Zhang et al. (2017) studied the toxicity of micro PVC on *Skeletonemacostatum* as shown in Figure 1.5. Initially, the cultivation of algae was carried out in the f/2 medium prepared using sterile seawater at 20 ± 1 °C under constant exposure to white fluorescence light (12 h of light and dark cycles). The algal sedimentation was prevented by shaking the mixture manually twice a day. In order to achieve the log phase of algae, the incubation was continued for 5–7 days. About 50×10^4 cell/mL algal concentration was employed initially for the experiment. An inverse relationship was observed between mPVC concentration and algal density. At 50 mg/L mPVC concentration, the inhibition in the algal density was 37.9%, 36.5%, 35.4%, and 39.7% at 24 h, 48 h, 72 h, and 96 h, respectively. The decrease in chlorophyll content was about 20%

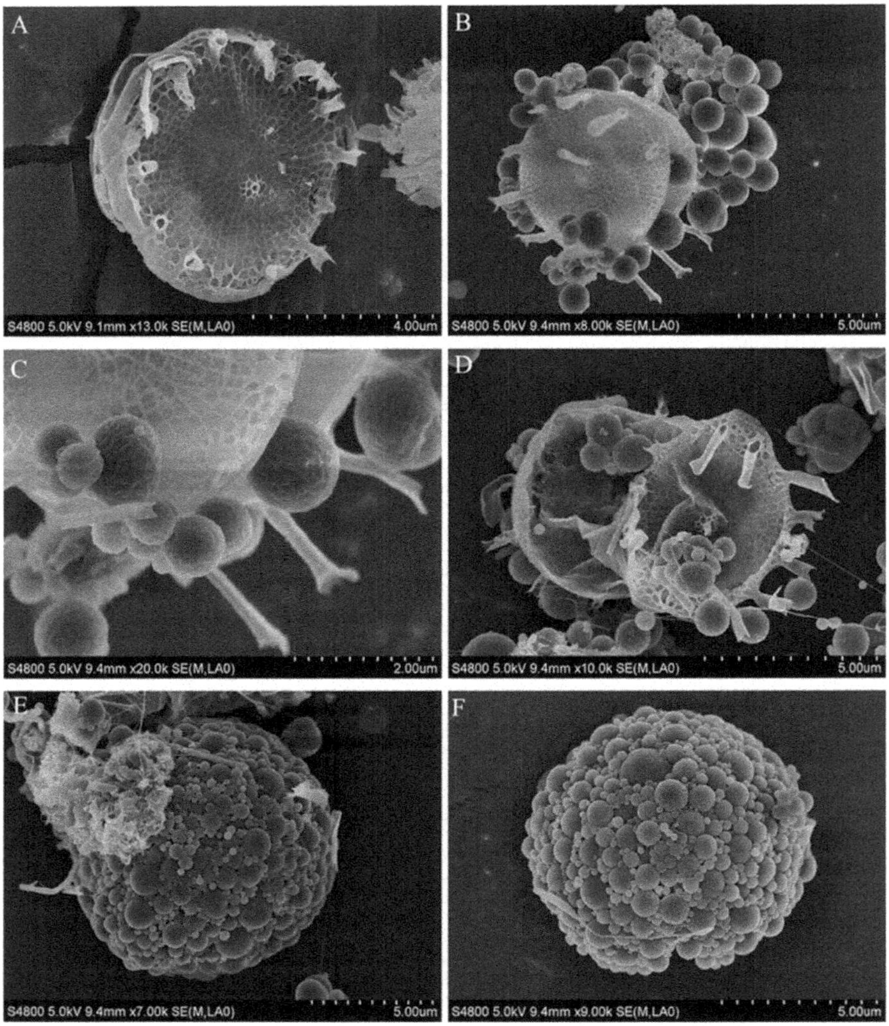

FIGURE 1.5 The SEM images of the interaction between mPVC and microalgae after 24 h exposure. A) control without mPVC; B) mPVC adsorbed on algal cells; C) mPVC embedded in algal cells; D) physical damage on algae by mPVC; E) algal fragments adsorbed on mPVC; F) aggregation of mPVC. Reprinted with permission from Zhang, C., Chen, X., Wang, J., Tan, L., 2017. Toxic effects of microplastic on marine microalgae *Skeletonemacostatum*: Interactions between microplastic and algae. Environ. Pollut. 220, 1282–1288. Copyright © 2016 Elsevier Ltd.

upon exposure to 50 mg/L mPVC from 24 to 96 h. In the case of PSII, a decrease of 32% was observed within 1 h of treatment with 50 mg/L mPVC. In the shading test, no significant effect on the algal growth and chlorophyll content was noted under the treatment with 50 mg/L mPVC except for the reduction in PSII after 1 h.

1.4.3 Polyethylene Terephthalate

Kim et al. (2020) studied the production of polyethylene terephthalate (PET) degrading enzyme, PETase by the *C. reinhardtii* strains CC-124 and CC-503. The strain CC-503 was the cell-wall less mutant of CC-125 and CC-124 was the normal wild-type strain comprising of mutation in *nit*1 and *nit*2 genes. Both the strains were transformed using pBR9_PETase_Cre plasmid followed by the spreading of strains on a zeocin-containing agar plate. Out of the antibiotic-resistant colonies formed on the plate, about 288 colonies were further inoculated. Afterwards, 61 and 17 clones of CC-124 and CC-503 strains were grown using a TAP medium supplemented with 10 mg/L zeocin. The cultivated cells were further transferred to 24 well and 12 well plates. About 11 CC-124 clones and 14 CC-503 clones were cultivated and transferred to 10 mL media. The mixture of PET powder and lysates was incubated for 2 weeks at 30 °C. About 9.12 mg TPA was produced from 30 mg PET powder which suggested a conversion rate of 35.17%. After 4 weeks of incubation with CC-124 strain, holes and dents were observed on the surface of PET, whereas no effect was observed in the case of CC-124 strain as evident from Figure 1.6.

Falah et al. (2020) evaluated the degradation of PET using the microalgae *Chlorella vulgaris* and also studied the effect of pretreatment. Initially, the substrate was prepared by cutting the small strips of PET bottles with the help of sterile scissors followed by their addition to the distilled water and 1 h of stirring. The pieces were further placed in 70% ethanol solution for 30 min under sterile conditions. Later, they were transferred to a Petri plate (sterile) for air drying. The pretreatment of the plastic films was carried out via two methods such as physical and chemical. In the physical method, UV and temperature (100 °C, 48 h) were used whereas, treatment with 90% concentrated nitric acid at ambient temperature for 5 days was carried out in the chemical method. The biological treatment involved selected microalgae. Two different setups were prepared for the treatment of plastic films with the microalgae. In the first setup, the films were pretreated with physical and chemical methods followed by biological treatment while, in another setup, the films were treated directly using the biological method. Around 2 mL of the alga was inoculated into the media followed by 1 month of incubation under light illumination at room temperature. No significant difference was observed in the plastic film degradation for the samples with (5.57%) and without pretreatment (5.45%). The pretreatment process aided in damaging the plastic films which further assisted the algal species to form biofilm as well as efficient degradation. FTIR spectra revealed the absorbance peak for primary amines, secondary amines, and carboxylic acids in the range of 2800 to 4000 cm^{-1}. The existence of aromatic anhydrides and conjugated acid halides was also confirmed from the peaks obtained in the range of 1500 to 2000 cm^{-1} for the pretreated sample. Scanning electron microscope (SEM) revealed the formation of cracks on the plastic films after pretreatment in which the growth of the algae was noted.

1.4.4 Polyethylene

Gowthami et al. (2023) analyzed the degradation ability of different microalgae like *Chloroidiumsaccharophilum*, *Picochlorum maculatum*, *Amphora* sp., *Hymenomonas*

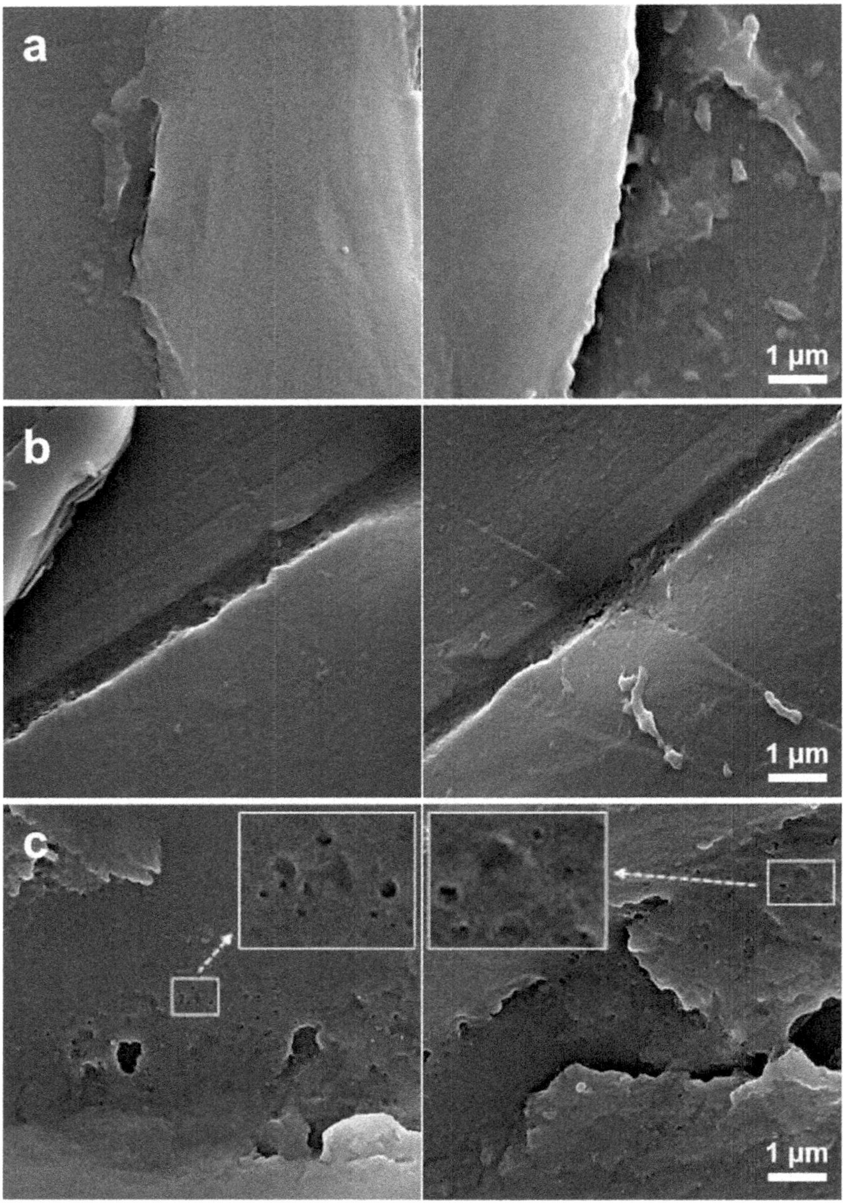

FIGURE 1.6 The results of scanning electron microscopy. a, c) PET surface after 4 weeks incubation with CC-124 wild type a) and CC-124_PETase #11; c) lysate. b PET surface after 2 weeks incubation with CC-124_PETase #11. The magnitudes were ×20,000 for all samples. Small and large red boxes represent original and zoomed pictures, respectively. Reprinted from Kim, J. W., Park, S. B., Tran, Q. G., Cho, D. H., Choi, D. Y., Lee, Y. J., Kim, H. S., 2020. Functional expression of polyethylene terephthalate-degrading enzyme (PETase) in green microalgae. Microb. Cell Fact. 19, 97.

globosa and *Limnospira indica* for low-density polyethylene (LDPE). The sheets were cut into smaller pieces followed by their sterilization using 70% ethanol and were further rinsed using water and dried at 60 °C overnight. Initially, the seawater was filtered (0.2 μm) and sterilized at 121 °C for 20 min. The salinity and pH of the seawater after sterilization was 27 ± 1 and 8 ± 0.5, respectively. The microalgal cells having a density equivalent to 10^5 cells/mL were inoculated in 300 mL sterile seawater comprising of nutrient medium. About 20 mg of the LDPE sheets were added to the reaction mixture followed by incubation under a light and dark cycle of 12:12 at 23 ± 2 °C for a period of 30 days. The highest reduction in weight of the LDPE sheet was noted upon treatment with *P. maculatum* (20.16 ± 0.148 %) followed by *H. globosa* (17.5 ± 0.047 %), *Amphora* sp. (16.66 ± 0.059 %), *L. indica* (13.66 ± 0.118 %), and *C. saccharophilum* (12 ± 0.054 %). The maximum rate of the LDPE reduction was noted for *P. maculatum* (0.005 g/day), which further resulted in a 50% reduction of LDPE sheet within 138 days. The microalgal colonization on the LDPE film was observed from the optical microscope analysis. The SEM images revealed the deteriorated surface of LDPE due to biofilm formation within 45 days of microalgal treatment. The surface was comprised of pits, grooves, and eroded regions. The FTIR analysis confirmed the successful degradation of the LDPE by microalgae from the development of new absorbance peaks at 1371, 1466, 3600, and 500–726 attributing to the CH_3 deformation, isopropyl groups, O-H stretching, and deformation of CH_2 group, respectively. The modification in the peaks present at 2928.83 and 2850.06 cm^{-1} represented the interactions of the microalgal enzymes and LDPE polymer. The crystallinities of LDPE were 64.13, 65.38, 67.2, 70.31, and 70.61% for *P. maculatum*, *Amphora* sp., *H. globosa*, *L. indica*, and *C. saccharophilum*, respectively, which was lower compared to the control (71.37%).

Bhuyar et al. (2021) studied the biodeterioration of polyethylene by the microalgae like *Chlorella*sp, *Cyanobacteria* sp. (blue-green algae), as well as their consortium. The organisms were isolated from the polyethylene bags and samples of water followed by their serial dilutions (10^{-1} to 10^{-10}). About 1 mL of the inoculum was added to the BBM and blue-green 11 (BG-11) specific for green microalgae and blue-green algae, respectively. The inoculated plates were further incubated for 12 h under illumination at 25 ± 2 °C. The degradation experiment was performed by pre-treating the LDPE sheets with UV before addition into the BG-11 medium. The medium was further inoculated with the selected microalgal species as well as their consortium separately. The growth of consortia of *Cyanobacteria* sp. and *Chlorella* sp. was higher compared to that of single species growth in the individual medium. The highest reduction in the LDPE concentration was 28% as noted during treatment with the consortia. SEM images of the *Cyanobacteria* sp. identified the formation of the network via EPS, which aided in the capitation, aggregation, as well as adhesion of the consortium on the microplastics. The EDX analysis identified a decrease in the amount of carbon from 92.30 ± 1.23% to 39.12 ± 1.08% and an increase in the amount of oxygen by 42.63 ± 0.94%. FTIR analysis indicated the successful degradation of the polymer from the development of an absorbance peak at 2839 cm^{-1} that was attributed to the carbonyl bond. The reduction in carbon weight percentage present in LDPE by 37.91% was observed during the degradation process.

Hadiyanto et al. (2022) evaluated the degradation of the oxidized high-density polyethylene (HDPE) and oxo-degradable oxium plastic employing microalgae, *Dunaliella salina*. The algal cell density of 1 ppm was used for the addition in the bioreactor containing saline water (30–35 ppt). The enrichment to the algal species was accomplished using 1 mL/L *D. salina* of Walne along with the mixture of 80 ppm urea, 36 ppm triple super phosphate (TSP), 40 ppm ammonium sulfate, 5 ppm ethylene diamine tetra acetic acid (EDTA), and 1 ppm $FeCl_3$. The samples of HDPE and oxium microplastics were prepared by oxidizing the dried samples of microplastics using 15 M hydrogen peroxide (H_2O_2) for 72 h at 25 °C under stirring conditions. A hetero-aggregation system comprising EPS, reducer microbes, and microplastics was developed in which the degradation was carried out by the conversion of microplastic into CO_2 by the microalgae which were further used as an energy source. The increase in the growth and density of microalgae was observed until 12 days of oxium treatment and 10 days of treatment with HDPE. A significant effect on the growth of microalgae was noted upon treatment with the oxidized HDPE microplastic compared to that of oxium treatment. SEM analysis revealed the changes in the surface morphology of the oxidized HDPE and oxium before and after microalgae treatment for 15 days. A softer surface along with smooth erosion and defects (caused by the oxidation process) were observed upon algal treatment of the oxidized HDPE. In the case of the oxium, larger patches were observed upon algal treatment after 15 days which suggested the effective formation of biofilm. FTIR analysis suggested the oxidation of alkyl groups of HDPE from the reduction in the absorbance peaks at 2914 and 2848 cm^{-1} upon microalgal treatment. Similarly, the reduction in peaks for CH (2914 and 2848 cm^{-1}), CH_2 bending (1468 cm^{-1}), and CH_2 rock (717 cm^{-1}) were observed in oxium microplastic upon microalgal treatment. The increase in the concentrations of carboxylic acid (1085 cm^{-1}) followed by the formation of a new alkene group (1645 cm^{-1}), and primary alcohol group (1045 cm^{-1}) were observed after the degradation of oxidized HDPE by microalgae. XRD analysis identified better crystalline properties of HDPE compared to oxium, thus suggesting greater difficulty in the natural degradation of HDPE. It was found that *D. salina* had better efficiency for HDPE degradation compared to the oxium microplastic.

Kumar et al. (2017) investigated polyethylene degradation via green photosynthetic microalgae such as *Scenedesmus dimorphus* (green microalga), *Anabaena spiroides* (blue-green alga), and *Naviculapupula* (diatom). Firstly, before the isolation of microalgae from the water samples collected from ponds, pools, etc., the samples were subjected to 5 min of centrifugation at 8000 rpm which resulted in the formation of a concentrated microalgae pellet. Simultaneously, the polyethylene bags containing the microalgae mass on their surface were wiped using a sterile blade followed by their dispersion in water. The samples of microalgae conserved from the above samples were serially diluted (10^{-1} to 10^{-10}) and were used for inoculating three different media like BBM specific for green microalgae, BG11 for blue-green algae, and diatom medium (DM) for diatoms. The plates were further incubated for 7 days under 12 h of illumination at 27 ± 2 °C. The polyethylene degradation experiment was carried out by inoculating 1% of microalgal species to their respective media comprising two polyethylene pieces followed by incubation at 27 ± 2 °C for

(a) *Scenedesmus dimorphus* (green algae)

(b) *Anabaena spiroides* (blue-green algae)

(c) *Navicula pupula* (Diatom)

FIGURE 1.7 The image showing the biological treatment of LD and HD polyethylene sheets with the three different groups of microalgae. Reprinted from Kumar, R. V., Kanna, G. R., Elumalai, S., 2017. Biodegradation of polyethylene by green photosynthetic microalgae. J. Bioremediat. Biodegrad. 8(1), 1000381.

the period of a month under 12 h light/dark conditions. The initial mean weight of low-density (LD) and high-density (HD) polyethylene pieces (1 sq cm) were 6.07 and 8.945 mg, respectively. The mean LD polyethylene degradation efficiency of 3.74, 8.18, and 4.44% were observed on treatment with *S. dismorphus*, *A. spiroides*, and *N. pupula*, respectively. SEM studies confirmed no visible impact of *S. dismorphus* adherence on the LD polyethylene (Figure 1.7), whereas colonization as well as a ruptured minute hole formation was demonstrated by *A. spiroides* and partial erosion of the LD polyethylene surface was observed in the case of *N. pupula* treatment.

Sanniyasi et al. (2021) reported the biodegradation of LDPE sheets by the use of microalga *Uronema africanum* Borge. About 100 mL of culture medium was supplemented with 10 LDPE sheets (1 cm^2) and the loopful of samples collected were inoculated into the medium followed by incubation under light and dark cycles of 12:12 h at 25 °C. In the biodegradation studies, the microalgae having the ability to colonize on the LDPE sheet surface were selected and about 1% inoculum was added in the 100 mL BBM culture medium along with 10 LDPE sheets. The reaction mixture was further subjected to incubation under 12:12 h of light and dark conditions for 30 days. After completion of incubation, the LDPE sheet fully colonized by microalga, as well as the protrusions of hair-like green structures from those sheets, were observed. The radial disc-like structures were observed on the surface of the sheet and after washing, the regions demonstrating erosions, grooves, abrasions, and ridges appeared. The dark field microscope analysis suggested the formation of erosions by radial discs and abrasions along with ridges and grooves by microalgal filaments. The gas-chromatography mass spectrometry (GC-MS) studies identified the presence of fatty acids methyl esters (FAME) in the control sample and a large number of hydrocarbons in the microalgae treated samples. The FTIR analysis revealed the absorbance bands attributing to the stretching vibrations of carboxylic O-H (3291.09 cm^{-1}), C=O (1745.03 cm^{-1}), -C=C- (1646.53 cm^{-1}), N-O (1545.2 cm^{-1}), C-N (1017.32 cm^{-1}), and -C≡C-H: C-H (667.31 cm^{-1}). SEM images confirmed the formation of erosion, abrasion, grooves, and ridges by the colonization as well as degradation of LDPE upon microalgal treatment. The size of the radial disc was around 20–30 μm in diameter and that of ridges and grooves were 10–15 μm in width.

1.4.5 Mixed

Lagarde et al. (2016) studied the interaction between *C. reinhardtii* and microplastics like polypropylene (PP) and HDPE. The samples were prepared initially by washing with water followed by cutting and the fragments were further formed by high energy milling that were sieved and isolated on the basis of the size in the range from 400 to 1000 μm. The microalgal species were initially cultivated in TAP medium at 24 ± 1 °C under cycles of 14 h of light and 10 h of darkness. The cultures were kept under shaking conditions at 100 rpm. The concentration of cells was around 6×10^6 cells/mL, which reduced to 4×10^6 cells/mL after 7 days. Even after 68 days of microplastic treatment with microalgal species no significant effect was observed whereas about 18% reduction in PP was observed after 78 days of treatment. Within 20 days of PP treatment with microalgae, the formation of hetero-aggregates began which reached up to a size of 4.5 mm at the end of the experiment. The aggregates were mostly ovoid in structure with the average size of 2.00 ± 0.48 mm and 1.49 ± 0.55 mm in length and width, respectively. In case of HDPE, no aggregates were formed. SEM images confirmed the microalgal colonization on the microplastics, which appeared homogeneous after 78 days of treatment. The density of aggregates was around 1.19. Upon mass calculation, the average composition of the aggregates was found to be around 52% in PP with respect to biomass (48%). About 95.5% plastic was recovered in the samples lacking microalgal species as well as in HDPE samples and the loss of 4.5% mainly occurred due to the adsorption of fragments on the surface of glass. In case of PP, the amount of plastic recovered was around 94.5%.

Song et al. (2020) analyzed the interactions of microalgal species like *Chlorella* sp. L38 and *Phaeodactylumtricornutum*MASCC-0025 with different microplastics with respect to their bioelimination. The microplastics utilized for the experiments were PE, PVC, PET, and PP at a concentration of 200 mg/L and the size of 74 μm. The microalgal species were cultivated in their respective medium like BG11 (*Chlorella* sp. L38) and modified f/2 (*P. tricornutum*) followed by the incubation under illumination of light at 30 ± 1 °C and 25 ± 1 °C for *Chlorella* sp. and *P. tricornutum*, respectively for 96 h. The promotion in the growth of *Chlorella* sp. was noted upon the addition of PET and PVC whereas, slight inhibition of growth was observed on treatment with PP. In case of *P. tricornutum*, the growth was inhibited upon treatment with all the microplastics. The highest specific growth rate of 0.066 d^{-1} was noted for *Chlorella* sp. in the presence of PET and PVC. PP demonstrated an inhibition of 25.8% on the third day of treatment. After 4 days of treatment, *P. tricornutum* with PP showed the highest inhibition of 21.2 %. Compared to the control group, about 1.4 times higher EPS production was noted in the case of *Chlorella* sp. and about 2.2 times in the case of *P. tricornutum*. Efficient production of antioxidant enzymes by the *Chlorella* sp. was noted, which was associated with the protection by scavenging of the ROS. SEM and TEM images identified the irregular morphology of PVC particles having rough porous surfaces along with the gathering of the large number of PVC microspheres on the surface. The cell of *Chlorella* sp. L38 was globular in shape with a smooth surface and size of 2–4 μm. The cell of *P. tricornutum* possessed a three-ray structure with the size of 10 μm, which was about 2–4 times larger than the size of *Chlorella* cells. The physical damage to the morphology of *P. tricornutum* cells was noted upon treatment with microplastic which resulted in the development of wrinkled and deformed surfaces with the aggregation of microplastics on their outer cell surface.

Cheng and Wang (2022) studied removal efficiencies of *Scenedesmus abundans* against different microplastics such as, PS, poly(methyl methacrylate) (PMMA), and polylactide (PLA). Initially, the selected microalgae was maintained on the BG-11 medium followed by their inoculation in the 75 mL liquid BG-11 medium in order to obtain the liquid culture. The culture medium was incubated under light illumination at 28 °C and 120 rpm along with the supplement of 5% carbon dioxide. The concentration of microplastics was around 90,000 number/mL. A rapid aggregation due to interaction between the microbe and the microplastic surface was noted that was largely dependent on the type of the latter. A shorter response time was seen in the case of PS and PMMA whereas, in the case of PLA it increased gradually with respect to contact time from 2 to 6 days. The zeta potential value of microalgal species showed a change from -25mV to -17 mV after 4 days while it was -18 mV after 6days. The maximum PMMA removal efficiency was 77 to 92% followed by PS and PLA having 50 to 71% and 31 to 53%, respectively by the un-pretreated microalgae. The lowest aggregation of the microalgal cells (1.48×10^5 to 2.05×10^5 number/mL) was induced by the PMMA. The removal efficiencies of microalgal species showed an enhancement after 6 days of pre-exposure to the microplastics. The removal efficiency increased from 59% to 84% for PS while it increased from 73% to 87% for PLA. Similarly, the removal efficiency for PMMA increased with the pre-exposure but was less significant as the removal rate by un-pretreated organism was already high. SEM images suggested a flocculent EPS formation upon exposure to PMMA and PLA whereas, a mucous membrane covered the aggregated cells on exposure to PS. Thus, the maximum removal of efficiency was for PMMA (98%) followed by PLA (87%) and PS (84%) upon treatment with the microalgae.

Khoironi et al. (2019) evaluated the interaction of PET and PP with *S. platensis* that was initially cultivated in the enrichment medium containing the mixture of TSP

TABLE 1.2
Microalgae with microplastic removal potential

Algae	Type of algae	Microplastic pollutant	Removal efficiency (%)	References
Chaetoceros-neogracile and *Rhodomonas salina*	microalgae	polystyrene	-	Long et al. (2015)
Chlamydomonas reinhardtii	microalgae	polystyrene	-	Li et al. (2020)
Chlorella vulgaris	microalgae	polystyrene	-	Tunali et al. (2020)
Fucus vesiculosus	-	polystyrene	94.5	Sundbaek et al. (2018)
Spirulina platensis	blue-green algae	styrofoam	-	Hadiyanto et al. (2021)
Tisochrysis lutea, *Heterocaspa triquetra*, and *Chaetoceros-neogracile*	microalgae	polystyrene	-	Long et al. (2017)
Chlorella sp. GEEL-08	microalgae	polyvinyl chloride	-	Manzi et al. (2022)
Chlorella vulgaris	microalgae	polyvinyl chloride	-	Fu et al. (2019)
Skeletonemacostatum	diatoms	polyvinyl chloride	-	Zhang et al. (2017)
Chlamydomonas reinhardtii	microalgae	polyethylene terephthalate	-	Kim et al. (2020)
Chlorella vulgaris	green microalgae	polyethylene terephthalate	5.57	Falah et al. (2020)
Chloroidium-saccharophilum, *Picochlorum maculatum*, *Amphora* sp., *Hymenomonas globosa* and *Limnospira indica*	microalgae	polyethylene	64.13	Gowthami et al. (2023)
Chlorella sp. *Cyanobacteria* sp. Consortium.	green microalgae blue-green algae	polyethylene	28	Bhuyar et al. (2021)
Dunaliella salina	microalgae	polyethylene and oxium	-	Hadiyanto et al. (2022)

TABLE 1.2 (Continued)
Microalgae with microplastic removal potential

Algae	Type of algae	Microplastic pollutant	Removal efficiency (%)	References
Scenedesmus dimorphus *Anabaena spiroides* *Naviculapupula*	green microalgae blue-green algae diatoms	polyethylene	3.74 (*S. dismorphus*), 8.18 (*A. spiroides*), and 4.44 (*N. pupula*)	Kumar et al. (2017)
Uronemaafricanum Borge	microalgae	polyethylene	-	Sanniyasi et al. (2021)
Chlamydomonas reinhardtii	microalgae	HDPE and polypropylene	-	Lagarde et al. (2016)
Chlorella sp. L38 and *Phaeodactylumtricornutum* MASCC-0025	microalgae	polyethylene terephthalate, polyvinyl chloride, polyethylene, and polypropylene	-	Song et al. (2020)
Scenedesmus abundans	Microalgae	polystyrene, poly(methyl methacrylate), and polylactide	84, 98, and 87	Cheng and Wang (2022)
Spirulina platensis	Microalgae	polyethylene terephthalate and polypropylene	-	Khoironi et al. (2019)

and urea (12.5 mg/250 mL) and $NaHCO_3$ (100 mg/250 mL) at 24–26 °C and pH 7–8 under light illumination and in the presence of oxygen. The samples of plastics were initially ethanol washed followed by 24 h of drying at room temperature, cutting into 1–2 μm sizes and mixing in the cyanobacterial culture (500 mL). At 150 mg/mL concentration of microplastics, the growth rate of microalgae was higher in the case of PET (0.3584 day^{-1}) compared to that of PP (0.2647 day^{-1}). However, with 200 and 275 mg/mL of microplastic the growth rates were 0.1229 and 0.907 day^{-1}, respectively in presence of PP and 0.1144 and 0.0832 day^{-1}, respectively in presence of PET. The reduction of the tensile strength of PET (density 1.37 kg/m^3) by 0.9939 MPa/day was more significant compared to that of PP (density 0.94 kg/m^3) that decreased by 0.1977 MPa/day after treatment with microalgae for 112 days. The higher hydrophilic nature of PET (higher polar C=O bond) attributed to more hetero-aggregation of EPS compared to that of PP. FTIR analysis revealed the reduction in the intensities of the absorbance peaks associated with the aromatic C=C (1613-1505 cm^{-1})

in PET after treatment with microalgae. Similarly, after the interaction of PP with the microalgae, the absorbance peaks appeared for amide C=O (1599 and 1534 cm^{-1}), ester and ketone C=O (1731.48 cm^{-1}). The presence of the microalgae was also confirmed from the new peak developed at 3343 cm^{-1}. SEM images showed the smooth surface of PET along with blisters and the coarser surface of PP having a textured defective surface before the microalgal treatment. The erosion of the surface along with the formation of pits and cavities was noted in the treated microplastics. EDX analysis suggested a reduction in the concentration of carbon by 48.61 and 36.7% in PET and PP, respectively.

1.5 SUMMARY AND CONCLUSION

Bioremediation is a sustainable approach for removal of several refractory pollutants from the environment and converting them to less toxic and easily degradable compounds. These techniques can be implemented for the removal of highly hazardous microplastics from the soil, water, and air. Hence, emerging strategies for bioremediation of the microplastics should be introduced and integrated with the conventional methods. Additionally, available infrastructure can be renovated for the establishment of innovative supplementary treatment systems that incorporate more effective and profitable microplastic bioremediation technologies. Various parameters such as initial microplastic concentration, inoculums load, contact time, media, pretreatment, temperature, pH and aeration should be carefully optimized in order to get maximum microplastic degradation in minimum time.

Genetic engineering techniques can be employed in order to develop efficient recombinant strains of bacteria, fungi or algae that are resistant to microplastics and can degrade them most efficiently. Integrated and interdisciplinary approaches can help in exploring the underlying mechanism of the microplastic degradation by the microbes. High throughput screening methods can help in isolation of novel microbes with microplastic degrading potential that can later be subjected to the strain improvement using genetic engineering techniques.

REFERENCES

Alimi, O.S., Farner Budarz, J , Hernandez, L.M., & Tufenkji, N. (2018). Microplastics and nanoplastics in aquatic environments: aggregation, deposition, and enhanced contaminant transport. *Environ. Sci. Technol.* 52(4), 1704–1724.

Barnes, D.K., Galgani, F., Thompson, R.C., & Barlaz, M. (2009). Accumulation and fragmentation of plastic debris in global environments. *Philos. Trans. R. Soc. Lond., B, Biol. Sci.* 364(1526), 1985–1998.

Bhuyar, P., Sundararaju, S., Feng, H.X., Rahim, M.H.A., Muniyasamy, S., Maniam, G.P., & Govindan, N. (2021). Evaluation of microalgae's plastic biodeterioration property by a consortium of *Chlorella* sp. and *Cyanobacteria* sp. *Environ. Res. Eng. Manag.* 77(3), 86–98.

Bläsing, M., & Amelung, W. (2018). Plastics in soil: Analytical methods and possible sources. *Sci. Total Environ.* 612, 422–435.

Browne, M.A., Underwood, A.J., Chapman, M.G., Williams, R., Thompson, R.C., & Van Franeker, J.A. (2015). Linking effects of anthropogenic debris to ecological impacts. *Philos. Trans. R. Soc. Lond., B, Biol. Sci.* 282(1807), 20142929.

Campanale, C., Massarelli, C., Savino, I., Locaputo, V., & Uricchio, V.F. (2020). A detailed review study on potential effects of microplastics and additives of concern on human health. *Int. J. Environ. Res. Public Health.* 17(4), 1212.

Cauwenberghe, L.V., Claessens, M., Vandegehuchte, M.B., & Janssen, C.R. (2015). Microplastics are taken up by mussels (*Mytilus edulis*) and lugworms (*Arenicola marina)* living in natural habitats. *Environ. Pollut.* 199,10–17.

Chaudhry, A.K., & Sachdeva, P. (2021). Microplastics' origin, distribution, and rising hazard to aquatic organisms and human health: socio-economic insinuations and management solutions. *Reg. Stud. Mar. Sci.* 48, 102018.

Chen, G., Feng, Q., & Wang, J. (2020). Mini-review of microplastics in the atmosphere and their risks to humans. *Sci. Total Environ.* 703, 135504.3.

Cole, M., Lindeque, P., Halsband, C., & Galloway, T.S. (2011). Microplastics as contaminants in the marine environment: a review. *Mar. Pollut. Bull.* 62(12), 2588–2597.

Cheng, Y.R., & Wang, H.Y. (2022). Highly effective removal of microplastics by microalgae *Scenedesmus abundans. Chem. Eng.* J. 435, 135079.

Duis, K., & Coors, A. (2016). Microplastics in the aquatic and terrestrial environment: sources (with a specific focus on personal care products), fate and effects. Environ. *Sci Europe.* 28(1), 1–25.4.

Eriksen, M., Lebreton, L.C., Carson, H.S., Thiel, M., Moore, C.J., Borerro, J.C., Galgani, F., Ryan, P.G., & Reisser, J. (2014). Plastic pollution in the world's oceans: more than 5 trillion plastic pieces weighing over 250,000 tons afloat at sea. *PloS one.* 9(12), 111913.

Fadare, O.O., & Okoffo, E.D. (2020). Covid-19 face masks: A potential source of microplastic fibers in the environment. *Sci. Total Environ.* 737, 140279.

Falah, W., Chen, F.J., Zeb, B.S., Hayat, M.T., Mahmood, Q., Ebadi, A., Toughani, M., & Li, E.Z. (2020). Polyethylene terephthalate degradation by microalga *Chlorella vulgaris* along with pretreatment. *Mater. Plast.* 57(3), 260–270.

Fasano, E., Bono-Blay, F., Cirillo, T., Montuori, P., & Lacorte, S. (2012). Migration of phthalates, alkylphenols, bisphenol A and di (2-ethylhexyl) adipate from food packaging. *Food Control.* 27(1), 132–138.

Fries, E., & Zarfl, C. (2012). Sorption of polycyclic aromatic hydrocarbons (PAHs) to low and high density polyethylene (PE). *Environ. Sci. Pollut. Res.* 19, 1296–1304.

Fu, D., Zhang, Q., Fan, Z., Qi, H., Wang, Z., Peng, L. (2019). Aged microplastics polyvinyl chloride interact with copper and cause oxidative stress towards microalgae *Chlorella vulgaris. Aquat. Toxicol.* 216, 105319.

Fuller, S., & Gautam, A. (2016). A procedure for measuring microplastics using pressurized fluid extraction. *Environ. Sci. Technol.* 50(11), 5774–5780.

Gowthami, A., Marjuk, M.S., Raju, P., Devi, K.N., Santhanam, P., Kumar, S.D., & Perumal, P. (2023). Biodegradation efficacy of selected marine microalgae against low-density polyethylene (LDPE): An environment friendly green approach. *Mar. Pollut. Bull.* 190, 114889.

Guzzetti, E., Sureda, A., Tejada, S., Faggio, C. (2018). Microplastic in marine organism: Environmental and toxicological effects. *Environ. Toxicol. Pharmacol.* 64, 164–171.

Habib, D., Locke, D.C., & Cannone, L.J. (1998). Synthetic fibers as indicators of municipal sewage sludge, sludge products, and sewage treatment plant effluents. *Water Air Soil Pollut.* 103, 1–8.

Hadiyanto, H., Haris, A., Muhammad, F., Afiati, N., & Khoironi, A. (2021). Interaction between styrofoam and microalgae *Spirulina platensis* in brackish water system. Toxics. 9(3), 43.8.

Hadiyanto, H., Khoironi, A., Dianratri, I., Huda, K., Suherman, S., & Muhammad, F. (2022). Biodegradation of oxidized high-density polyethylene and oxo-degradable plastic using microalgae *Dunaliella salina. Environ. Pollut. Bioavailab.* 34(1), 469–481.

Haq, I., Kumar, S., Kumar, V., Singh, S.K., & Raj, A. (2016a). Evaluation of bioremediation potentiality of ligninolytic Serratia liquefaciens for detoxification of pulp and paper mill effluent. *J Hazard Mater* 305: 190–199.

Haq, I., Kumari, V., Kumar, S., Raj, A., Lohani, M., & Bhargava, R.N. (2016b). Evaluation of the phytotoxic and genotoxic potential of pulp and paper mill effluent using Vigna radiata and Allium cepa. *Adv Biol* 8065736.

Haq, I., Kumar, S., Raj, A., Lohani, M., & Satyanarayana, G.N.V. (2017). Genotoxicity assessment of pulp and paper mill effluent before and after bacterial degradation using Allium cepa test. *Chemosphere* 169: 642–650.

Haq, I., Raj, A., & Markandeya (2018). Biodegradation of Azure-B dye by Serratia liquefaciens and its validation by phytotoxicity, genotoxicity and cytotoxicity studies. *Chemosphere* 196, 58–68.

Haq, I., Mazumder, P., & Kalamdhad, A.S. (2020). Recent advances in removal of lignin from paper industry wastewater and its industrial applications – A review. *Bioresour. Technol* 312: 123636.

Haq, I., & Kalamdhad, A.S. (2021). Phytotoxicity and cyto-genotoxicity evaluation of organic and inorganic pollutants containing petroleum refinery wastewater using plant bioassay. *Environ. Technol. Innov* 23: 101651.

Haq, I., Kalamdhad, A.S., & Pandey, A. (2022a). Genotoxicity evaluation of paper industry wastewater prior and post-treatment with laccase producing Pseudomonas putida MTCC 7525. *J. Clean. Prod* 342: 130981.

Haq, I., Singh, A., & Kalamdhad, A.S. (2022b) Integrated technologies for the remediation of paper industry waste in a circular bioeconomy, Editor(s): Sunita Varjani, Ashok Pandey, Mohammad J. Taherzadeh, Huu Hao Ngo, & R.D. Tyagi, *Biomass, Biofuels, Biochemicals*, Elsevier, Pages 351–363, ISBN 9780323885119.

Haq, I., Roy, S., & Kalamdhad, A.S. (2022c). Characterization of Pulp and Paper Mill Wastewater and Its Toxicity Analysis Using Vigna Radiata. In *Environmental Degradation: Monitoring, Assessment and Treatment Technologies*; Springer International Publishing: Cham, Switzerland, pp. 261–269. ISBN 9783030941475.

Haq, I., & Kalamdhad, A.S. (2022). Advanced and Ecofriendly Technologies for the Treatment of Industrial Wastewater to Constrain Environmental Pollution. In: Arora, S., Kumar, A., Ogita, S., Yau, Y.Y. (eds) *Innovations in Environmental Biotechnology*. Springer, Singapore. https://doi.org/10.1007/978-981-16-4445-0_22.

Haq, I., & Kalamdhad, A.S. (2023). Enhanced biodegradation of toxic pollutants from paper industry wastewater using Pseudomonas sp. immobilized in composite biocarriers and its toxicity evaluation. *Bioresour. Technol. Repor* 24:101674.

Haq, I., Kalamdhad, A.S., & Malik, A. (2024). Bioremediation of Petroleum Refinery Wastewater Using Bacillus subtilis IH-1 and Assessment of Its Toxicity. *Arch. Microbiol*, 206, 296.

Hidalgo-Ruz, V., Gutow, L., Thompson, R.C., & Thiel, M. (2012). Microplastics in the marine environment: a review of the methods used for identification and quantification. *Environ. Sci. Technol.* 46(6), 3060–3075.

Hirai, H., Takada, H., Ogata, Y., Yamashita, R., Mizukawa, K., Saha, M., Kwan, C., Moore, C., Gray, H., Laursen, D., & Zettler, E.R. (2011). Organic micropollutants in marine plastics debris from the open ocean and remote and urban beaches. *Mar. Pollut. Bull.* 62(8), 1683–1692.

Hurley, R., Woodward, J., & Rothwell, J.J. (2018). Microplastic contamination of river beds significantly reduced by catchment-wide flooding. *Nat. Geosci.* 11(4), 251–257.

Imhof, H.K., Ivleva, N.P., Schmid, J., Niessner, R., & Laforsch, C. (2013). Contamination of beach sediments of a subalpine lake with microplastic particles. *Curr. Biol.* 23(19), R867–R868.

Jambeck, J.R., Geyer, R., Wilcox, C., Siegler, T.R., Perryman, M., Andrady, A., Narayan, R., & Law, K.L. (2015). Plastic waste inputs from land into the ocean. *Science*. 347(6223), 768–771.

Jung, Y.S., Sampath, V., Prunicki, M., Aguilera, J., Allen, H., LaBeaud, D., Veidis, E., Barry, M., Erny, B., Patel, L., & Akdis, C. (2022). Characterization and regulation of microplastic pollution for protecting planetary and human health. *Environ. Pollut*. 315, 120442.

Karbalaei, S., Hanachi, P., Walker, T.R., & Cole, M. (2018). Occurrence, sources, human health impacts and mitigation of microplastic pollution. *Environ. Sci. Pollut. Res*. 25, 36046–36063.

Khoironi, A., Anggoro, S., & Sudarno. (2019). Evaluation of the interaction among microalgae *Spirulina* sp, plastics polyethylene terephthalate and polypropylene in freshwater environment. J. *Ecol. Eng*. 20(6), 161–173.

Kim, J.W., Park, S. B., Tran, Q.G., Cho, D.H., Choi, D.Y., Lee, Y.J., & Kim, H.S. (2020). Functional expression of polyethylene terephthalate-degrading enzyme (PETase) in green microalgae. *Microb. Cell Fact*. 19, 97.

Kumar, R.V., Kanna, G.R., & Elumalai, S. (2017). Biodegradation of polyethylene by green photosynthetic microalgae. *J. Bioremediat. Biodegrad*. 8(1), 1000381.

Lagarde, F., Olivier, O., Zanella, M., Daniel, P., Hiard, S., & Caruso, A. (2016). Microplastic interactions with freshwater microalgae: Hetero-aggregation and changes in plastic density appear strongly dependent on polymer type. *Environ. Pollut*. 215, 331–339.

Lee, H., Shim, W.J., & Kwon, J.H. (2014). Sorption capacity of plastic debris for hydrophobic organic chemicals. *Sci. Total Environ*. 470, 1545–1552.

Li, S., Wang, P., Zhang, C., Zhou, X., Yin, Z., Hu, T., Hu, D., Liu, C., & Zhu, L. (2020). Influence of polystyrene microplastics on the growth, photosynthetic efficiency and aggregation of freshwater microalgae *Chlamydomonas reinhardtii. Sci. Total Environ*. 714, 136767.

Lithner, D., Larsson, Å., & Dave, G. (2011). Environmental and health hazard ranking and assessment of plastic polymers based on chemical composition. *Sci. Total Environ*. 409(18), 3309–3324.

Liu, Z., Yu, P., Cai, M., Wu, D., Zhang, M., Chen, M., & Zhao, Y. (2019). Effects of microplastics on the innate immunity and intestinal microflora of juvenile *Eriocheir sinensis. Sci. Total Environ*. 685, 836–846.

Long, M., Moriceau, B., Gallinari, M., Lambert, C., Huvet, A., Raffray, J., & Soudant, P. (2015). Interactions between microplastics and phytoplankton aggregates: Impact on their respective fates. *Mar. Chem*. 175, 39–46.

Long, M., Paul-Pont, I., Hegaret, H., Moriceau, B., Lambert, C., Huvet, A., & Soudant, P. (2017). Interactions between polystyrene microplastics and marine phytoplankton lead to species-specific hetero-aggregation. *Environ. Pollut*. 228, 454–463.

Manzi, H.P., Zhang, M., & Salama, E.S. (2022). Extensive investigation and beyond the removal of micro-polyvinyl chloride by microalgae to promote environmental health. *Chemosphere*. 300, 134530.

McDevitt, J.P., Criddle, C.S., Morse, M., Hale, R.C., Bott, C.B., & Rochman, C.M. (2017). Addressing the issue of microplastics in the wake of the microbead-free waters act. A new standard can facilitate improved policy. *Environ. Sci. Technol*. 51(12), 6611–6617.

Piarulli, S., Scapinello, S., Comandini, P., Magnusson, K., Granberg, M., Wong, J.X., Sciutto, G., Prati, S., Mazzeo, R., Booth, A.M., & Airoldi, L. (2019). Microplastic in wild populations of the omnivorous crab *Carcinusaestuarii*: A review and a regional-scale test of extraction methods, including microfibres. *Environ. Pollut*. 251, 117–127.

Rowdhwal, S.S.S., & Chen, J. (2018). Toxic effects of di-2-ethylhexyl phthalate: an overview. *Biomed Res. Int.* 2018, Article ID 1750368.

Sanniyasi, E., Gopal, R.K., Gunasekar, D.K., & Raj, P.P. (2021). Biodegradation of low-density polyethylene (LDPE) sheet by microalga, *Uronemaafricanum Borge. Sci Rep.* 11, 17233.

Sharma, R., & Kaushik, H. (2021). Micro-plastics: An invisible danger to human health. *CGC Int. J. Contemp. Technol.* 3(2),1–5.

Song, C., Liu, Z., Wang, C., Li, S., & Kitamura, Y. (2020). Different interaction performance between microplastics and microalgae: The bio-elimination potential of *Chlorella* sp. L38 and *Phaeodactylumtricornutum* MASCC-0025. *Sci. Total Environ.* 723, 138146.

Sundbæk, K.B., Koch, I.D. W., Villaro, C.G., Rasmussen, N.S., Holdt, S.L., & Hartmann, N.B. (2018). Sorption of fluorescent polystyrene microplastic particles to edible seaweed *Fucus Vesiculosus. J. Appl. Phycol.* 30, 2923–2927.

Teuten, E.L., Saquing, J.M., Knappe, D.R., Barlaz, M.A., Jonsson, S., Björn, A., Rowland, S.J., Thompson, R.C., Galloway, T.S., Yamashita, R., Ochi, D., (2009). Transport and release of chemicals from plastics to the environment and to wildlife. *Philos. Trans. R. Soc. Lond., B, Biol. Sci.* 364(1526), 2027–2045.

Ustabasi, G.S., & Baysal, A. (2020). Bacterial interactions of microplastics extracted from toothpaste under controlled conditions and the influence of seawater. . 703,135024.

Thompson, R.C., Moore, C.J., Vom Saal, F.S., & Swan, S.H. (2009). Plastics, the environment and human health: current consensus and future trends. Philos. Trans. R. Soc. Lond., *B, Biol. Sci.* 364(1526), 2153–2166.

Tunali, M., Uzoefuna, E.N., Tunali, M.M., & Yenigun, O. (2020). Effect of microplastics and microplastic-metal combinations on growth and chlorophyll a concentration of *Chlorella vulgaris. Sci. Total Environ.* 743, 140479.

Vianello, A., Jensen, R.L., Liu, L., & Vollertsen, J. (2019). Simulating human exposure to indoor airborne microplastics using a breathing thermal manikin. *Sci. Rep.* 9(1), 1–11.36.

Wang, J., Tan, Z., Peng, J., Qiu, Q., & Li, M. (2016). The behaviors of microplastics in the marine environment. *Mar. Environ. Res.* 113, 7–17.

Zhang, C., Chen, X., Wang, J., Tan, L., (2017). Toxic effects of microplastic on marine microalgae *Skeletonemacostatum*: Interactions between microplastic and algae. *Environ. Pollut.* 220, 1282–1288.

2 Advanced Oxidation Processes for Degrading Microplastics in Aqueous Media

Wendy Quilumbaquin, Katherine Vaca-Escobar, and Patricio J. Espinoza-Montero

2.1 INTRODUCTION

Plastics are widely used in a variety of products owing to their versatility, ease of processing, and low cost. They are found in short-term items such as consumer goods packaging, as well as products that require extended outdoor durability and performance over decades (Jambeck et al., 2015; Liu et al., 2022; Zinke, 2020). The chemical composition of the base polymer, along with the effects of additives, impurities, and processing methods, ultimately determine the product's service life in any environment (Baruah et al., 2022; Eriksen et al., 2014).

Plastic pollution has received increasing global attention because of its presence in various species and environments. Marine and freshwater environments are the primary areas affected by plastic pollution, with approximately 322 metric tons of plastic waste discharged into the sea annually. By 2050, it is estimated that this figure will reach 33 billion tons. As plastic degrades, it forms microplastics (MPs), a process exacerbated by mechanical wear from weathering. This raises concerns regarding MPs' ecotoxicological risks to aquatic environments (Hu et al., 2021). MPs are defined as plastic particles smaller than 5 mm, characterized by high molecular weight and hydrophobicity. They can exist in the environment for extended periods and have the capacity to adsorb heavy metals and organic pollutants, such as polychlorinated biphenyls and polycyclic aromatic hydrocarbons. This adsorption ability makes MPs a pathway for transporting such contaminants (Shen et al., 2022).

Consequently, MPs can induce diverse toxic effects and damage to aquatic organisms, primarily through bioaccumulation in the digestive system; plastic additives and other adsorbed toxic compounds act as endocrine disruptors, affecting metabolic processes and causing reproductive and DNA damage (Pizzichetti et al., 2021). For humans, the primary source of contamination is through the ingestion of contaminated food (from trophic chains) or drinking water containing MPs.

DOI: 10.1201/9781003379386-2

The accumulation of MPs in the body can lead to adverse effects such as oxidative stress, metabolic impairment, and disruption of immune function (Koelmans et al., 2022).

Traditional treatments for addressing plastic pollution include filtration, sedimentation, and bioremediation, among others (Hu et al., 2021). However, these methods do not contribute to the degradation or fragmentation of MPs; they merely transfer the pollutants from one medium to another. Therefore, it is imperative to apply new environmentally friendly, cost-effective, and highly efficient technologies to remove these pollutants (Eriksen et al., 2014).

Currently, advanced oxidation processes (AOPs) are appealing technologies for the effective degradation of persistent organic pollutants in wastewater (Jiang et al., 2021). AOPs are characterized by their generation of highly reactive oxidizing species (HROS), including reactive oxygen species (ROS), capable of mineralizing recalcitrant compounds, including MPs. Depending on the electrolytic medium, different HROS can be generated, such as hydroxyl ($^{\bullet}OH$), superoxide $\left(O_2^{\bullet-}\right)$, hydroperoxyl $\left(HO_2^{\bullet}\right)$, alkoxyl ($RO^{\bullet}$), sulfate $\left(SO_4^{\bullet-}\right)$, and chlorine ($Cl^{\bullet}$) radicals (Xu et al., 2021).

This chapter presents recent advances in the degradation of MPs by AOPs and discusses the advantages and disadvantages of each technology. Initially, the sources and types of MPs, their properties, and their feasibility for degradation by different treatments are briefly reviewed. Subsequently, the characteristics that inform the selection of appropriate technologies are discussed. Additionally, sampling methods are outlined, which are crucial to future research on the identification and quantification of environmental MPs prior to remediation. The effectiveness of each method is also explained. Finally, perspectives and guidelines for future studies on MP degradation are provided, emphasizing the goal of using AOPs to clean up highly polluted waters rather than simply shifting the pollutant phase, as occurs with physical processes. From a circular economy perspective, transforming MPs into valuable products is essential.

2.2 ADVANCED OXIDATION PROCESSES

2.2.1 General Fundamentals

AOPs were first proposed in the 1980s and involve the generation and use of highly oxidizing ROS, primarily the hydroxyl radical ($^{\bullet}OH$) (Hodges et al., 2018). $^{\bullet}OH$, is the second strongest oxidant in nature after fluorine, has a redox potential of 2.8 V *vs.* NHE (normal hydrogen electrode) and a very short lifetime constituting nanoseconds. In addition to $^{\bullet}OH$, other highly oxidizing radicals can be generated from the oxidation of electrolytes, such as sulfate radicals when the electrolyte is sulfate, resulting in a synergistic effect on the degradation of organic matter. Table 2.1 presents various oxidizing species involved in AOPs (Luo et al., 2021).

Depending on the target pollutant, total oxidation products include water, carbon dioxide, and inorganic salts; however, total mineralization is not always achieved, resulting in small biodegradable and non-hazardous molecules such as formic acid and acetic acid (Shen et al., 2022). Therefore, AOPs can be applied as a stand-alone

TABLE 2.1
Redox potentials relative to the normal hydrogen electrode (NHE)

Redox potentials	Species E^0 (V, 25 °C)
Fluorine	3.03
Hydroxyl radical	2.80
Sulphate radical	2.5–3.1
Ozone	2.07
Hydrogen peroxide	1.78
Perhydroxyl radical	1.70
Permanganate	1.68

Source: Feilizadeh et al., 2023.

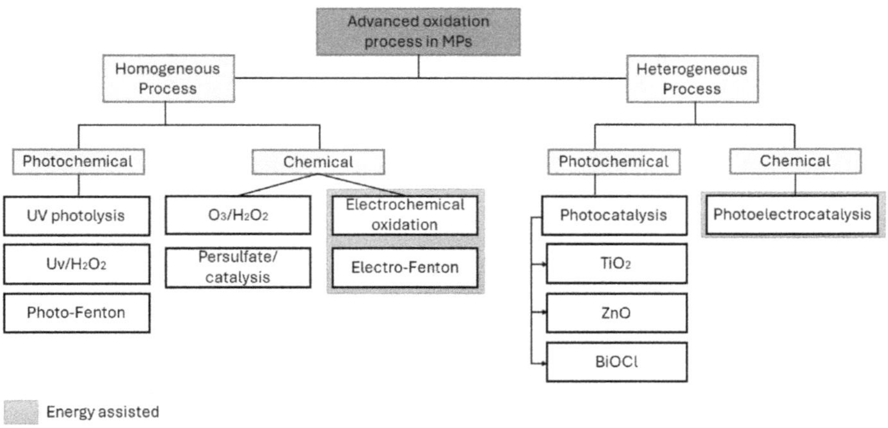

FIGURE 2.1 Classification of various advanced oxidation processes used for microplastic degradation.

water treatment technology or in combination with other techniques, depending on the contaminant concentration level (< 5 mg/L) (Amor et al., 2019; Luo et al., 2021). AOPs can be homogeneous or heterogeneous, as illustrated in Figure 2.1. In both cases, light (natural or artificial) can be applied to enhance the production of hydroxyl radicals and improve degradation (Cai et al., 2019).

2.2.2 Application of Advanced Oxidation Process Treatments for Microplastics

According to recent literature, various AOPs have been employed to degrade MPs in aqueous media, such as UV/H_2O_2, ozone, UV/catalysts, heat/persulfate, electro-Fenton, electrochemical oxidation, and photoelectrocatalysis. The primary objective of these AOPs is to generate $^{\bullet}OH$; however, in cases where H_2O_2, O_3, or other oxidants

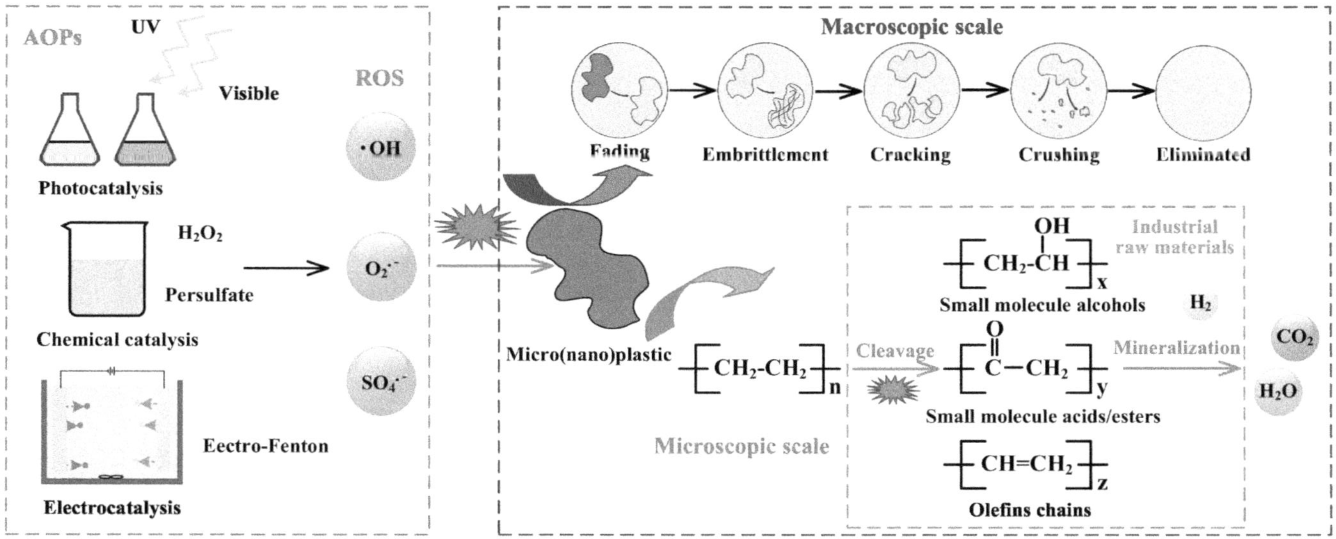

FIGURE 2.2 Mechanisms for eliminating microplastics via advanced oxidation processes (AOPs) with representations at the macro and microscopic scales (Shen et al., 2022).

are used, chemical oxidation can occur, which, in the presence of light, enhances oxidizing power by generating $^{\bullet}OH$ and other HROS such as superoxide ion radicals ($O_2^{\bullet-}$) and their conjugate acid form, the hydroperoxyl radical (Lu et al., 2022).

Generally, MP degradation via AOPs begins with the attack of the $^{\bullet}OH$ on the surface of the MPs, forming (R–$^{\bullet}OH$) (**Reaction 1**) and continues with degradation-fragmentation until complete destruction and mineralization are achieved (**Reaction 2**) (Feilizadeh et al., 2023).

$$\text{Organic pollutants} + \text{oxidizing} \rightarrow \text{oxidation intermediates} \quad (1)$$

$$\text{Intermediates} \rightarrow H_2O + CO_2 + \text{other minerals} \quad (2)$$

A more detailed representation of the MP degradation process is shown in Figure 2.2, encompassing direct photocatalysis, chemical catalysis, and electrocatalysis. Although AOPs can effectively degrade MPs, the process can be time-consuming and costly. To improve efficiency, AOPs can be combined with other treatment methods such as adsorption and biodegradation (Kim et al., 2020).

2.2.3 Degradation of Microplastics via Homogeneous Advanced Oxidation Processes

2.2.3.1 UV Photolysis and UV/H_2O_2

Photolysis, or the photodegradation of polymers, serves as the foundation for all photo-driven AOPs, as it involves physical and chemical changes in polymer structure caused by light (Zhu et al., 2020). Many organic polymers are susceptible to visible, particularly (ultraviolet) UV irradiation, as UV light has higher energy levels (Fechine et al., 2002). One widely used AOP involves H_2O_2 in the presence of UV radiation (H_2O_2/UV), in which light induces the generation of $^{\bullet}OH$ according to **Reactions 3 and 4**. This process accelerates and enhances the degradation of organic matter, as $^{\bullet}OH$ is the second strongest oxidant in nature (C. Wu & Linden, 2008).

$$H_2O_2 + h\nu \rightarrow 2^{\bullet}OH \quad (3)$$

$$HO_2^- + h\nu \rightarrow \cdot O + \cdot OH \quad (4)$$

Photolysis is primarily used as a pretreatment prior to MPs biodegradation to reduce the hydrophobicity of the MPs surface. Since most bacteria are hydrophilic, this treatment is necessary to facilitate biodegradation. Photolysis promotes surface oxidation, chain scission, and the formation of several polar functional groups on the surface of the MPs, thereby decreasing their average molecular weight and hydrophobicity (Suhrhoff & Scholz-Böttcher, 2016).

Complete decomposition of MPs by photolysis alone can be time-consuming, making it necessary to combine methods such as photolysis and biological treatments to accelerate MP degradation. The use of H_2O_2/UV has been found to be a more

TABLE 2.2
UV photolysis and UV/H_2O_2 of different types of MPs in aqueous media

Technology	MPs type	Reaction time	Evaluation method	Efficiency	Reference
UV photolysis	PET, LDPE, PS, and PVC	57 days	GC-FID/MS	PET: dibutyl phthalate leached LDPE: 4.9% PS: 0.07% PVC: 53 mg bisphenol A leached/g PVC	Suhrhoff & Scholz-Böttcher (2016)
UV and visible photolysis	PS	56 days	Nanoparticle tracking analysis	N/A	Lambert & Wagner (2016)
UV photolysis	PE, PS, PVC, and PET	50 h	ATR-FTIR, SEM, XPS and GC-MS	Drastic decrease in contact angle (> 20°)	Lin et al. (2020)
UV/H_2O_2 UV/Cl_2 UV/water	PS	5 days	ATR-FTIR DSC	Increased crack formation, crystallinity, CI index, average molecular weight.	X. Liu et al. (2021)
Solar photolysis	PS	60 days 150 days	ATR-FTIR, SEM, XPS and gel permeation chromatography, EPR spectrometry	Average size decreased to 93 µm (60 days), and 52 µm (150 days)	Zhu et al. (2020)

Note: PS: polystyrene; PP: polypropylene; LDPE: low-density polyethylene; PE: polyethylene; PVC: polyvinyl chloride; PET: polyethylene terephthalate; N/A: not applicable; MP: microplastic; GC-FID/MS: Gas chromatograph-flame ionization/mass spectrometer detector; ATR-FTIR: attenuated total reflectance- Fourier transform infrared; SEM: Scanning electron microscopy; XPS: X-Ray photoelectron spectroscopy; DSC: differential scanning calorimetry; EPR: electron paramagnetic resonance.

powerful method for MP degradation, as summarized in Table 2.2. It is important to highlight that this topic is relatively new, and there are few reports available at present.

MPs degradation is normally assessed by examining changes in surface morphology, brittleness, the carbonyl index (CI), crystallinity, glass transition temperatures, hydrophobicity, and average molecular weight (X. Liu et al., 2021). Table 2.2 summarizes the most representative studies on the degradation of MPs by photolysis in aqueous media under different conditions.

2.2.3.2 Ozone Oxidation

Ozone (O_3) is a strong oxidant (2.07 V *vs.* NHE) used for water decontamination processes through direct chemical oxidation. O_3 can easily react with a variety of polymeric materials, particularly polymers with unsaturated bonds and aromatic rings of polymers (Hidayaturrahman & Lee, 2019). In addition, O_3can induce the formation of free radicals (mainly $^{\bullet}OH$) when the treatment is performed in an alkaline medium, as shown in **Reaction 5** (Hu et al., 2021). O_3has been used to degrade MPs and oxidize their surfaces, thereby improving the biodegradation efficiency of polymers (Derco et al., 2015).

$$O_3 + H_2O \rightarrow \frac{1}{2}O_2 + HO_2^{\bullet} + {}^{\bullet}OH \tag{5}$$

The combination of ozonation and UV irradiation, along with the addition of chemicals like H_2O_2, represents one of the most advanced treatment processes applied in water treatment plants. Adding H_2O_2 to the water in UV systems enhances the breakdown kinetics of target contaminants. When OH- radicals are not dominant, H_2O_2 can act as an extra oxidant (Derco et al., 2015; Kurian, 2021). Under UV irradiation, H_2O_2 can be transformed into $^{\bullet}OH$, accelerating the degradation of contaminants in water (Hidayaturrahman & Lee, 2019). The peroxone process is a specific application of ozonation with the addition of H_2O_2. In this process, H_2O_2 is added to the water treatment system, generating the necessary $^{\bullet}OH$ radicals. The combination of $O_3$and H_2O_2 in the peroxone process further enhances oxidation capabilities and improves overall treatment efficiency (Bilińska et al., 2016).

These AOPs, such as ozonation with the addition of H_2O_2 and UV irradiation (**Reactions 6–7**), are effective in removing a wide range of contaminants from water, including organic compounds, pesticides, and pharmaceuticals. They offer advantages in terms of their ability to degrade complex and persistent contaminants, as well as their potential for disinfection (Bilińska et al., 2016).

$$O_3 + H_2O + h\nu \rightarrow H_2O_2 + O_2 \tag{6}$$

$$2O_3 + H_2O_2 \rightarrow 3O_2 + 2\,{}^{\bullet}OH \tag{7}$$

Although ozonation is an effective process for the degradation or removal of MPs, it has certain disadvantages. First, O_3 has an active half-life in water of approximately 25 minutes. Second, O_3 cannot be transported, and therefore ozonation must be produced in situ. Third, the equipment used for ozonation can be corroded by nitric acid or nitric oxide, which may be produced during the process. Fourth, special measures must be taken when working with O_3 to avoid exposure. Finally, the oxidizing effects of O_3 can affect the eyes and lungs (Bui et al., 2016; Derco et al., 2015).

2.2.3.3 Electrochemical Oxidation

Electrochemical oxidation is defined as the heterogeneous catalysis of an electrochemical reaction occurring at the electrode–electrolyte interface, where the electrode functions as both an electron donor/acceptor and a catalyst (C. Zhou et al., 2019). An electrode acts as a catalyst for these reactions by either providing sites for the adsorption of intermediates (internal transfer mechanism) or acting as an electron carrier that transfers charge by tunneling to species at the interface (external transfer mechanism) (Sen et al., 2022). Electrochemical reactions occur at the same overpotential value but at distinct rates depending on the electrode material. This variation in reaction rates is due to catalysis, the difference in the work function of the material, and the interaction energy between the electrode and the compound to be transformed. This is quantified by the heterogeneous electronic transfer constant (k°) (Castillo-Cabrera et al., 2024; Martínez-Huitle et al., 2023; Ochoa-Chavez et al., 2024).

In recent decades, electrocatalysis has grown exponentially owing to its wide range of applications, including in the environmental field. For example, electrochemical technologies are considered chemically green or environmentally friendly, as the main reagent is the electron (Oturan et al., 2015). Consequently, various reports have emphasized the elimination of organic and inorganic pollutants in wastewater and drinking water using electrochemical techniques (Cisneros-León et al., 2023; Rajeshwar et al., 1994). In addition, oxidizing species key to pollutant treatment can be electrogenerated without adding an oxidant to the effluent under treatment. The main electrochemical technologies used for wastewater remediation include electrocoagulation, electroflotation, electrodialysis, electroreduction, electrochemical oxidation, and advanced electrochemical oxidation (AEO) (L. Zhang et al., 2020).

In AEO, oxidation occurs both directly and indirectly. Indirect processes commonly generate highly oxidizing species derived from the oxidation of the medium, such as $^{\bullet}OH$ or by the oxidation of a precursor present in the solution being treated (e.g., $SO_4^{2-} \rightarrow S_2O_8^{2-}$, etc.) (Sen et al., 2022). While $^{\bullet}OH$ can be considered a direct process (vide infra), owing to its high reactivity, the oxidations it induces on organic molecules occur within the interface, and thus this radical does not react within the solution like other highly oxidizing species (Priyadarshini et al., 2022). Moreover, it has a half-life of approximately 1×10^{-9} s and is generally physiosorbed on the electrode surface(Rajeshwar et al., 1994). However, other highly oxidizing species, such as O_3, H_2O_2, and $S_2O_8^{2-}$, can diffuse and react in the solution, constituting an indirect oxidation process(Drogui et al., 2007).Thus, the electrocatalytic process can be categorized as direct or indirect, both of which rely on the situ production of hydroxyl radicals (Figure 2.3).

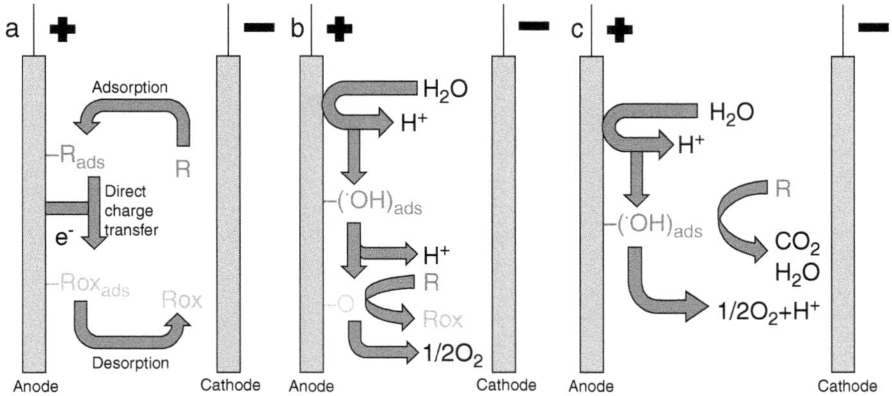

FIGURE 2.3 (a) Direct electrochemical oxidation and the mechanism of active oxygen species on the surface of the (b) active anode and (c) inactive anode in the electrochemical oxidation system (Garcia-Segura et al., 2018).

Both direct and indirect anodic oxidation, methods of electrolysis, are used for MPs degradation and fragmentation (Kiendrebeogo et al., 2021). This approach allows organic contaminants to be oxidized at specific potentials that are more negative than that of the oxygen evolution reaction (Oturan et al., 2015). These electrochemical processes offer possible solutions for the removal and treatment of MPs in water systems (Ganiyu et al., 2021). Table 2.3 summarizes the most representative studies on the electrochemical degradation of MPs in aqueous media that applied semiconductors prepared by various synthesis methods.

2.2.3.4 Persulfate Treatment

AOPs based on sulfate radicals ($E^0 = 3.1$ V), a powerful oxidant, have also been shown to be effective for the degradation of MPs. These radicals are generally obtained through the physical activation of persulfate ($S_2O_8^{2-}$) **(Reaction 8)** and peroxymonosulfate (HSO_5^-) **(Reaction 9)** via heat, photo-irradiation, or ultrasound techniques (Lee et al., 2021; Moreno-Andrés et al., 2020).

$$S_2O_8^{2-} \rightarrow 2SO_4^{\bullet-} \tag{8}$$

$$HSO_5^- \rightarrow SO_4^{\bullet-} + {}^{\bullet}OH \tag{9}$$

Another method to activate persulfate is through chemical activation with addition of transition metals, which can produce sufficient $SO_4^{\bullet-}$ **(Reactions 10–11)** (Y. Zhang et al., 2020).

$$S_2O_8^{2-} + Fe^{2+} \rightarrow SO_4^{\bullet-} + SO_4^{2-} + Fe^{3+} \tag{10}$$

$$HSO_5^- + Fe^{2+} \rightarrow SO_4^{\bullet-} + OH^{\bullet} + Fe^{3+} \tag{11}$$

TABLE 2.3
Representative studies on electrochemical oxidation of different MPs in aqueous media

Electrochemical Oxidation	MP type	Reaction Time	Evaluation method	Effciency	Reference
Direct	PS	6 h	Weight loss and size distribution	89% degraded	Kiendrebeogo et al. (2021)
Indirect	-	6 h	Carbonyl index	79.7% degraded	Durán et al. (2018)

Note: MP: microplastic; PS: Polystyrene.

Persulfate treatments offer several benefits over $^{\bullet}OH$ formation procedures. First, $SO_4^{\bullet -}$ has a substantially longer lifetime than $^{\bullet}OH$ and hence has a greater chance of reacting with MPs. Furthermore, its oxidizing ability is much less affected by pH fluctuations compared to that of $^{\bullet}OH$ (D. Zhou et al., 2015). Table 2.4 presents treatments using persulfate under different conditions.

2.2.3.5 Photo-Fenton Process

The use of UV or visible light in conjunction with the traditional Fenton method can improve the catalytic capacity of the catalyst, boost the degradation efficiency of the organic pollutant, and minimize iron sludge generation. The photo-Fenton method is designed to expedite the reduction of Fe^{3+} to Fe^{2+} (Figure 2.4) by utilizing the energy provided by light (Babuponnusami & Muthukumar, 2014; Kalal et al., 2014).

$$\left[Fe(OH)\right]^{2+} + h\nu \rightarrow Fe^{2+} + {}^{\bullet}OH \tag{12}$$

$$H_2O_2 + h\nu \rightarrow 2\,{}^{\bullet}OH \tag{13}$$

The reaction mechanism for the photo-Fenton reaction process begins with Fe^{2+} reacting with H_2O_2 to generate Fe^{3+} or $\left[Fe(OH)\right]^{2+}$, which, under light irradiation, produces $^{\bullet}OH$ that degrades MPs (**Reactions 12–13**) (Babuponnusami & Muthukumar, 2014). Table 2.5 presents applications of the photo-Fenton process.

2.2.3.6 Electro-Fenton Process

The electro-Fenton (EF) technique involves electrochemically producing H_2O_2, which then reacts with Fe^{2+} to generate radical species capable of degrading MPs (Ahmed et al., 2022). The EF approach can also be utilized to completely degrade and mineralize persistent organic pollutants such as personal care products, pharmaceuticals, landfill leachate, and pesticides. A primary limitation of this method is its requirement for an acidic medium (pH = 2.8–3.5), which can harm the environment (Ganiyu et al., 2021). To address this issue, heterogeneous EF processes involving solid catalysts have been proposed (Miao et al., 2020).

$$O_2 + 2e^- + 2H^+ \rightarrow H_2O_2 \tag{14}$$

$$Fe^{2+} + H_2O_2 + H^+ \rightarrow Fe^{3+} + H_2O + {}^{\bullet}OH \tag{15}$$

$$Fe^{3+} + e^- \rightarrow Fe^{2+} \tag{16}$$

Catalysts can substantially broaden the pH range within which the EF reaction can occur. Indeed, the EF process is a popular subgroup of electrochemical AOPs, although few studies have comprehensively examined factors influencing the reaction, the synthesis of intermediate products and their toxicity, and the formation of

TABLE 2.4
Persulfate treatments of different MPs in aqueous media

Treatment	MP type	Reaction time	Evaluation method	Efficiency	Reference
Persulfate treatment and Fenton process	PS and HDPE	10 days	FTIR	-	Y. Zhang et al. (2020)
Persulfate treatment	PS y PVC	30 days	-Contact angle -FTIR	Initial: 12.20° Final: 0.13°	P. Liu et al. (2019)

Note: PS: polystyrene; HDPE: high-density polyethylene; PVC: polyvinyl chloride; FTIR: Fourier transform infrared.

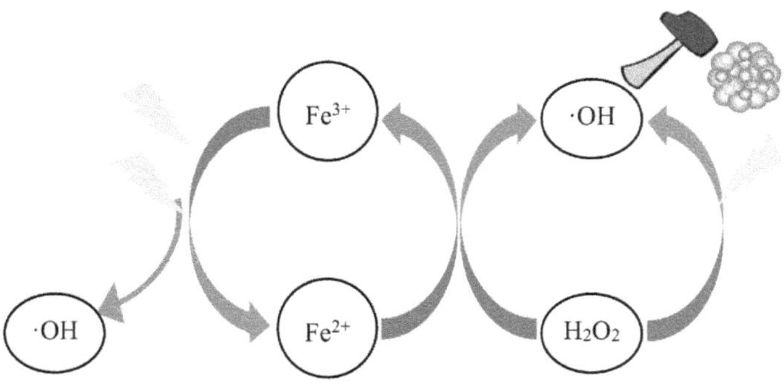

FIGURE 2.4 Reaction mechanism for the photo-Fenton process (M. Zhang et al., 2019).

TABLE 2.5
Applications of the photo-Fenton process to degrade different MPs in aqueous media

Photo-Fenton	MP type	Reaction time	Evaluation method	Efficiency	Reference
Fe^{2+}/H_2O_2	PS	250 min	Size distribution	99 %	Feng et al. (2011)
TiO2-GOx	PE	6 h	FTIR SEM EDX	-	Ghatge et al. (2022)

Note: PS: polystyrene; PE: polyethylene; MP: microplastic; FTIR: Fourier transform infrared; SEM: Scanning electron microscopy; EDX: energy dispersive X-ray.

secondary MPs (Y. Liu et al., 2021). A primary objective of this approach is to identify effective catalysts capable of enhancing H_2O_2 production (**Reactions 14–16**) (Li et al., 2022). Table 2.6 summarizes studies on the EF process, demonstrating its potential for degrading MPs and achieving up to 75% dichlorination and 56% weight loss using ZnO NRs/SnOx (Miao et al., 2020). Nonetheless, further research is needed to optimize the process conditions, understand its mechanisms, and assess feasibility on a larger scale.

2.3.1 Degradation of Microplastics via Heterogeneous Advanced Oxidation Processes

2.3.1.1 Photocatalysis

Heterogeneous photocatalysis involves the absorption of light by the crystal lattice of a photocatalyst dispersed in an aqueous medium (where light energy > bandgap), resulting in the generation of electron-hole pairs (e^-_{CB} / h^+_{VB}) within the semiconductor (CB = conduction band, VB = valence band). Subsequently, oxidation reactions mediated by the generated h^+_{VB} occur. h^+_{VB} is expected to possess sufficient energy

TABLE 2.6
Studies on the electro-Fenton process to degrade different MPs in aqueous media

Electro-Fenton process	MP type	Reaction time	Evaluation method	Efficiency	Reference
Fe^{2+}/H_2O_2 or PDS	PS and PE MPs	30 days	Size distribution	Percentage of small MPs increased	Y. Liu et al. (2018)
Mn@NCNTs/ PMS	PE from cosmetic products	8 h	Weight loss	44%	Kang et al. (2019)
ZnO NRs/ SnOx/ Fe^0 nanoparticles under visible-light lamp	PP and PVC	6 h	Weight loss	75% dichlorination and 56% weight loss of MPs	Miao et al. (2020)

Note: PS: polystyrene; PP: polypropylene; PE: polyethylene; PVC: polyvinyl chloride; MP: microplastic; PDS: peroxydisulfate; Mn@NCNTs/PMS: manganese carbide nanoparticles/ peroxymonosulfate; NRs: nanorods.

to either oxidize water to hydroxyl radicals ($^{\bullet}OH$) or directly oxidize organic matter (Ameta et al., 2018; Priyadarshini et al., 2022).

In part (b) inside a circle is the band gap with the transport of electrons that goes from the valence band to the conduction band, on the outside reduction reactions on the left side with O_2 and the polymer chain in light conditions that are represented with yellow lines; On the right side, the Fe^{3+} 3d defect states oxidation reactions with $H_2O/^{\bullet}OH$ and $^{\bullet}OH$ with the product of the polymer chain (CO_2+H_2O).

Figure 2.5 depicts the mechanism of photocatalysis, illustrating how the photocatalyst absorbs light energy in the valence band, where electrons are excited to the conduction band (e^-_{CB}), leaving behind holes in the valence band$\left(h^+_{VB}\right)$. These holes then react with the organic matter to initiate the oxidization process (C. Zhou et al., 2018). The importance of photocatalysis lies in its ability to act as an AOP capable of degrading pollutants in aqueous media through the generation of active species ($^{\bullet}OH$, $O_2^{\bullet -}$, and holes (h^+)) under light irradiation (UV, sunlight) (Zhou et al., 2018).

Furthermore, semiconductor photocatalyst technology promotes high degradation efficiency while employing nanoscale metal oxides and metal oxide complexes, with TiO_2, BiOCl, and ZnO demonstrating the best photocatalytic performance under UV light (λ< 387 nm). Various important active species are formed during heterogeneous photocatalysis (L. Cai et al., 2018; Ebrahimbabaie et al., 2022; Wang et al., 2020).

The difference in energy between the valence (HOMO) and conduction bands (LUMO) is known as the bandgap (E_g) (see Table 2.7).

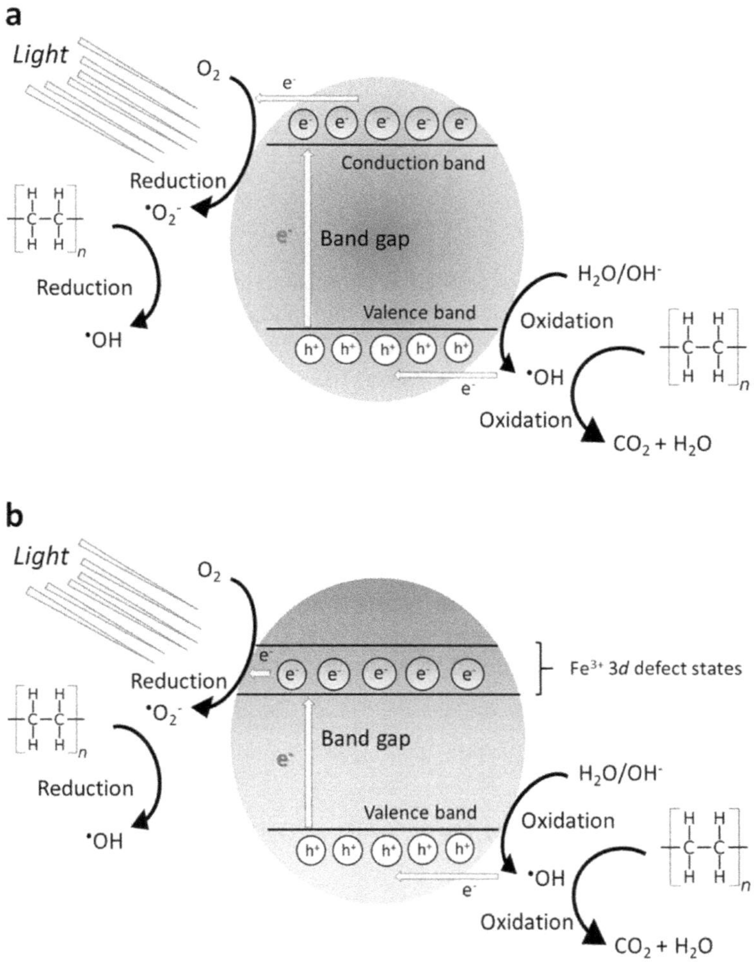

FIGURE 2.5 Mode of action of a photocatalyst for polyethylene chain decomposition: (a) single catalyst such as ZnO and (b) Fe-ZnO photocatalyst (Ebrahimbabaie et al., 2022).

TABLE 2.7
Bandgap energy of different photocatalysts and wavelengths

Semiconductor E_g< 1.5–3.0 eV	Bandgap energy (eV)	Wavelength (nm)
$TiO_{2\ (anatase)}$	3.2	413
$TiO_{2\ (rutile)}$	3.1	388
ZnO	3.37	388
BiOCl	3.55	310

presents the bandgap energy of different semiconductors and the corresponding wavelengths required for their activation under light irradiation. The following subsection discusses the potential of different photocatalysts for MP removal (Ameta et al., 2018).

2.3.1.1.1 Semiconductors

- **TiO_2**

The semiconductor TiO_2, when applied in photocatalytic MP degradation in aqueous systems, is a highly effective option given its high photostability and oxidative, hydrophilic, and non-toxic properties. Depending on its crystalline structure, TiO_2 can exist in anatase, rutile, or brookite forms; however, TiO_2–anatase (3.2 eV) is commonly used as an oxidative photocatalyst and is well-known for its ability to catalyze the degradation of organic compounds under UV irradiation (Denisse Vital-Grappin et al., 2021).

The mechanics of TiO_2–MWCNTs-PE, as described by An et al. (2014), begin with the absorption of UV light by semiconducting particles, which generates mobile electrons and holes in the conduction and valance bands (**Reaction 17**). This is followed by the creation of an electron–hole pair, initiating a sequence of events (as shown in **Reactions 18–22**) that produces hydroxyl radicals with a high redox potential. These active oxygen species then attack nearby polymer chains, starting the breakdown process (**Reaction 23**). As the ROS penetrate the polymer matrix, the polymer ultimately degrades.

$$TiO_2 + hv \rightarrow e^- + h^+ \tag{17}$$

$$H_2O + h^+ \rightarrow H^+ + {}^{\bullet}OH \tag{18}$$

$$O_2 + e^- \rightarrow O_2^{\bullet -} \tag{19}$$

$$O_2 + H^+ \rightarrow HOO^{\bullet} \tag{20}$$

$$2HOO^{\bullet} \rightarrow H_2O_2 + O_2 \tag{21}$$

$$H_2O_2 + e^- \rightarrow OH^- + {}^{\bullet}OH \tag{22}$$

$$-\left(CH_2CH_2\right)- + {}^{\bullet}OH \rightarrow \text{Degradation products} \tag{23}$$

- **ZnO**

Zinc oxide nanoparticles (ZnO-NPs) are one of the most frequently used engineered nanomaterials in consumer and industrial products and possess exceptional physicochemical qualities (Wigger et al., 2020). Widely produced and extensively applied, ZnO-NPs can be discharged into and accumulated in a variety of aquatic environments.

Additionally, given its high redox potential, ease of synthesis, non-toxicity, and high electron mobility, ZnO is a promising photocatalyst for MPs degradation (Ahmed et al., 2022).

The mechanics described by Tong et al. (2022) indicate that MPs can considerably increase Zn^{2+} release by promoting both h^+–dependent ZnO oxidative dissolution and proton-dependent ZnO dissolution reactions (**Reactions 24–28**).

$$ZnO + 2H^+ \rightarrow Zn^{2+} + H_2O \tag{24}$$

$$2ZnO + 4H^+ \rightarrow 2Zn^{2+} + O_2 \tag{25}$$

$$e^- + MPs \rightarrow MPs^- + O_2 \rightarrow O_2^{\bullet -} \tag{26}$$

$$O_2^{\bullet -} + h^+ \rightarrow O_2 \tag{27}$$

$$OH^- + h^+ \rightarrow {}^{\bullet}OH \tag{28}$$

- **Bismuth-based Oxides**

BiOCl is p-type semiconductor known for its chemical stability and capacity to adsorb both visible and UV light, with a wide bandgap of 3.55 eV (Castillo-Cabrera et al., 2022). The abundant surface hydroxyl groups of BiOCl in water contribute to its strong photocatalytic activity for MP degradation. This is because the OH^- species benefit from photoinduced oxygen vacancies and can regenerate in a water-rich atmosphere (S. Wu et al., 2017).

The general mechanics described by Jiang et al. (2021) are summarized in **Reactions 29–33**. Initially, the process begins with the reaction of holes and free radicals. However, the photocatalytic performance is slightly reduced by the trapping of holes, indicating that holes are not the primary free radicals driving the degradation process.

$$BiOCl + h\nu \rightarrow e^- + h^+ \tag{29}$$

$$e^- + O_2 \rightarrow O_2^{\bullet -} \tag{30}$$

$$h^+ + OH^- \rightarrow {}^{\bullet}OH \tag{31}$$

$$BiOCl - OH^- + H_2O \rightarrow BiOCl - OH^- + H^+ + {}^{\bullet}OH \tag{32}$$

$$O_2^{\bullet -} + {}^{\bullet}OH \rightarrow (MPs) + H_2O + CO_2 \tag{33}$$

Table 2.8 summarizes the most representative studies on the photocatalysis degradation of MPs in aqueous media, using semiconductors prepared by various synthesis methods. It includes the evaluation methods used and their respective efficiencies.

TABLE 2.8
Photocatalytic degradation of different MPs in aqueous media

Semiconductor	MP type	Reaction time	Evaluation method	Efficiency	Reference
TiO_2 and ZnO	PVC and PVLC/PVC copolymer	10–30 days	Pseudo first-order kinetics of Cl– production	TiO_2: 0.4×10^{-4}/h ZnO: 3.6×10^{-4}/h	Hidaka et al. (1996)
ZnO-Pt	LDPE film fragments	175 h	CI and VI	33.9% increase in CI; 34.5% increase in VI	Tofa et al. (2019)
C,N TiO_2	HDPE MPs from cosmetic products	50 h	Weight loss	71.77 ± 1.88%	Denisse Vital-Grappin et al. (2021)
ZnO nanorods	PP	2 weeks	Volume reduction	65%	Uheida et al. (2021)
BiOCl	PE	24 h	Weight loss	5.38%	Jiang et al. (2021)
TiO_2/Pt	PE and PVC	10 h	H_2 production efficiency	PE: 113 and 620 µmol H $gsub^{-1}$ in a neutral and 5 M NaOH environment, respectively; PVC: 133 and 300 µmol H_2 $gsub^{-1}$ in a neutral and 5 M NaOH environment, respectively	Kawai and Sakata (1981)
CdS/CdOx quantum dots	PE, PVC, PLA, PET, and PUR (10 M NaOH pretreated)	22 h	H_2 production efficiency	PLA: 3.09 ± 0.15 mmol H_2 $gsub^{-1}$ PET: 0.21 ± 0.04 mmolH_2 gsub–1 PUR: 0.04 ± 0.01 mmol H_2 $gsub^{-1}$	Uekert et al. (2018)
CNx-Ni_2P	PET, PLA, and polyester microfiber (2 M KOH pretreated)	50 h	H_2 production efficiency	82.5 ± 7.3 and 178 ± 12 µmol H_2 $gsub^{-1}$ from PET and PLA, respectively.	Uekert et al. (2019)
Nb_2O_5 atomic layers	PE, PP, and PVC	40–90 h	Carbon conversion rate	PE: 100% in 40 h PP: 100% in 60 h PVC: 100% in 90 h	

Note: PS: polystyrene; PP: polypropylene; HDPE: high-density polyethylene; PE: polyethylene; PVC: polyvinyl chloride; PVLC: polyvinylidene chloride; PET: polyethylene terephthalate; PLA: polylactic acid; PUR: polyurethane; MP: microplastic.

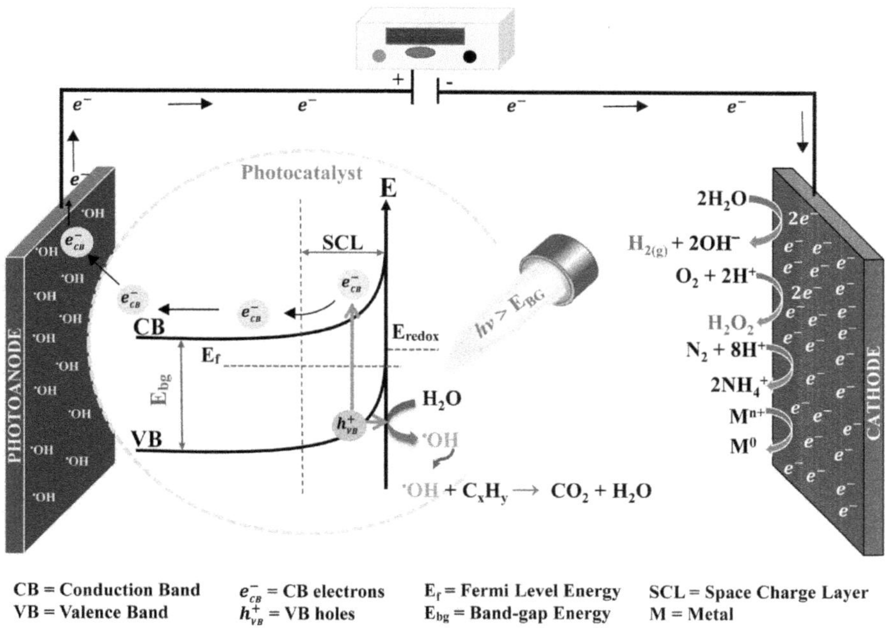

FIGURE 2.6 Photoelectrocatalytic mechanism for the degradation of organic pollutants in aqueous media (Alulema-Pullupaxi, Espinoza-Montero, et al., 2021).

2.3.1.2 Photoelectrocatalysis

Among AOPs, photoelectrocatalysis (PEC) is recognized for its environmental friendliness and efficiency. It utilizes photons and electrical energy to generate powerful oxidants by combining heterogeneous photocatalysis and positive polarization to the photoanode (Amelia et al., 2021; Espinoza-Montero et al., 2022). The PEC mechanism, as shown in Figure 2.6, involves applying an external anodic polarization potential to a semiconductor film (photocatalyst) supported on a conductive substrate (electrode), which enhances the photocatalytic process (Collivignarelli et al., 2020; Rajput et al., 2021) .When light with energy higher than that of the forbidden band ($h\nu > E_g$) is incident, it promotes the transport of electrons from the valence band ($\mathbf{e}^-_{\mathbf{Bv}}$) to the conduction band $\left(\mathbf{e}^-_{\mathbf{CB}}\right)$, thus generating electron-hole pairs ($\mathbf{e}^-_{\mathbf{CB}}/\mathbf{h}^+_{\mathbf{VB}}$) (Alulema-Pullupaxi, Fernández, et al., 2021). With the anodic potential applied, the recombination of the photogenerated e^-_{CB}/h^+_{VB} is minimized due to electrical polarization, facilitating the mobilization of electrons toward the external circuit. This increases the availability of h^+_{VB} and promotes the formation of •OH through the oxidation of water (Castillo-Cabrera et al., 2022). However, the effectiveness of PEC in degrading organic compounds depends on the photoanode's ability to generate ROS (Zarei & Ojani, 2017).

The application of PEC for the degradation of MPs is relatively recent. One study reported generating value-added compounds using solar energy and biocatalytic artificial photosynthesis, which combines photocatalysis with redox biocatalysis (J.

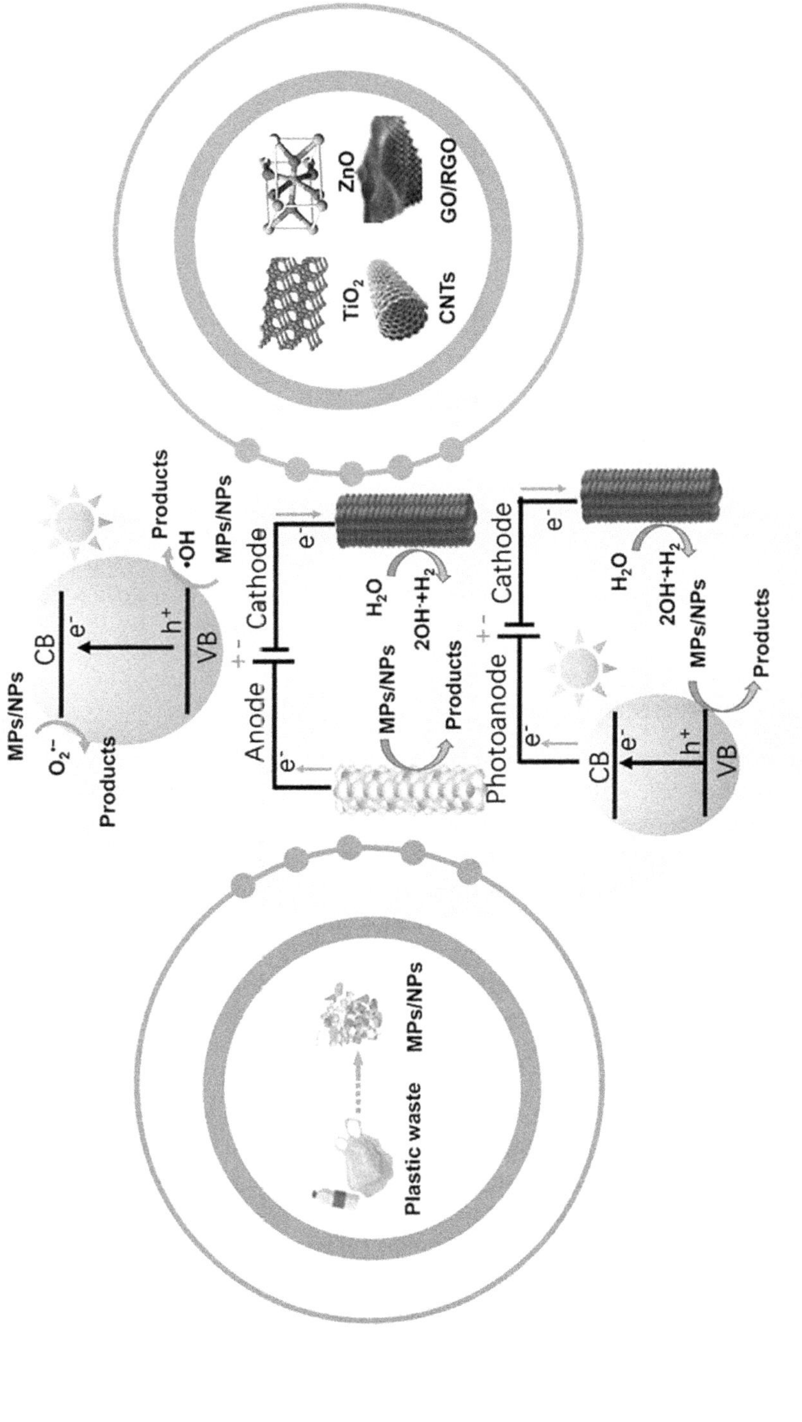

FIGURE 2.7 Overview of nanomaterial-induced MP degradation using heterogeneous AOPs (Du et al., 2022).

Kim et al., 2022). However, this nature-inspired technique (Figure 2.7) has certain limitations such as slow reaction rates due to difficult water oxidation kinetics and the need for non-recyclable real-world materials for photoelectrochemical biosynthetic processes. Quilumbaquin et al. (2024) used a TiO_2-modified boron-doped-diamond photoanode constructed through electrophoretic deposition, achieving a degradation efficiency of 89.91 ± 0.08% of HDPE-MPs (polyethylene microplastics) in 10 h, with a higher efficiency at a lower current density (6.89 $mAcm^{-2}$).

2.4 SUMMARY AND CONCLUSIONS

MPs are recognized as emerging environmental contaminants with persistent risks, exacerbated by their long-term accumulation. Conventional methods are often inadequate at completely removing MPs from water, necessitating the development of environmentally friendly, effective, and low-cost alternatives. Bioremediation offers promise in meeting these criteria, despite its extended treatment durations. AOPs present a viable solution, capable of mineralizing MPs into benign water and CO_2 molecules, primarily via the generation of highly reactive radicals such as $^{\bullet}OH$. However, the efficacy of AOPs is contingent upon various reaction conditions, including pH, temperature, UV irradiation, and reaction time. Previous studies have demonstrated the benefits and drawbacks of homogeneous and heterogeneous AOPs, such as UV photolysis, O_3, UV/visible light-induced photocatalysis, and heat activation, in effectively degrading diverse types of MPs, albeit typically limited to surface layers. Nevertheless, although current research has indicated AOPs' potential for complete MPs removal, specific studies remain scarce, highlighting an urgent need for further investigation and development of strategies for the total elimination of MPs in the environment.

To address MPs pollution at its source, substantial efforts are underway, including the creation of biodegradable plastics and improvement of recycling capacity. Legislation and agreements have also been pivotal in combating MPs pollution. However, scaling up AOPs for industrial-level treatment of MPs requires comprehensive data on design and operational parameters from laboratory- or pilot-scale research.

BIBLIOGRAPHY

Ahmed, R., Hamid, A. K., Krebsbach, S. A., He, J., & Wang, D. (2022). Critical review of microplastics removal from the environment. *Chemosphere*, *293*, 133557. https://doi.org/https://doi.org/10.1016/j.chemosphere.2022.133557

Alulema-Pullupaxi, P., Espinoza-Montero, P. J., Sigcha-Pallo, C., Vargas, R., Fernández, L., Peralta-Hernández, J. M., & Paz, J. L. (2021). Fundamentals and applications of photoelectrocatalysis as an efficient process to remove pollutants from water: A review. *Chemosphere*, *281* (December 2020). https://doi.org/10.1016/j.chemosphere.2021.130821

Alulema-Pullupaxi, P., Fernández, L., Debut, A., Santacruz, C. P., Villacis, W., Fierro, C., & Espinoza-Montero, P. J. (2021). Photoelectrocatalytic degradation of glyphosate on titanium dioxide synthesized by sol-gel/spin-coating on boron doped diamond (TiO2/

BDD) as a photoanode. *Chemosphere*, *278*, 130488. https://doi.org/10.1016/j.chemosphere.2021.130488

Amelia, D., Fathul Karamah, E., Mahardika, M., Syafri, E., Mavinkere Rangappa, S., Siengchin, S., & Asrofi, M. (2021). Effect of advanced oxidation process for chemical structure changes of polyethylene microplastics. *Materials Today: Proceedings*, *xxxx*, 10–13. https://doi.org/10.1016/j.matpr.2021.10.438

Ameta, R., Solanki, M. S., Benjamin, S., & Ameta, S. C. (2018). Photocatalysis. In *Advanced Oxidation Processes for Wastewater Treatment: Emerging Green Chemical Technology* (pp. 135–175). Elsevier Inc. https://doi.org/10.1016/B978-0-12-810499-6.00006-1

Amor, C., Marchão, L., Lucas, M. S., & Peres, J. A. (2019). Application of advanced oxidation processes for the treatment of recalcitrant agro-industrial wastewater: A review. *Water (Switzerland)*, *11*(2). https://doi.org/10.3390/w11020205

An, Y., Hou, J., Liu, Z., & Peng, B. (2014). Enhanced solid-phase photocatalytic degradation of polyethylene by TiO2-MWCNTs nanocomposites. *Materials Chemistry and Physics*, *148*(1–2), 387–394. https://doi.org/10.1016/j.matchemphys.2014.08.001

Babuponnusami, A., & Muthukumar, K. (2014). A review on Fenton and improvements to the Fenton process for wastewater treatment. *Journal of Environmental Chemical Engineering*, *2*(1), 557–572. https://doi.org/10.1016/J.JECE.2013.10.011

Baruah, A., Sharma, A., Sharma, S., & Nagraik, R. (2022). An insight into different microplastic detection methods. In *International Journal of Environmental Science and Technology* (Vol. 19, Issue 6, pp. 5721–5730). Springer Science and Business Media Deutschland GmbH. https://doi.org/10.1007/s13762-021-03384-1

Bilińska, L., Gmurek, M., & Ledakowicz, S. (2016). Comparison between industrial and simulated textile wastewater treatment by AOPs – Biodegradability, toxicity and cost assessment. *Chemical Engineering Journal*, *306*, 550–559. https://doi.org/10.1016/j.cej.2016.07.100

Bui, X. T., Vo, T. P. T., Ngo, H. H., Guo, W. S., & Nguyen, T. T. (2016). Multicriteria assessment of advanced treatment technologies for micropollutants removal at large-scale applications. *Science of The Total Environment*, *563–564*, 1050–1067. https://doi.org/10.1016/J.SCITOTENV.2016.04.191

Cai, J., Niu, T., Shi, P., & Zhao, G. (2019). Boron-Doped Diamond for Hydroxyl Radical and Sulfate Radical Anion Electrogeneration, Transformation, and Voltage-Free Sustainable Oxidation. *Small*, *15*(48). https://doi.org/10.1002/smll.201900153

Cai, L., Wang, J., Peng, J., Wu, Z., &Tan, X. (2018). Observation of the degradation of three types of plastic pellets exposed to UV irradiation in three different environments. *Science of The Total Environment*, *628–629*, 740–747. https://doi.org/10.1016/J.SCITOTENV.2018.02.079

Castillo-Cabrera, G. X., Espinoza-Montero, P. J., Alulema-Pullupaxi, P., Mora, J. R., & Villacís-García, M. H. (2022). Bismuth Oxyhalide-Based Materials (BiOX: X = Cl, Br, I) and Their Application in Photoelectrocatalytic Degradation of Organic Pollutants in Water: A Review. In *Frontiers in Chemistry* (Vol. 10). Frontiers Media S.A. https://doi.org/10.3389/fchem.2022 900622

Castillo-Cabrera, G. X., Pliego-Cerdán, C. I., Méndez, E., & Espinoza-Montero, P. J. (2024). Step-by-step guide for electrochemical generation of highly oxidizing reactive species on BDD for beginners. *Frontiers in Chemistry*, *11*. https://doi.org/10.3389/fchem.2023.1298630

Cisneros-León, D. G., Espinoza-Montero, P. J., Bolaños-Mendez, D., Alvarez-Paguay, J., Fernández, L., Saavedra-Alulema, P. F., Lopez, K., Astorga, D., & Piñeiros, J. L. (2023). Electrochemical degradation of surfactants in domestic wastewater using a DiaClean® cell equipped with a boron-doped diamond electrode. *Frontiers in Chemistry*, *11*. https://doi.org/10.3389/fchem.2023.900670

Collivignarelli, M. C., Abbà, A., Carnevale Miino, M., Arab, H., Bestetti, M., & Franz, S. (2020). Decolorization and biodegradability of a real pharmaceutical wastewater treated by H2O2-assisted photoelectrocatalysis on TiO2 meshes. *Journal of Hazardous Materials, 387*, 121668. https://doi.org/10.1016/J.JHAZMAT.2019.121668

Denisse Vital-Grappin, A., Ariza-Tarazona, M. C., Montserrat Luna-Hernández, V., Francisco Villarreal-Chiu, J., Manuel Hernández-López, J., Siligardi, C., Cedillo-González, E. I., & Mx, J. M. H. (2021). *The Role of the Reactive Species Involved in the Photocatalytic Degradation of HDPE Microplastics Using C,N-TiO 2 Powders*. https://doi.org/10.3390/polym13070999

Derco, J., Dudáš, J., Valičková, M., Šimovičová, K., & Kecskés, J. (2015). Removal of micropollutants by ozone based processes. *Chemical Engineering and Processing – Process Intensification, 94*, 78–84. https://doi.org/10.1016/J.CEP.2015.03.014

Drogui, P., Blais, J.-F., & Mercier, G. (2007). Review of Electrochemical Technologies for Environmental Applications. In *Recent Patents on Engineering* (Vol. 1). https://doi.org/10.2174/187221207782411629

Du, H., Wang, Q., Chen, G., & Wang, J. (2022). Photo/electro-catalytic degradation of micro- and nano-plastics by nanomaterials and corresponding degradation mechanism. In *TrAC– Trends in Analytical Chemistry* (Vol. 157). Elsevier B.V. https://doi.org/10.1016/j.trac.2022.116815

Durán, F. E., de Araújo, D. M., do Nascimento Brito, C., Santos, E. V., Ganiyu, S. O., & Martínez-Huitle, C. A. (2018). Electrochemical technology for the treatment of real washing machine effluent at pre-pilot plant scale by using active and non-active anodes. *Journal of Electroanalytical Chemistry, 818*, 216–222. https://doi.org/10.1016/J.JELECHEM.2018.04.029

Ebrahimbabaie, P., Yousefi, K., & Pichtel, J. (2022). Photocatalytic and biological technologies for elimination of microplastics in water: Current status. *Science of The Total Environment, 806*, 150603. https://doi.org/10.1016/J.SCITOTENV.2021.150603

Eriksen, M., Lebreton, L. C. M., Carson, H. S., Thiel, M., Moore, C. J., Borerro, J. C., Galgani, F., Ryan, P. G., & Reisser, J. (2014). Plastic Pollution in the World's Oceans: More than 5 Trillion Plastic Pieces Weighing over 250,000 Tons Afloat at Sea. *PLoS ONE, 9*(12). https://doi.org/10.1371/journal.pone.0111913

Espinoza-Montero, P. J., Vargas, R., Alulema-Pullupaxi, P., & Fernández, L. (2022). Photoelectrocatalysis: Principles and Applications. In *Advanced Oxidation Processes for Wastewater Treatment* (pp. 53–68). CRC Press. https://doi.org/10.1201/9781003165958-5

Fechine, G. J. M., Souto-Maior, R. M., & Rabello, M. S. (2002). Structural changes during photodegradation of poly(ethylene terephthalate). *Journal of Materials Science 37, 4979–4984*. https://doi.org/https://doi.org/10.1023/A:1021067027612

Feilizadeh, M., Karimi Estahbanati, M. R., Nayernia, Z., Avazpour, S., Drogui, P., & Tyagi, R. D. (2023). Chemical degradation of microplastics and nanoplastics in water and wastewater. In *Current Developments in Biotechnology and Bioengineering* (pp. 315–332). Elsevier. https://doi.org/10.1016/b978-0-323-99908-3.00014-2

Feng, H. M., Zheng, J. C., Lei, N. Y., Yu, L., Kong, K. H. K., Yu, H. Q., Lau, T. C., & Lam, M. H. W. (2011). Photoassistedfenton degradation of polystyrene. *Environmental Science and Technology, 45*(2), 744–750. https://doi.org/10.1021/es102182g

Ganiyu, S. O., Martínez-Huitle, C. A., & Oturan, M. A. (2021). Electrochemical advanced oxidation processes for wastewater treatment: Advances in formation and detection of reactive species and mechanisms. In *Current Opinion in Electrochemistry* (Vol. 27). Elsevier B.V. https://doi.org/10.1016/j.coelec.2020.100678

Garcia-Segura, S., Ocon, J. D., & Chong, M. N. (2018). Electrochemical oxidation remediation of real wastewater effluents — A review. *Process Safety and Environmental Protection, 113*, 48–67. https://doi.org/10.1016/J.PSEP.2017.09.014

Ghatge, S., Yang, Y., Ko, Y., Yoon, Y., Ahn, J. H., Kim, J. J., & Hur, H. G. (2022). Degradation of sulfonated polyethylene by a bio-photo-fenton approach using glucose oxidase immobilized on titanium dioxide. *Journal of Hazardous Materials, 423*, 127067.https://doi.org/10.1016/J.JHAZMAT.2021.127067

Hidaka, H., Suzuki, Y., Nohara, K., Horikoshi, S., Hisamatsu, Y., Pelizzetti, E., & Serpone, N. (1996). Photocatalyzed degradation of polymers in aqueous semiconductor suspensions. I. Photooxidation of solid particles of polyvinylchloride. *Journal of Polymer Science, Part A: Polymer Chemistry, 34*(7), 1311–1316. https://doi.org/10.1002/(SICI)1099-0518(199605)34:7<1311:AID-POLA18>3.0.CO;2-C

Hidayaturrahman, H., & Lee, T. G. (2019). A study on characteristics of microplastic in wastewater of South Korea: Identification, quantification, and fate of microplastics during treatment process. *Marine Pollution Bulletin, 146*, 696–702. https://doi.org/10.1016/J.MARPOLBUL.2019.06.071

Hodges, B. C., Cates, E. L., & Kim, J. H. (2018). Challenges and prospects of advanced oxidation water treatment processes using catalytic nanomaterials. *Nature Nanotechnology, 13*(8), 642–650. https://doi.org/10.1038/s41565-018-0216-x

Hu, K., Tian, W., Yang, Y., Nie, G., Zhou, P., Wang, Y., Duan, X., & Wang, S. (2021). Microplastics remediation in aqueous systems: Strategies and technologies. In *Water Research* (Vol. 198). Elsevier Ltd. https://doi.org/10.1016/j.watres.2021.117144

Jambeck, J. R., Geyer, R., Wilcox, C., Siegler, T. R., Perryman, M., Andrady, A., Narayan, R., & Law, K. L. (2015). Plastic waste inputs from land into the ocean. *Science, 347*(6223), 768–771. https://doi.org/10.1126/science.1260352

Jiang, R., Lu, G., Yan, Z., Liu, J., Wu, D., & Wang, Y. (2021). Microplastic degradation by hydroxy-rich bismuth oxychloride. *Journal of Hazardous Materials, 405*. https://doi.org/10.1016/j.jhazmat.2020.124247

Kalal, S., Singh Chauhan, N. P., Ameta, N., Ameta, R., Kumar, S., & Punjabi, P. B. (2014). Role of copper pyrovanadate as heterogeneous photo-Fenton like catalyst for the degradation of neutral red and azure-B: An eco-friendly approach. *Korean Journal of Chemical Engineering, 31*(12), 2183–2191. https://doi.org/10.1007/s11814-014-0142-z

Kang, J., Zhou, L., Duan, X., Sun, H., Ao, Z., & Wang, S. (2019). Degradation of cosmetic microplastics via functionalized carbon nanosprings. *Matter, 1*(3), 745–758.

Kiendrebeogo, M., Karimi Estahbanati, M. R., Khosravanipour Mostafazadeh, A., Drogui, P., & Tyagi, R. D. (2021). Treatment of microplastics in water by anodic oxidation: A case study for polystyrene. *Environmental Pollution, 269*. https://doi.org/10.1016/j.envpol.2020.116168

Kim, J., Jang, J., Hilberath, T., Hollmann, F., & Park, C. B. (2022). Photoelectrocatalytic biosynthesis fuelled by microplastics. *Nature Synthesis, 1*(10), 776–786. https://doi.org/10.1038/s44160-022-00153-x

Kim, S., Seo, A. Y., & Lee, T. G. (2020). Functionalized cellulose to remove surfactants from cosmetic products in wastewater. *Carbohydrate Polymers, 236*, 116010. https://doi.org/https://doi.org/10.1016/j.carbpol.2020.116010

Koelmans, A. A., Redondo-Hasselerharm, P. E., Nor, N. H. M., de Ruijter, V. N., Mintenig, S. M., & Kooi, M. (2022). Risk assessment of microplastic particles. In *Nature Reviews Materials* (Vol. 7, Issue 2, pp. 138–152). Nature Research. https://doi.org/10.1038/s41578-021-00411-y

Kurian, M. (2021). Advanced oxidation processes and nanomaterials -a review. *Cleaner Engineering and Technology*, *2*, 100090.https://doi.org/10.1016/J.CLET.2021.100090

Lambert, S., & Wagner, M. (2016). Characterisation of nanoplastics during the degradation of polystyrene. *Chemosphere*, *145*, 265–268. https://doi.org/10.1016/J.CHEMOSPHERE.2015.11.078

Lee, Y., Lee, S., Cui, M., Ren, Y., Park, B., Ma, J., Han, Z., & Khim, J. (2021). Activation of peroxodisulfate and peroxymonosulfate by ultrasound with different frequencies: Impact on ibuprofen removal efficient, cost estimation and energy analysis. *Chemical Engineering Journal*, *413*, 127487.https://doi.org/10.1016/J.CEJ.2020.127487

Li, Y., Wang, C., Pan, S., Zhao, X., & Liu, N. (2022). Mn doping improves in-situ H2O2 generation and activation in electro-Fenton process by Fe/Mn@CC cathode using high-temperature shock technique. *Chemosphere*, *307*, 136074. https://doi.org/10.1016/J.CHEMOSPHERE.2022.136074

Lin, J., Yan, D., Fu, J., Chen, Y., & Ou, H. (2020). Ultraviolet-C and vacuum ultraviolet inducing surface degradation of microplastics. *Water Research*, *186*, 116360. https://doi.org/10.1016/J.WATRES.2020.116360

Liu, L., Xu, M., Ye, Y., & Zhang, B. (2022). On the degradation of (micro)plastics: Degradation methods, influencing factors, environmental impacts. In *Science of the Total Environment* (Vol. 806). Elsevier B.V. https://doi.org/10.1016/j.scitotenv.2021.151312

Liu, P., Qian, L., Wang, H., Zhan, X., Lu, K., Gu, C., & Gao, S. (2019). New Insights into the Aging Behavior of Microplastics Accelerated by Advanced Oxidation Processes. *Environmental Science and Technology*, *53*(7), 3579–3588. https://doi.org/10.1021/acs.est.9b00493

Liu, X., Sun, P., Qu, G., Jing, J., Zhang, T., Shi, H., & Zhao, Y. (2021). Insight into the characteristics and sorption behaviors of aged polystyrene microplastics through three type of accelerated oxidation processes. *Journal of Hazardous Materials*, *407*, 124836. https://doi.org/10.1016/J.JHAZMAT.2020.124836

Liu, Y., Fan, Q., Liu, Y., & Wang, J. (2018). Fenton-like oxidation of 4-chlorophenol using H2O2 in situ generated by Zn-Fe-CNTs composite. *Journal of Environmental Management*, *214*, 252–260. https://doi.org/10.1016/J.JENVMAN.2018.03.024

Lu, J., Hou, R., Wang, Y., Zhou, L., & Yuan, Y. (2022). Surfactant-sodium dodecyl sulfate enhanced degradation of polystyrene microplastics with an energy-saving electrochemical advanced oxidation process (EAOP) strategy. *Water Research*, *226*. https://doi.org/10.1016/j.watres.2022.119277

Luo, H., Zeng, Y., Zhao, Y., Xiang, Y., Li, Y., & Pan, X. (2021). Effects of advanced oxidation processes on leachates and properties of microplastics. *Journal of Hazardous Materials*, *413*. https://doi.org/10.1016/j.jhazmat.2021.125342

Martínez-Huitle, C. A., Rodrigo, M. A., Sirés, I., & Scialdone, O. (2023). A critical review on latest innovations and future challenges of electrochemical technology for the abatement of organics in water. In *Applied Catalysis B: Environmental* (Vol. 328). Elsevier B.V. https://doi.org/10.1016/j.apcatb.2023.122430

Miao, F., Liu, Y., Gao, M., Yu, X., Xiao, P., Wang, M., Wang, S., & Wang, X. (2020). Degradation of polyvinyl chloride microplastics via an electro-Fenton-like system with a TiO2/graphite cathode. *Journal of Hazardous Materials*, *399*, 123023. https://doi.org/10.1016/J.JHAZMAT.2020.123023

Moreno-Andrés, J., Rueda-Márquez, J. J., Homola, T., Vielma, J., Moríñigo, M. Á., Mikola, A., Sillanpää, M., Acevedo-Merino, A., Nebot, E., & Levchuk, I. (2020). A comparison of photolytic, photochemical and photocatalytic processes for disinfection of recirculation aquaculture systems (RAS) streams. *Water Research*, *181*, 115928.https://doi.org/10.1016/J.WATRES.2020.115928

Ochoa-Chavez, S. A., Montenegro-Rosero, K. L., Espinoza-Montero, P. J., Yar-Hernández, A. V., & Frontana-Uribe, B. A. (2024). Advanced oxidation processes for removal of pharmaceuticals and personal care products. *Development in Wastewater Treatment Research and Processes: Emerging Technologies for Removal of Pharmaceuticals and Personal Care Products: State of the Art, Challenges and Future Perspectives*, 255–286. https://doi.org/10.1016/B978-0-443-19207-4.00014-8

Oturan, N., Van Hullebusch, E. D., Zhang, H., Mazeas, L., Budzinski, H., Le Menach, K., & Oturan, M. A. (2015). Occurrence and Removal of Organic Micropollutants in Landfill Leachates Treated by Electrochemical Advanced Oxidation Processes. *Environmental Science and Technology*, *49*(20), 12187–12196. https://doi.org/10.1021/acs.est.5b02809

Pizzichetti, A. R. P., Pablos, C., Álvarez-Fernández, C., Reynolds, K., Stanley, S., & Marugán, J. (2021). Evaluation of membranes performance for microplastic removal in a simple and low-cost filtration system. *Case Studies in Chemical and Environmental Engineering*, *3*, 100075. https://doi.org/10.1016/J.CSCEE.2020.100075

Priyadarshini, M., Das, I., Ghangrekar, M. M., & Blaney, L. (2022). Advanced oxidation processes: Performance, advantages, and scale-up of emerging technologies. *Journal of Environmental Management*, *316*, 115295. https://doi.org/10.1016/J.JENVMAN.2022.115295

Quilumbaquin, W., Castillo-Cabrera, G. X., Borrero-González, L. J., Mora, J. R., Valle, V., Debut, A., Loor-Urgilés, L. D., & Espinoza-Montero, P. J. (2024). Photoelectrocatalytic degradation of high-density polyethylene microplastics on TiO2-modified boron-doped diamond photoanode. *IScience*, *27*(3). https://doi.org/10.1016/j.isci.2024.109192

Rajeshwar, K., Ibanez, J. G., & Swain, G. M. (1994). Electrochemistry and the environment. In *Journal of Applied Electrochemistry* (Vol. 24). https://doi.org/https://doi.org/10.1007/BF00242533

Rajput, H., Kwon, E. E., Younis, S. A., Weon, S., Jeon, T. H., Choi, W., & Kim, K. H. (2021). Photoelectrocatalysis as a high-efficiency platform for pulping wastewater treatment and energy production. *Chemical Engineering Journal*, *412*, 128612. https://doi.org/10.1016/J.CEJ.2021.128612

Sen, R., Das, S., Nath, A., Maharana, P., Kar, P., Verpoort, F., Liang, P., & Roy, S. (2022). Electrocatalytic Water Oxidation: An Overview With an Example of Translation From Lab to Market. In *Frontiers in Chemistry* (Vol. 10). Frontiers Media S.A. https://doi.org/10.3389/fchem.2022.861604

Shen, M., Song, B., Zhou, C., Hu, T., Zeng, G., & Zhang, Y. (2022). Advanced oxidation processes for the elimination of microplastics from aqueous systems: Assessment of efficiency, perspectives and limitations. In *Science of the Total Environment* (Vol. 842). Elsevier B.V. https://doi.org/10.1016/j.scitotenv.2022.156723

Suhrhoff, T. J., & Scholz-Böttcher, B. M. (2016). Qualitative impact of salinity, UV radiation and turbulence on leaching of organic plastic additives from four common plastics – A lab experiment. *Marine Pollution Bulletin*, *102*(1), 84–94. https://doi.org/10.1016/j.marpolbul.2015.11.054

Tofa, T. S., Kunjali, K. L., Paul, S., & Dutta, J. (2019). Visible light photocatalytic degradation of microplastic residues with zinc oxide nanorods. *Environmental Chemistry Letters*, *17*(3), 1341–1346. https://doi.org/10.1007/s10311-019-00859-z

Tong, L., Song, K., Wang, Y., Yang, J., ji, J., Lu, J., Chen, Z., & Zhang, W. (2022). Zinc oxide nanoparticles dissolution and toxicity enhancement by polystyrene microplastics under sunlight irradiation. *Chemosphere*, *299*, 134421. https://doi.org/10.1016/J.CHEMOSPHERE.2022.134421

Uheida, A., Mejía, H. G., Abdel-Rehim, M., Hamd, W., & Dutta, J. (2021). Visible light photocatalytic degradation of polypropylene microplastics in a continuous water flow system. *Journal of Hazardous Materials*, *406*, 124299. https://doi.org/10.1016/j.jhazmat.2020.124299

Uekert, T., Kuehnel, M. F., Wakerley, D. W., & Reisner, E. (2018). *Energy & Environmental Science*, *11*, 2853–2857.

Uekert, T., Kasap, H., & Reisner, E. (2019). *Journal of the American Chemical Society, 141*, 15201–15210.

Wang, Z., Xin, X., Shi, X., & Zhang, Y. (2020). A polystyrene-degrading Acinetobacter bacterium isolated from the larvae of Triboliumcastaneum. *Science of the Total Environment*, 726. https://doi.org/10.1016/j.scitotenv.2020.138564

Wigger, H., Kägi, R., Wiesner, M., & Nowack, B. (2020). Exposure and Possible Risks of Engineered Nanomaterials in the Environment—Current Knowledge and Directions for the Future. In *Reviews of Geophysics* (Vol. 58, Issue 4). Blackwell Publishing Ltd. https://doi.org/10.1029/2020RG000710

Wu, C., & Linden, K. G. (2008). Degradation and byproduct formation of parathion in aqueous solutions by UV and UV/H2O2 treatment. *Water Research*, *42*(19), 4780–4790. https://doi.org/10.1016/j.watres.2008.08.023

Wu, S., Xiong, J., Sun, J., Hood, Z. D., Zeng, W., Yang, Z., Gu, L., Zhang, X., & Yang, S. Z. (2017). Hydroxyl-Dependent Evolution of Oxygen Vacancies Enables the Regeneration of BiOCl Photocatalyst. *ACS Applied Materials and Interfaces*, *9*(19), 16620–16626. https://doi.org/10.1021/acsami.7b01701

Xu, Q., Huang, Q.-S., Luo, T.-Y., Wu, R.-L., Wei, W., & Ni, B.-J. (2021). Coagulation removal and photocatalytic degradation of microplastics in urban waters. *Chemical Engineering Journal*, *416*, 129123. https://doi.org/10.1016/j.cej.2021.129123

Zarei, E., & Ojani, R. (2017). Fundamentals and some applications of photoelectrocatalysis and effective factors on its efficiency: a review. In *Journal of Solid State Electrochemistry* (Vol. 21, Issue 2, pp. 305–336). Springer New York LLC. https://doi.org/10.1007/s10008-016-3385-2

Zhang, L., Zhao, H., Wilkinson, D. P., Sun, X., & Zhang, J. (2020). *Electrochemical Water Electrolysis; Fundamentals and Technologies* (illustrated). CRC Press.

Zhang, M., Dong, H., Zhao, L., Wang, D., & Meng, D. (2019). A review on Fenton process for organic wastewater treatment based on optimization perspective. *Science of The Total Environment*, *670*, 110–121. https://doi.org/https://doi.org/10.1016/j.scitotenv.2019.03.180

Zhou, C., Lai, C., Xu, P., Zeng, G., Huang, D., Zhang, C., Cheng, M., Hu, L., Wan, J., Liu, Y., Xiong, W., Deng, Y., & Wen, M. (2018). In Situ Grown AgI/Bi12O17Cl2 Heterojunction Photocatalysts for Visible Light Degradation of Sulfamethazine: Efficiency, Pathway, and Mechanism. *ACS Sustainable Chemistry and Engineering*, *6*(3), 4174–4184. https://doi.org/10.1021/acssuschemeng.7b04584

Zhou, C., Tao, L., Yang, F., Wang, B., Wan, X., Jin, Y., Yu, H., & Yang, Y. (2019). Application of electrochemical methods in heterogeneous catalysis. *Current Opinion in Chemical Engineering*, *26*, 88–95. https://doi.org/https://doi.org/10.1016/j.coche.2019.09.007

Zhou, D., Zhang, H., &Chen, L. (2015). Sulfur-replaced Fenton systems: Can sulfate radical substitute hydroxyl radical for advanced oxidation technologies? *Journal of*

Chemical Technology and Biotechnology, *90*(5), 775–779. https://doi.org/10.1002/jctb.4525

Zhu, K., Jia, H., Sun, Y., Dai, Y., Zhang, C., Guo, X., Wang, T., & Zhu, L. (2020). Long-term phototransformation of microplastics under simulated sunlight irradiation in aquatic environments: Roles of reactive oxygen species. *Water Research*, *173*, 115564.https://doi.org/10.1016/J.WATRES.2020.115564

Zinke, L. (2020). Moving marine microplastics. In *Nature Reviews Earth and Environment* (Vol. 1, Issue 4, p. 186). Springer Nature. https://doi.org/10.1038/s43017-020-0047-9

3 Role of Microbes in Microplastic Removal and Its Effect on Human Health

Saman Aftab and Hamid Mukhtar

3.1 INTRODUCTION

An African scientist in the 1990s used the term microplastic for the first time in an article titled "Plastic and other artifact on South African beaches: temporal trends in abundance and composition." Since then, microplastic has been used as a term for describing tiny particles of plastic. Although tiny plastic particles, having a size between 100 and 5 mm are categorized as microplastics, their characteristics and properties are still in the phase of exploration. As far as the classification of plastic particles on the basis of their size is generally concerned, this includes macroplastics that are > 25 mm in size, mesoplastics that are 5–25 mm in size and nanoplastics that are < 100 nm in size.

The United Nations Environment Program (UNEP, 2014), has listed microplastic pollution arising from the environmental degradation of plastic waste among the top environmental issues in the world that need attention. It is evident that being present in the environment, microplastics adversely affect vast flora and fauna species by causing numerous harmful effects on them. Having a large surface area attributed to smaller size, the microplastics have enhanced capability to absorb into biological cells. Other contaminants present in the environment such as toxicants, flame retardants, plasticizers, and heavy metals get attracted by the hydrophobic surface of microplastics eventually entering the biological organisms and their systems. Thus, microplastics make their way to the food web and being recalcitrant to degradation in the environment, get entered into the food chain, where the final consumers are humans.

For the removal of microplastics in the environment, bioremediation has of lately received significant interest from researchers across the world. Since 1970, several research studies have been done for the screening of potent microorganisms having the ability to degrade plastic polymers. Microorganisms are capable of undergoing biodegradation of plastics through enzymatic hydrolysis after colonizing the plastic surface. Unlike composting, the conventional synthetic plastics can be used as a carbon source in the process of biodegradation and valorization. The initial step of

DOI: 10.1201/9781003379386-3

valorization of microplastics (enzymes acting on microplastics to result in depolymerization) includes the screening of microorganisms capable of degrading plastics for further identification and extraction of potent enzymes involved in plastic biodegradation. The depolymerization products formed during the process of microbial valorization of microplastics wastes, could be utilized in the concluding phase (e.g., biofermentation of products formed as a result of depolymerization). After the segregation of products, the final step would use these products via different metabolic pathways in the biosynthesis of value-added chemical products. The abovementioned bioremediation approaches include screening microbes (isolated from the open environmental sources including waste from landfills, marine water and mulch films) and potent biocatalysts for the removal of microplastics through the process of biodegradation. For better understanding of entire process, the knowledge of steps involved in biodegradation holds significance.

3.2 MICROPLASTIC BIODEGRADATION PROCESS

The complex process of biodegradation involves a number of key steps. The process starts with bio-deterioration in which the biotic and abiotic factors act in combination to result in the breakdown of larger plastic particles into smaller fragments. The next step is the defragmentation also widely known as depolymerization; this involves the release of microbial enzymes that possess the ability to break long chain polymer into smaller subunits such as oligomers, dimers, and monomers. Extracellular and intracellular depolymerases are the two significant classes of enzymes which participate actively in the biodegradation of various plastics. Consequently, polymer of lower molecular weight is obtained due to cleavage of long chain polymer and having fragile physical properties; this smaller fragmented polymer is readily accessible for further process of microbial assimilation.

The next significant step in the process of biodegradation is known as assimilation, in which some of the degraded molecules are recognized by the microbial cell receptors, thus resulting in the intake of those molecules across the plasma membrane. The last step of biodegradation is known as mineralization, which typically involves the release of molecules such as carbon dioxide, methane, nitrogen, and water from intracellular metabolites into the environment. Biodegradation of plastic resulting through aerobic metabolism of microorganisms releases $CO_2$and water as end products. Whereas the anaerobic metabolism leads to the production of carbon dioxide, methane and water as end products of plastic degradation.

Several abiotic factors cause alterations in the properties of polymer. For example, light, temperature, pressure, moisture, and chemical conditions along with biological processes have caused various modifications which are commonly categorized as polymer degradation. It has been reported that biofilm formation and cells' interactions are defined by varying differences in the size of plastic particles. The microbial colonization of plastics through biofilm formation on the surface of macroplastics has been evidenced by the imaging of plastic particles using Scanning electron microscopy (SEM). As the degradation of plastic particles proceeds further, the polymer particles become smaller in size causing the cells on the polymer surface to embed in the microplastics or adhere within the cracks. In anaerobic conditions, pyrolysis has been documented as an efficient method to result in the formation of low molecular weight aliphatic hydrocarbons. The monomers formed through pyrolysis are subjected to a terminal oxidation process for further degradation. This process is similar to the degradation pathway of n-alkane by an enzyme known as alkane hydroxylase resulting in the formation of primary alcohol. Thus, microorganisms and their associated enzymes involved in the degradation of plastics including polyethylene (PE), polyvinyl chloride(PVC), polystyrene (PS) and polypropylene (PP) have been isolated and identified in the literature. Moreover, the capacity of these microbes to bioremediate plastic pollution has been documented on the basis of the characterization of biofilm produced on polymers, surface erosion, and alterations in the physical and chemical properties of plastics. Although the majority of such studies investigate the degradation of macroplastic films, it has been assumed that interactions between microorganisms and the plastic can be applied to microplastics as well.

3.3 MICROBIAL COMMUNITIES INVOLVED IN MICROPLASTICS BIODEGRADATION

3.3.1 Bacteria Capable of Degrading Plastics

Bacterial biodegradation of synthetic polymer has been widely reported (Table 3.1 and Figure 3.1) in the literature. The most relevant abundance was documented for *Proteobacteria* followed by *Actinobacteria*, *Firmicutes*, *Bacteroidetes,* and *Acidobacteria* in degrading polymers. The low density polyethylene (LDPE) has been documented to be most susceptible to degradation by microbial attack (Gravouil et al., 2017; Raddadi and Fava, 2019; Liet al., 2020). The degradation of LDPE ensues via cleavage of the PE chains following the enzymatic action of oxidases resulting in the formation of low molecular weight compounds. These compounds are then assimilated within the cell where alkane biodegradation pathway results in complete mineralization of these compounds (Gravouil et al., 2017).

Polyethylene has also been degraded by actinomycete *Rhodococcus ruber* (Sivan et al., 2006) and *Lysinibacillus xylanilyticus* (Esmaeili et al., 2013). In addition, Skariyachan et al. (2016) reported that members of *Enterobacter* sp. *and Pantoea* sp. are capable of degrading polyethylene. Likewise, the polyethylene degradation was also shown by the microbial strains from the genera *Comamonas, Delftia and Stenotrophomonas* (Peixoto et al., 2017). Moreover, Vimala et al. (2016) reported that polyethylene can be degraded by the microbial action of *Bacillus subtilis* both in the presence and absence of bio-surfactants. Likewise, Skariyachan et al. (2018) reported the potential of microbial strains (*Brevibacillus* sp. and *Aneurinibacillus* sp.)isolated from sewage treatment plants and landfills to degrade polyethylene. Recently, Park and Kim (2019) demonstrated the capability of mixed bacterial cultures (composed of *Bacillus* sp. and *Paenibacillus* sp.) to result in biodegradation of PE.

TABLE 3.1
Polymer degrading bacteria

BACTERIA	SOURCE	POLYMER	POLYMER BIODEGRADATION ASSAY	BIODEGRADATION ASSESSMENT	REFERENCES
Rhodococcus ruber	Commercially available strain	PE	Under aerobic conditions, shaken flasks were incubated for a period of 56 days	Gravimetric weight loss of 7.5% was observed in PE films	Sivan et al. (2006)
Pseudomonas putida	Commercially available strain	PS	One pyrolysis and four fermentation runs in shake flasks were conducted	About 6.4 g of PHA was produced by utilizing 64 g of polystyrene.	Ward et al. (2006)
Pure microbial cultures	Commercially available strain	PS	shaken flasks were incubated at 28°C for a period of 56 days	Gravimetric weight loss of 0.8% was observed in PS films	Mor and Sivan(2008)
Bacillus flexus	Soil (Plastic dumping site)	PP (UV pretreated)	365 days of incubation	Gravimetric weight loss of 2.5% was observed in PP films	Arkatkar et al. (2010)
Streptomyces sps	Soil (Garbage)	PE	180 days of incubation in liquid culture	Gravimetric weight loss of 46.16% and 35.78% loss was observed in PE bag and plastic cups respectively	Usha et al. (2011)
Pseudomonas aeruginosa	Soil (Petroleum oil fields)	PVC	shaken flasks were incubated at 37°C for a period of 180 days	Reduction of 53% in tensile strength of PVC was observed along with FTIR and SEM analyses	Das et al. (2012)
Pseudomonas aeruginosa, Bacillus subtilis	Soil (Garden, Mangrove, Forest, Petrol Pump, Garbage)	PET, PS	30 days of incubation	*Bacillus subtilis* contributed to highest % loss in the weight of PS whereas in case of PET, % loss in the weight was maximum by *Pseudomonas aeruginosa* (in Bushnell Hass broth), and *Bacillus subtilis* (in Nutrient broth)	Asmita et al. (2015)

Stenotrophomonas panacihumi	Soil (Municipal solid waste)	PP	under composting conditions at 37°C, an incubation period of 90 days was observed	Plastic molecular weight was measured by using GPC	Jeon and Kim(2016)
Cupriavidus necator	Commercially available strain	PS	Oxidative degradation of PS flakes, pyrolysis and fermentation was run to produce PHAs	The strain was able to utilize prodegraded PS as carbon source consequently producing PHAs	Johnston et al. (2018)
Mixed bacterial consortium (*Brevibacillus sps*, *Aneurinibacillus sp.*)	sewage treatment plants and landfills	HDPE, LDPE, PP films and pellets	under aerobic conditions, the test tubes were incubated for a period of 4 months at 50 °C	Gravimetric weight reduction for LDPE(58.21 ± 2) HDPE (46.6 ± 3) and PP strips (56.3 ± 2%) were observed; those for LDPE, HDPE and PP pellets, losses in the weight were 45.7 ± 3, 37.2 ± 3 and 44.2 ± 3% respectively. GC-MS analysis was also performed	Skariyachan et al. (2018)
Mixed bacterial consortium (*Bacillus sp. Paenibacillus sp.*)	Soil (Landfill site)	PE	The mixed bacterial consortium was used to inoculate PE microplastic granules in mineral culture medium, subjected to an incubation period of 60 days at 30°C.	The loss of 14.7% in the weight and 22.8% loss in the mean particle diameter of granules were reported as analyzed by field emission SEM, GC-MS, TGA and FTIR analyses	Park and Kim (2019)
Pseudomonas citronellolis, *Bacillus flexus*	Commercially available strains	PVC	45 and 30 days of incubation	At the end of incubation period of 45 days, a decrease in average molecular weight of 10% and 7% for plastic incubated with *Pseudomonas citronellolis* and *Bacillus flexus* respectively were confirmed using GPC analysis. After an incubation period of a month, Gravimetric weight loss of 19% was observed in waste PVC film incubated with *P. citronellolis*	Giacomucci et al. (2019)

(continued)

TABLE 3.1 (Continued)
Polymer degrading bacteria

BACTERIA	SOURCE	POLYMER	POLYMER BIODEGRADATION ASSAY	BIODEGRADATION ASSESSMENT	REFERENCES
Bacillus sp.and *Pseudomonas* sp.	Soil (Waste dumpsite)	HDPE	30 days of incubation period	Formation of new functional groups on the polymer film structure was confirmed by using FTIR analysis.	Devi et al. (2019)
Periphyton biofilm (Deinococcus-thermus >Proteobacteria > Cyanobacteria)	Lake	PE, PP and PET (Microplastics)	Two months of incubation period with the addition of glucose and peptone as sources of carbon	The biofilm formation increased for all the polymers due to the addition of glucose. An improvement in the degradation rate from 9.52%–18.02%, 5.95%–14.02% and 13.24–19.72% for PP, PE and PET respectively was also observed. Biodegradation was also confirmed by GPC, SEM, and FTIR analyses.	Shabbir et al. (2020)
Pseudomonas aeruginosa	Soil (Waste dumpsite)	LDPE	Tween 80 and bleach were used as pretreatments on the plastic films. An incubation period of 2 months was given for degradation of LDPE by the bacterial strain at 37°C in Bushnell Hass broth and Minimal Salt medium	Surface modifications and increased carbonyl index of the LDPE was observed by Field emission SEM and FTIR analyses	Gupta and Devi(2020)
Actinobacteria belonging to genera of *Streptomyces, Nocardia*, and *Rhodococcus*	Soil (Landfill site)	LDPE	An incubation period of 60 days was provided for individual bacterial strains for degradation of LDPE films	Bacterial Strains of *Streptomyces* sp. showed greater plastic degrading potential as observed by maximum loss of plastic weight and modifications in the structure of LDPE film observed in FTIR. Whereas, *Rhodococcus* sp. demonstrated maximum reduction in the tensile strength of plastic	Soleimani et al. (2021)

3.3.2 Fungi Capable of Degrading Plastics

Similar to bacteria, many fungal strains (Table 3.2) also possess the ability to participate in the removal of plastic pollution through the process of biodegradation. Fungal isolates including *Aspergillus niger* and *Aspergillus clavatus* are well reported to degrade polymers under certain laboratory conditions. Genera in the phylum *Ascomycota* (*e.g.* genus of *Aspergillus* and *Fusarium*), followed by *Basidiomycota* and *Zygomycota* are most efficiently related to microplastics degradation. Muhonja et al. (2018) reported great potential of *Aspergillus* sp. to result in the degradation of polyethylene. Polyurethane (PU), is a polymer commonly used in the manufacture of tires such as skateboard wheels, high-resilience foam seating and rigid foam insulation panels. This highly recalcitrant plastic was subjected to degradation by an endophytic fungal species as reported by US researchers in 2011 (Russell et al., 2011). So far the evidence of biodegradation of polypropylene polymers is scarce. However, the potential of microbial communities to degrade polypropylene was reported for the very first time by Cacciari et al. (1993). Few other researches concerned with the degradation of polypropylene have also been recorded in the literature, including the research conducted by Fontanella et al. (2010), Arkatkar et al. (2010), and Skariyachan et al. (2018).

Mostly pure microbial cultures have been investigated for their ability to perform Polystyrene biodegradation in various research studies (Mor and Sivan, 2008; Tian et al., 2017). In addition, *Pseudomonas aeruginosa* (Das et al., 2012) and bacterial strains of *Achromobacter* sp., demonstrated the ability to cause degradation of polyvinylchloride (Giacomucci et al., 2019a). Whereas, fungi (Webb et al., 2000) capable of degrading polyvinylchloride included *Aspergillus niger, Phanerochaete chrysosporium, Lentinus tigrinus,* and *Aspergillus sydowii* as well (Ali et al., 2014).

The microhabitats of caves in particular have been reported to host a distinctive variety of chemoheterotrophs that are culturable and could be explored for plastic biodegradation studies. Living under low nutrient habitats or stressful conditions may result in the development of an ability in the microorganisms to utilize any available nutrient for their survival. Likewise, there is extensive literature available on this pretext which reports the biodeteriorative capability of several microorganisms living in the stressful cave environment.

The microbial biodegradation is widely accepted and is still underway for its enhanced efficiency. One effective approach to deal with plastic pollution could be to change the polymer manufacturing process and new formulations should be made with special considerations on mechanism for their microbial degradation. These considerations could include manipulating the factors that can accelerate the rate of degradation such as temperature, pH, morphology and chemical structure of polymer, presence or absence of plasticizers and, most importantly, the type of organisms which can be involved in the biodegradation process. In addition, there is a need for the exploration of novel biocatalysts/enzymes and optimization of conditions for better degradation outcomes concerning all synthetic polymers.

TABLE 3.2
Polymer degrading fungi

FUNGI	SOURCE	POLYMER	POLYMER BIODEGRADATION ASSAY	BIODEGRADATION ASSESSMENT	REFERENCES
Phanerochaete chrysosporium, *Engyodontium album*	Commercially available strains	PP (UV pretreated)	365 days of incubation in shaken flasks	% decreases of 18.8% and 9.42% were observed in the weight of polymers incubated with *P.chrysosporium* and *E.album* respectively	Jeyakumar et al. (2013)
Aspergillus niger, *Aspergillus sydowii Phanerochaete chrysosporium*, *Lentinus tigrinus*,	Sterilized soil soaked with sewage sludge for 10 months	PVC	49 days of incubation in mineral salt medium	Decrease in the molecular weight of plastic; alterations on plastic surface and modifications in FTIR profile were monitored in the presence of *Phanerochaete chrysosporium*	Ali et al. (2014)
Aspergillus niger	Soil (Garden, Mangrove, Forest, Petrol Pump, Garbage)	PET	30 days of incubation	% decrease in weight of PET was highest by *Aspergillus niger* in Rose Bengal broth	Asmita et al. (2015)
Fungal consortium (*Curvularia lunat*, *Penicillium* simplicissimum, Alternaria *alternate*, and *Fusarium* sp.)	Soil (Plastic dumping site)	PE	Polyethylene discs were inoculated with fungal isolates as well as fungal consortium separately in mineral salt medium for an incubation period of 90 days	The decrease in the weight of polymer attributed to individual fungal strains, i.e. *Curvularia lunata* (1.2%), *Penicillium simplicissimum* (7.7%), *Alternaria alternata* (0.8%) and *Fusarium* sp. (0.7%) was less when compared with the fungal consortium (27%). Plastic degradation was also indicated by FTIR and SEM analyses	Sowmya et al. (2015)

Aspergillus sp., Paecilomyces lilacinusI, Lasiodiplodia theobromae	Endemic plants(*Psychotria flavida* and *Humboldtia brunonis*)	LDPE, PP (both gamma irradiated)	An incubation period of 3 months in Rose Bengal Broth culture was given	SEM, Differential scanning calorimetric and FTIR analyses were done. Decrease in intrinsic viscosity and average molecular weight of the polymers were observed.	Sheik et al. (2015)
Aspergillus oryzae	Soil (Plastic dumping site)	PE	60 days of incubation in soil culture medium	Gravimetric weight loss of 30% was observed in PE strips	Indumathi and Gayathri (2016)
Aspergillus clavatus	Soil (Landfill site)	LDPE	shaken flasks were incubated for a period of 90 days	Gravimetric weight loss of 35% was observed while Strum test revealed the production of 2.32 g l-1 CO2. Moreover, surface alterations were revealed by FTIR and microscopic analyses	Gajendiran et al. (2016)
Aspergillus sp.	Floating plastic debris in the shoreline of a lake	PE	120 days of incubation	Mean weight reduction of 36.4±5.53% in PE was reported along with the detection of new functional groups in FTIR analysis of the PE	Muhonja et al. (2018)
Aspergillus niger	Soil (Garden)	LDPE	shaken flasks containing mineral salt medium were incubated for a month	When % decrease in the weight of plastic was evaluated, LDPE film was more degraded (up to 2.46%) as compared to LDPE sheet (up to 1.53%). *Aspergillus niger* was the most efficient in degrading LDPE	Singh et al. (2018)
Aspergillus terreus, Aspergillus sydowii	Rhizosphere Soil	PE	60 days of incubation with continuous shaking at ambient temperature and varied pH (3.5, 7 and 9.5)	*Aspergillus terreus* (resulting in 58.51 ± 8.14 % weight loss of the plastic at pH 7) and *Aspergillus sydowii* (resulting in 94.44 ± 2.40% loss in	Sangale et al. (2019)

(*continued*)

TABLE 3.2 (Continued)
Polymer degrading fungi

FUNGI	SOURCE	POLYMER	POLYMER BIODEGRADATION ASSAY	BIODEGRADATION ASSESSMENT	REFERENCES
				tensile strength of the plastic at pH 3.5) were considered as the most efficient polyethylene degrading fungal isolates. Results were substantiated with FTIR and SEM analyses	
Bjerkandera adusta	Wild fungus (Mountains)	HDPE	Under lignocellulose substrate treatment, the fungus was investigated for its ability to degrade HDPE. It was suggested that secretion of fungal enzyme lac case was stimulated by ash wood chips. Therefore, an incubation period of three months was observed by using ash wood chips with the fungal inoculation	At the end of incubation, the plastic surface erosion was reported while degradation of the amorphous structure of HDPE due to enzymatic action was confirmed using Raman analysis	Kang et al. (2019)
Aspergillus sp, *Penicillium* sp.	Commercially available strains	PP (Neat), PP (blend)	30 days of incubation in petri dish at 29°C	It was concluded that addition of PBAT/TPS to PP had no significant effect on the polymer mass loss, however, slight degradation was observed in FTIR spectra, microscopic images and contact angle analysis	De Oliveira et al. (2020)

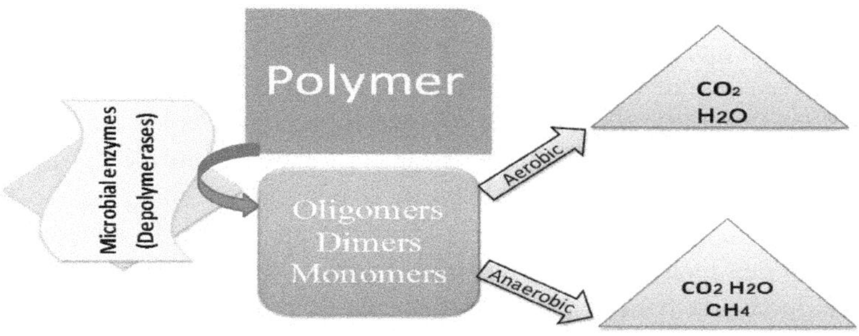

FIGURE 3.1 Schematic illustration of polymer biodegradation.

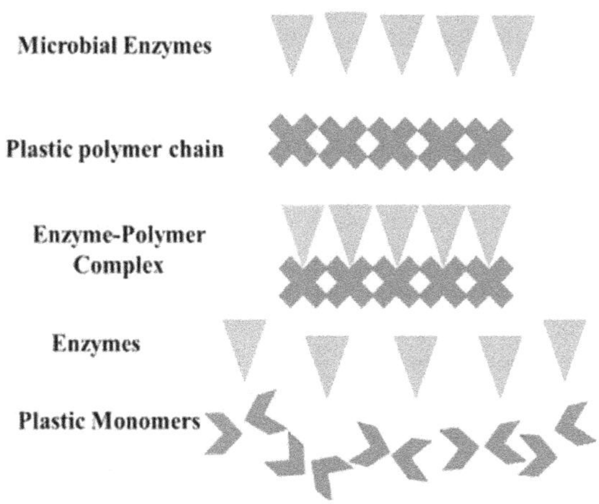

FIGURE 3.2 Pictorial illustration of enzymes degrading polymers into monomers.

3.3.3 Microbial Enzymes Capable of Degrading Microplastics

Microbial enzymes are well known for their role in the functioning and regulation of cell, but they are also involved in the biodegradation of environmental pollutants emerging from anthropogenic activities including microplastics. In this regard, the microbial enzyme capable of degrading plastic can target the polymer structure of microplastic through its enzymatic action and degrade it into its monomers. The monomeric material will be later used as a carbon source in the energy production cycle of microorganism (Figure 3.2).

Extracellular enzymes of microorganisms such as esterases, lignin peroxides, lipases, manganese peroxides, and laccases are reported to enhance the microbial adherence to the plastic with the conversion of functional groups (such as carbonyl and alcohol) present in the polymer structure thereby increasing its hydrophilicity for the process of degradation to proceed further. Hydrolases including lipases, cutinases, poly (3-hydroxybutyrate) depolymerases, and esterases break down plastic into smaller particles by acting on the polymer surface. The polymer chain cleavage is enhanced during biodegradation when these enzymes attach chemical groups or bonds within the side chain of plastics. However, their size does not allow them to diffuse into the structure of polymer; therefore, degradation occurs on the plastic surface, resulting in surface erosion evident by the formation of holes and cracks.

Unique interaction mechanisms exist for each enzyme involved in degradation of microplastics. Generally, the mechanism is initiated by changing the polymer surface due to enzymatic attack. This is known as enzyme surface modification in which hydrolases such as lipases, cutinases, carboxylesterases, and proteases were reported to modify polymer surfaces of microplastic prone to the process of biodegradation. This mechanism was comprehensively reviewed by Kawai et al. (2019) who suggested that specific enzymes (hydrolases) only act as a surface modifier to change the polymer surface of a microplastic. As suggested by the author, such biocatalyst is known as a surface modifying enzyme. Therefore, this enzyme is only known to enhance the surface hydrophilicity of microplastic and does not take part in the further process of degradation of microplastic (Kawai et al., 2019). The Elemental spectroscopy chemical analysis (ESCA) has been used to explain the unique interaction between the enzyme and surface of microplastic. When the enzyme-treated microplastic was compared with the control, modifications in the C–O bonding were clearly observed.

From this analysis, it was confirmed that the change in C–O bonding was attributed to enzyme interaction with the polymer surface. As a result of the plastic surface modification, only a few cutinases were reported to be capable of degrading the building blocks of the microplastic. Extensive research aimed at finding a novel enzyme that possesses the ability to degrade microplastics has been done in the last ten years. Various enzymes are claimed to have the capacity to conduct biodegradation of plastics by degrading polymers into their monomeric forms. Among these, hydrolases, laccases, oxidases, peroxidases, and amidases, have been reported in the literature to be potent enzymes involved in polymer degradation (Gómez-Méndez et al.,2018).

Laccase and alkane hydrolase are found to be Polyethylene (PE) degraders. However, Laccase is most reportedly associated with the high density polyethylene (HDPE) degradation. Being an oxidase group enzyme, laccase is known to provide carbonyl region within the polymer chain by depolymerizing the polymer via oxidative cleavage of the amorphous region of HDPE (Ghatge et al.,2020). Moreover, many enzymes are known to cause microplastics degradation either directly or indirectly. For example, fungal Manganese peroxidase isolated from *Phanerochaete chrysosporium* was capable of reducing and decreasing the tensile strength as well as the total molecular weight of polyethylene. Moreover, the soybean peroxidase has been reported to decrease the hydrophobicity of the polyethylene surface, particularly in the presence of hydrogen peroxide. While the enzymatic potential of specific enzymes for the removal of microplastics has been documented, the detailed mechanism of action is still unexplored. Polyethyleneterephthalate (PET) degradation into its monomers by the two enzymes PETase and MHETase is most extensively studied in literature. Although PET degradation involves tricarboxylic acid cycle and β-ketoadipate pathway, its complete mineralization is dependent on the metabolic substrates obtained as a result of degradation. Other than PETase and MHETase, few other enzymes such as lipase and esterase are also capable of degrading PET but the mechanism of their action is still unexplored.

The first enzyme capable of degrading polystyrene (PS) was reported in 1997. The enzyme hydroquinone peroxidase from *Azotobacter beijerinckii* is known to conduct biodegradation of polystyrene. For the conversion of water insoluble polystyrene into a water soluble polystyrene, an appropriate amount of Dichloromethane is used. Then the hydroquinone peroxidase enzyme carries out the degradation process of water-soluble polystyrene (Nakamiya et al.,1997). Lately, serine hydrolase has also been reported to carry out the enzymatic degradation of polystyrene. The mineralization of PS monomer will result in the production of acetyl-coenzyme A (CoA) through β-oxidation. Next, in the microbial metabolism, the Acetyl-CoA will be utilized in the tricarboxylic acid cycle.

Polyvinyl acetate/alcohol (PVA) degradation is associated with PVA oxidase in the literature. The gram-negative bacteria, including *P. vesicularis* and the bacterial strains of *Pseudomonas* sp. have been reported to produce this oxidase enzyme (Wilkes and Aristilde, 2017). In addition, activity of PVA oxidase is in correlation with the activity of another enzyme known as PVA dehydrogenase. First, PVA dehydrogenase acts to introduce β-diketone group into the PVA polymer, resulting in the formation of a product which then initiates the oxidation of PVA molecule by PVA oxidase enzyme. PVA oxidase carries out the oxidation of the PVA via serine hydrolase active site. After that, PVA oxidase hydrolyzes the integrated β-diketone group within the PVA molecule to produce PVA monomer (Shimao et al., 2000). In contrast to other microplastics, literature on polypropylene biodegradation and isolated enzymes involved in its degradation is still scarce. In the end, it can be stated that although enzymatic degradation of some microplastics such as PE, PET, and PS is comprehensively known and understood, the information based on evidence on enzymatic degradation of other microplastics such as PP and PVC is still unknown. Therefore, further research to comprehend the biodegradation of these microplastics through enzymatic action is needed. Next, the methods used to analyze polymer biodegradation are discussed briefly for better understanding of microplastic degradation employing biotechnological strategies.

3.4 BIOTECHNOLOGICAL ASSESSMENT OF MICROPLASTIC BIODEGRADATION

Evaluation and analysis of plastic biodegradation can be done using standardized analytical methods. Physical changes in the polymer pertaining to surface erosion as an evidence of biodegradation can be observed either through the naked eye or the use of microscope. These observable changes include roughened surface of polymer, defragmentation, appearance of cracks or holes, discoloration, or bio-film appearance on the plastic surface. All of these are strong indications of plastic degradation. However, advanced microscopic techniques such as scanning electron microscopy (SEM) or atomic force microscopy (AFM) could provide detailed observation of the phenomenon. Thus, SEM and AFM have been employed in most of the research studies concerning biodegradation of plastics to observe eroded polymers after microbial degradation. Other techniques, such as fourier transform infrared spectroscopy

(FTIR), x-ray photoelectron spectroscopy (XPS) and x-ray diffraction (XRD) are also employed extensively for characterizing the degraded polymer.

3.4.1 Gravimetric Weight Loss Assessment

Another important technique used in various studies of polymeric biodegradation is the measure of loss of mass. Being a very useful and convenient technique, it is widely used in the area of research. However, this method does not provide a strong evidence of biodegradation when used alone in the assessment of degradation outcome. Particularly for plastics, where additives are added in the formulations, the gravimetric weigh loss analysis alone may provide misleading results. Furthermore, owing to very slow process of degradation, the loss in mass is usually very low and inaccuracy might affect the data as well. Therefore, the gravimetric weigh loss technique should always be used with other techniques for the assessment and characterization of biodegradation activities accurately.

3.4.2 Evaluation of Change in Mechanical Properties

Change in the physical properties of a polymer such as density, contact angle, viscosity, molecular weight distribution (using GPC), melting temperature (Tm), glass transition temperature (Tg) can be evaluated using Thermogravimetric analysis (TGA) and Differential scanning calorimetric (DSC) analysis whereas changes in the crystalline and amorphous regions are observed by using X-ray diffraction, Small angle X-ray scattering (SAXS) and Wide angle X-ray scattering (WAXS) techniques (Witt et al., 2001). Specific alterations in the polymer including tensile strength reduction, modification in chain elongation, and change of polymer molar mass can also be attributed as an evidence of biodegradation.

Such techniques are helpful in the analysis of biodegradation where breaking of bonds result in the reduction of molar mass, with small change in the total mass of the plastic, referring to the structural damage of the degraded polymer.

3.4.3 Evaluation of Change in Chemical Properties

Changes in the chemical properties of the plastic as a result of degradation by microbial attack may include the appearance or removal of functional groups in the polymer structure. These alterations can be easily observed by using Fourier transform infrared spectroscopy (FTIR). The chemical modifications of the polymer structure are widely analyzed with the help of FTIR in most of the research studies concerning bioremediation of plastics. Similarly, the use of micro-Fourier transform infrared (FTIR) has recently gained traction. It made the collection and observance of spectra in reflectance or transmittance and attainment of attenuated total reflectance mode possible. It also facilitated mapping as well as characterization of multiple distinctive polymers. In addition, micro-FTIR helped in the identification of microplastics having irregular shapes. The only drawback associated with this technique is the procedural requirement of linking the FTIR to the microscope, which renders it expensive and time-consuming. However, this method has been accepted widely as a reliable method used for the characterization of microplastics. For example, Horton et al. (2017) explored the utilization of micro-FTIR and suggested that the molecular mapping ability of this technique is very promising. Other methods applied for the identification of microplastics include pyrolysis-gas chromatography-mass spectrometry and Scanning electron microscope-energy dispersive spectroscopy (SEM-EDS). The most relevant techniques used for the analysis of molecular weight and molecular weight distribution of the degraded products or intermediates include Gas chromatography mass spectrometry (GCMS) and Nuclear magnetic resonance spectroscopy (NMR); whereas, the average molecular weight (Mn %) and the molecular weight distribution of the plastic material can be observed using Gel permeation chromatography (GPC). A reduction in the average molecular weight of the polymer is an evidence of chain cleavage and, in turn, of microbial degradation (Ali et al., 2014; Yang et al., 2015; Tian et al., 2017; Giacomucci et al., 2019). However, this technique may not detect early biodegradation of plastic since such degradation starts with the polymer surface modification whereas GPC is performed on the bulk polymer. Such drawback of this analytical technique renders it not highly sensitive. Therefore, GPC could be helpful when applied with other techniques for the analysis of recalcitrant plastic degradation.

3.4.4 Standardized Methods

For the analysis of the biodegradation of plastics under both aerobic and anaerobic conditions, various organizations, including the International Organization for Standardization (ISO), European Committee for Standardization, and American Society for Testing and Materials (ASTM), have developed different standard tests.

In case of aerobic conditions, microorganisms utilize oxygen to form carbon dioxide through oxidation of carbon. Those microorganisms then release carbon dioxide as one of the major metabolic end products. Accordingly, the consumption of oxygen, known as the Respirometric Test, or the formation of carbon dioxide called the Sturm Test, are good indicators for degradation of polymers. These tests are commonly used to analyze biodegradation under laboratory conditions. However, in the case of petroleum-based plastics this method could not be reliable as during biodegradation only a small amount of gas could be formed. Thereby, due to low sensitivity, this method should be used in combination with other analytical methods.

3.4.5 Radiolabeling

Radiolabeling is a technique which consists of labelling the carbon in the polymer with carbon isotope 14C. The labelled carbon would be utilized as substrate for the growth of microorganisms during the process of biodegradation. The radioactive gas produced ($14CO_2$, $14CH_4$) in the process is then measured as a confirmation of complete mineralization. (Tian et al., 2017). The availability of 14C labeled plastic material makes the measurements and the interpretation of the outcome straightforward. The additives or other impurities in the polymer do not interfere with the results obtained through this method. Moreover, the technique shows a high degree of precision and accuracy. Yet, the drawback linked with this method is its high cost as labeled materials are costly and are not readily accessible. Moreover, the issues associated with radioactive waste disposal may also be a drawback in this regard.

3.4.6 Advanced Analytical Techniques

Besides the above-mentioned analytical methods, other techniques have also been reported to evaluate the degradation of a few biodegradable polymers. But these techniques have not been standardized for common use in the evaluation of degradation of petroleum-derived conventional plastics through microbial action. The techniques are as follows:

Reflectometric interference spectroscopy (RIfS) is used for the monitoring of alteration in the physical thickness of a biodegradable polymer. This has been employed for analyzing the enzymatic degradation of thin polycaprolactone (PCL) polymer film. However, this technique has not been applied frequently in the case of synthetic non-biodegradable plastics.

Elemental analyser/isotope ratio mass spectrometry (EA/IRMS) is another advanced technique based on the quantification of carbon stable isotopes (δ13C), which may indicate degradation of plastic as demonstrated by an increase of δ13C values.

3.5 BIODEGRADATION STUDIES OF FOUR MAJOR SYNTHETIC POLYMERS (PE, PVC, PS, PP)

Numerous environmental hazards arise from physical and chemical degradation of plastic waste. Conversely, biodegradation of plastics by employing various microorganisms seems to be a feasible option. Considering the abundance of plastic degrading microbes in the environment and their specificity in attacking plastics, bioremediation is the most effective and ecofriendly approach for dealing with plastic pollution.

3.5.1 Polyethylene Biodegradation

Polyethylene films have been most commonly used to study the degradation of conventional plastics and polymers by using pure as well as mixed cultures under aerobic conditions (Roy et al., 2011; Restrepo-Flórez et al., 2014). Several bacteria have demonstrated the ability to degrade polyethylene (Tribedi and Dey, 2013) such as actinomycete *Rhodococcus ruber* (Sivan et al., 2006) and members of *Enterobacter, Pantoea* (Skariyachan et al., 2016), *Bacillus, Streptococcus, Staphylococcus, Moraxella and Pseudomonas, Cellulosimicrobium,, Brevibacillus,* and *Lysinibacillus* (Singh et al., 2018; Muhonja et al., 2018). In addition, the microbial strains capable of producing antibiotic were isolated from a limestone cave and investigated for their potential to degrade LDPE. The microbial consortium (*Bacillus licheniformis, Serrentia* sp, *Bacillus* sp, and *Stenotrophomonas* sp), after a month of incubation at 37°C resulted in polymer defragmentation as analyzed by SEM. Furthermore, the formation of new functional groups was also observed by using the FTIR analysis. According to that study, the use of calcium as a co metabolite in the culture medium also enhanced the biodegradation process (Jamil et al., 2015).

Singh et al. (2016) isolated and studied the potential of *Pseudomonas* sp., *Staphylococcus* sp., and *Bacillus* sp. to degrade PE. The plastic strips were pretreated with Tween 80 and bleach. The microorganisms were able to degrade plastic when incubated with pretreated PE strips for a period of 40 days. The study reported *Bacillus* sp. causing a highest 42.5% loss in plastic weight. Similarly, *Enterobacter* spp and *Pontoea* spp were isolated from plastic waste processing sites and investigated in consortium for their biodegrading ability. The novel bacterial consortia demonstrated enhanced LDPE degradation over the incubation period of 120 days. The microbial action of consortia contributed to 81% and 38% of weight reduction for LDPE strips and LDPE pellets, respectively (Skariyachan et al., 2016). However, no GPC analysis was done and information about polymer additives was not provided by the authors. The bacterial strains (from the genera *Comamonas, Delftia, and Stenotrophomonas*) isolated from plastic debris also demonstrated PE degradation after 90 days of incubation as revealed by SEM and attenuated total reflectance (ATR)-FTIR spectroscopic analyses. Likewise, a reduction of 46.7% in the viscous area of defragmented PE was observed by using Phase imaging as a tool to assess degradation (Peixoto et al., 2017).

Although there is limited evidence of biodegradation of Polyethylene terephthalate (PET) or its consumption to facilitate microbial growth, Yoshida et al. (2016) conducted a successful study and isolated a novel bacterium (namely *Ideonella sakaiensis*) from sludge samples collected near PET bottle recycling site. PET film was treated with the bacterial inoculum following an incubation period of 42 days at 30°C temperature. At the end of incubation, SEM analysis of the film revealed complete PET biodegradation. The researchers suggested hydrolytic activity of two enzymes (PETases) responsible for its degradation. Recently, Sarkhel et al. (2020) studied PET biodegrading ability of isolated strains of *Vibrio* sp. and *Aspergillus* sp. and reported plastic biodegradation of 35% by bacterial attack and 22% by fungal attack after an incubation period of 6 weeks. This outcome regarded bacteria as better degraders of PET as compared to fungi. Likewise, Eashur and Jasim (2020) collected soil samples from plastic garbage dumping site and isolated bacteria for investigating their PE degrading potential under laboratory conditions. The incubation period comprised of 40 days in a carbon deficient minimal salt medium. At the end of incubation, PE was subjected to gravimetric weight loss analysis for the assessment of degradation. The results indicated that 30.5% weight loss of the polymer was due to microbial action of *Pseudomonas fluorescens*, whereas19% reduction of in the weight of PE was due to the action of bacterial strain belonging to *Streptomyces* sp. Similarly, in a recent study conducted by Wu et al. (2023), it was confirmed that *Bacillus paramycoides*, isolated from a recycling plant dealing with waste mulch was able to degrade PE film. The study reported a 12% decrease in the initial weight of the film within 45 days after co-incubation with TW-2. While the surface tension of PE film increased from 9.755 to 31.013, a decrease in the surface hydrophobicity of film was observed. The biodegradation of PE was further confirmed using SEM, FTIR, and X-ray diffraction (XRD) analyses. Moreover, GPC analysis was done to substantiate the claim of plastic degradation, which showed a decrease in the molecular weight, whereas an increase in the number of fragments having low molecular weight was also recorded. These results strongly suggested that the isolated strain of bacterium was able to degrade PE to some extent under laboratory conditions.

Similar to bacterial degradation, polyethylene is well documented to be degraded by several fungal strains as well. A study investigating PE degradability of fungal consortia (Alternaria *alternate*, *Curvularia lunat*, *Penicillium* simplicissimum, and *Fusarium* sp.) obtained from a waste dumping site reported greater reduction in weight of plastic by fungal consortium (27%) as compared to individual fungal strains, confirmed by FTIR and SEM analyses. Moreover, the enzymes responsible for degradation were identified as laccase and manganese peroxidase by the researchers (Sowmya et al , 2015). *Aspergillus oryzae* was found to degrade PE strips in 60 days of incubation resulting in 30% of plastic weight loss (Indumathi and Gayathri, 2016), while after 120 days of incubation, the strain contributed to 36.4±5.53% mean weight reduction of PE film (Muhonja et al., 2018). The FTIR analysis of PE showed the appearance of aldehyde, carboxyl, and ether functional groups, and suggested fungi as better at degrading PE when compared to bacteria. In another study, a fungal species *Cladosporium halotolerans* was isolated from the digestive tract of *Galleria mellonella* larvae. This fungal species was able to degrade HDPE polymer by interacting with polymeric material using its hyphae. Moreover, secretion of enzymes involved in degradation of polyethylene helped in the process of biodegradation (Di Napoli et al., 2023).

Bacteria and fungi are also used in consortium for investigating plastic degrading ability. *Lysinibacillus xylanilyticus* and *Aspergillus niger* isolated from landfill soil samples were reported to colonize and modify LDPE (UV pretreated and non-UV irradiated) films. After 126 days of incubation with developed consortium, about 29.5% biodegradation occurred in the UV irradiated film while 15.8% biodegradation occurred in non-UV irradiated film. The findings were confirmed using FTIR, SEM, and XRD analyses (Esmaeili et al., 2013). Likewise, Padmanabhan et al. (2020) isolated microorganisms from grease, petroleum products as well as cooking oil to investigate their potential for degrading polyethylene. After screening of several microorganisms, two bacterial species (*Sphingomonas* and *Pseudomonas aeuroginosa*) and three fungal species (including *Aspergillus niger*, *Aspergillus flavus*) were investigated for plastic degradation by giving an incubation period of 35 days with the plastic strips. During the process of biodegradation, an alteration of pH from 7 to 8 was observed. At the end of the study, it was reported that *Sphingomonas* resulted in the highest plastic degradation while an unknown fungal strain also caused maximum plastic breakdown as revealed by the weight loss of plastic strips as well as modifications in the polymer structure.

3.5.2 Polyvinylchloride Biodegradation

A few studies have been conducted on PVC biodegradation. The bacterial strains isolated from soil samples contaminated with petroleum oil were investigated for their potential to biodegrade PVC. After an incubation period of 180,the tensile strength of PVC was evaluated as a tool to assess biodegradation. It was reported that a 53% reduction in tensile strength of PVC was attributed to *Pseudomonas aeruginosa*, whereas *Achromobacter sp.* acted to result in 43% reduction. Moreover, deterioration of polymer surface and structural modifications were also reported using SEM and FTIR analyses (Das et al., 2012). Whereas a developed bacterial consortium (*Pseudomonas otitidis, Bacillus aerius, Acanthopleuribacter pedis* and *Bacillus cereus*) better degraded PVC than LDPE film in natural conditions in a given 90 days of incubation period as analyzed through FT-IR spectroscopy and SEM (Anwar et al., 2013). Biodegradation of PVC by *Pseudomonas citronellolis* and *Bacillus flexus* has also been reported. The reduction in average molecular weight of 10% for polyvinylchloride incubated with *Pseudomonas citronellolis* was observed in the GPC analysis performed at the end of incubation period of 45 days. Furthermore, the researchers set up waste PVC film biodegradation by employing this strain for an incubation period of 30 days. This resulted in 19% reduction in the weight of PVC at the end of incubation, which reiterated the potential of the strain to degrade PVC plastics (Giacomucci et al., 2019).

An interesting comparison between biodegradation and photocatalysis of composite film containing PVC, polycaprolactone (PCL), and Zinc Oxide was drawn by Nag et al. (2020). The samples of soil were collected from municipal solid waste dumping site from which PVC degrading strain of *Bacillus altitudinis* was isolated. It was revealed in the study that photocatalysis resulted in maximum degradation of 17.8% in 2 hours while biodegradation took 25 days to result in 10% weight reduction of the plastic film. In a most recent study, the researchers conducted biodegradation of additive-free PVC by bacterial consortium obtained from the gut of *Tenebrio molitor* larvae. The plastic films were inoculated with bacterial consortium for about a month. After the period of incubation ended, the films were analyzed by various analytical methods including atomic force microscopy, SEM, water contact angle and time-of-flight secondary ion mass spectrometry. In addition, the films were also subjected to evaluation by FTIR, differential scanning calorimetry, thermogravimetric analysis technique, and ion chromatography. It was observed that biofilms were formed on bio-treated PVC films and there was an increase in the functional groups while chlorine contents were decreased. Moreover, polymer surface modifications were visible with the appearance of erosions and cracks on the surfaces of PVC films. Meanwhile, an increase in the hydrophilicity of bio-treated films was also observed with the decrease in their thermal stability. Biodegradation was also confirmed with the help of GPC. Furthermore, Gas chromatography-mass spectrometry analysis revealed the presence of three medium-chain aliphatic primary alcohols and their corresponding fatty acids identified as PVC degradation intermediates (Xu et al., 2023).

Fungal strains have also been reported to degrade PVC. In particular, the capability of several fungi to biodegrade PVC plastic was investigated by Webb et al. (2000). The strains included *Penicillium, Thanatephorous, Emericella, Aureobasidium, Cladosporium, Paecilomyces, Alternaria, Aspergillus, Phaeococcomyces sp.*, Epicoccum, *Taphrina, Kluyveromyces, Rhodotorula*. Isolation of the strains involved in the study was done from the surface of plasticized PVC films exposed to the atmosphere for up to 95 weeks. One month of incubation at 25°C was observed for all the fungal isolates for biodegradation of PVC film pieces placed on agar surface. At the end of incubation, the researchers reported a reduction of 7% gravimetric weight of the plastic film. In another study, an experimental setup was arranged placing thin pure PVC films in the sterilized soil and soaking with sewage sludge for a long incubation period of 10 months. When the incubation ended, fungal strains including *Aspergillus niger, Aspergillus sydowii, Lentinus tigrinus,* and *Phanerochaete chrysosporium,* were isolated. Furthermore, it was observed that microbial isolates were able to degrade PVC after 49 days of incubation. Specifically, decreased polymer molecular weight; polymer surface alterations and FTIR profile were monitored in the presence of *Phanerochaete chrysosporium* (Ali et al.,2014).

3.5.3 Polystyrene Biodegradation

The biodegradation of PS was studied (Kaplan et al., 1979) employing seventeen fungal species demonstrating the potential to degrade plastics. The study revealed a very low (< 1%) rate of polystyrene biodegradation after the incubation of PS in the soils containing fungi, invertebrates, and several microbes was conducted for a period of three months. Further, it was reported that the rate of plastic degradation was not improved with the addition of minerals and cellulose. Mostly pure microbial cultures have been employed in various studies concerning the biodegradation of polystyrene (Raddadi and Fava, 2019). A study confirmed the ability of actinomycete *Rhodococcus ruber* strain to form biofilm on pure PS flakes within a few hours. However, only a small reduction in the weight of PS, i.e., 0.8% of gravimetric weight loss was observed due to microbial action within an incubation period of 56 days. In addition, the study confirmed an increase in the formation of biofilm particularly attributed to the addition of mineral oil (Mor and Sivan,2008).

In another study, microbial strains of *Pseudomonas aeruginosa, Staphylococcus aureus, Streptococcus pyogenes, Bacillus subtilis*, and *Aspergillus niger* were isolated from the soil samples after giving an incubation period of four months. These strains were then subjected to an investigation of their ability to degrade PET and PS strips separately. After one month of incubation, it was observed that *Bacillus subtilis* resulted in the maximum percentage loss in the weight of PS and demonstrated greater potential for degrading PS (Asmita et al.,2015). In a relatively different kind of a study conducted by Ward et al. in 2006, polystyrene was used for the synthesis of biodegradable plastic. First, polystyrene was converted to styrene oil through pyrolysis, then *Pseudomonas putida* was employed by the researchers to convert styrene oil to polyhydroxyalkanoate (PHA). By using styrene oil as a carbon source, the bacterial strain was converted 1g of styrene oil into 62.5 mg of polyhydroxyalkanoatein shake flasks while 250 mg of biomass was generated in the process. In the end, it was concluded, 64 g of polystyrene was utilized to produce 6.4 g of PHA in an overall procedural requirement of a pyrolysis and four fermentation runs.

Recently, Johnston et al. (2018) conducted a similar study employing *Cupriavidus necator* for the conversion of the pretreated polystyrene waste to PHAs. Oxidative degradation was given as a pretreatment to high impact and general purpose PS flakes used as plastic waste in the study. The plastics were exposed to ozone and defragmented at varied high temperatures. For fermentation, the bacterial strain demonstrated the potential to grow in the presence of oxidized Mn-myristicate and Co-stearate having waste PS as carbon source and produce biodegradable plastic (PHAs).

Sincemonooxygenases, alkane hydroxylases, andcytochrome P4500sare capable of breaking C–C bonds, these enzymes have been ranked as the top potent enzyme classes that can degrade PS(Hou and Majumder, 2021).Certain enzymes such as Ring-hydroxylating dioxygenases may be able to cleave the polymer chain of polystyrene and carry out the oxidation of aromatic ring compounds produced from degradation of plastic. These specific enzymes were abundant in Proteobacteria, Actinobacteria, Bacteroidetes, and Firmicutes, signifying a promising potential for degradation of polystyrene in different environments. Most recently, in a study the diversified polystyrene-degrading bacterial community was isolated from PS waste dumping sites from a subtropical mangrove area. The degradation consortia was dominated by bacteria of Sphingomonadaceae, Rhodanobacteraceae, Rhizobiaceae, Dermacoccaceae, Rhodocyclaceae, Hyphomicrobiaceae, and Methyloligellaceae sharing similar microbial community structure. The members belonging to the genera of Alcanivorax, Flexivirga, Mesobacillus, Cytobacillus, Thioclava, Novosphingobium, Gordonia, Stappia, and Thalassospira demonstrated their ability to degrade PS. Furthermore, degradation of polystyrene by Gordonia sp. and *Novosphingobium sp.* was observed by weight reduction, SEM, ATR-FTIR, and contact angle analyses, all of which revealed the morphological and structural modifications of the PS films attributed to microbial degradation. The appearance of new functional groups containing oxygen suggested the oxidative degradation. Although, after 30 days of incubation under laboratory conditions, the degradation rates ranging from 2.7% to 7.7% were low but their role in PS degradation is still promising (Liu et al., 2023).

3.5.4 Polypropylene Biodegradation

Biodegradation of polypropylene was investigated by Arkatkar et al. (2010) using unpretreated and pretreated PP films. The chemically pretreated PP films were given aquaregia and fenton as pretreatments and physically pretreated PP films were exposed to thermal and short UV radiations as pretreatment. The bacterial isolates were *Pseudomonas azotoformans*, *Pseudomonas stutzeri*, *Bacillus flexus*, and *Bacillus subtilis*, all of which were isolated from plastic garbage dumping sites. For the study of biodegradation, plastic films were incubated in minimal media, inoculated separately with each bacterial isolate for up to one year. The formation of biofilm by all four microorganisms was observed in the study but maximum degradation was observed mainly by *B. flexus* on the UV pretreated polymer which resulted in 2.5% weight loss of the plastic film. When a similar kind of study (Jeyakumar et al., 2013) on UV pretreated pro-oxidant blended PP was done employing *Engyodontium album*and *Phanerochaete chrysosporium*, the researchers reported that after365 days of incubation, about 18.8% and 9.42% gravimetric weight losses in the plastic were attributed to microbial attack by *Phanerochaete chrysosporium* and *Engyodontium album* respectively. Moreover, the synergistic effect of polymer blending as well as pretreatment on the rate of degradation was also suggested by the researchers (Wang et al., 2022).

Fontanella et al. (2013) compared the degradation of additives free PP films and PP polymer having Mn+Fe or Mn additive by the microbial action of *Rhodococcus Rhodochrous*. After six months of incubation, it was evident that the microbial strain was able to grow and biodegrade additives containing PP, whereas no changes in the additives free PP films were observed. The degradation assessment was done by analyzing ADP/ATP ratio, ATP content measurement, and cell viability corresponding to the metabolic activity of microorganisms. For the biodegradation study of additive free PP, Jeon and Kim (2016) investigated the PP biodegrading potential of Stenotrophomonas panacihumi under composting conditions by giving three months of incubation at 37°C. The polymer biodegradation was confirmed by GPC analysis. In a study conducted by Jeon et al. (2021), *Lysinibacillus sp.*, isolated and identified as a novel strain, was demonstrated to degrade polyethylene and polypropylene in 26 days by resulting in approximately 4 and 9% weight reduction respectively. During the biodegradation process, a roughened surface referring to surface erosion was observed by SEM. Furthermore, the biodegradation of polyethylene and polypropyleneal so resulted in the production of various oxidation products containing CH2 groups as observed in GC-MS analysis. Recently, in a study, two strains of *Pseudomonas aeruginosa*, were isolated on the basis of their potential to utilize PP as a carbon source. Afterwards, the degradation by the bacterial strains was confirmed using diverse analytical methods such as weight loss, SEM, contact angle, FTIR, Thermal Gravimetric Analysis, and Differential Scanning Calorimetry. Consequently, the strains were able to degrade PP particles causing a weight reduction of 9.35% ± 2.22 and 17.2% ± 1.56, respectively after 40 days of incubation. Moreover, PP surface showed signs of surface erosion and formation of functional group containing oxygen. Whereas, the hydrophobicity and thermal stability of plastic were also decreased suggesting biodegradation.

In another study, *Aspergillus sp.* and *Penicillium sp.* fungi were employed for the degradation of neat PP and Polypropylene/poly (butyleneadipate-co terephthalate)/ thermoplastic starch blends. The polymers were incubated with the microbial strains separately for a month at 29°C. At the end of incubation, no significant changes in the mass loss of polymer were observed with the addition of PBAT/TPS to the PP. However, the results based on FTIR spectra, contact angle analysis and microscopic images indicated a slight degradation, which was attributed to improved fungal attack in PP blends (De Oliveira et al.,2020). Most recently, in a study the researchers isolated seven bacterial as well as seven fungal strains with the ability to degrade polyethylene. The bacterial isolates included three *Priestia megaterium strains, Klebsiella pneumoniae, Pseudomonas fluorescens, Enterobacter ludwigii, Chryseobacterium sp.*, while fungal isolates included four *Fusarium spp., two Lecanicillium spp. and Trichoderma sp.* In addition, microbial isolates having the ability to degrade polypropylene were also isolated. These included seven bacterial strains (five *Serratia marcescens and two Enterobacter spp.)* and six fungal strains (four *Aspergillus spp., Fusarium oxysporum and Penicillium granulatum).* The formation of a biofilm as well as and surface alterations in both PE and PP pellets were confirmed in the SEM analysis. However, the most potent modifications(microcracks and corrugations) strongly suggesting degradation were obtained for plastic treated with bacteria. FTIR analysis revealed the polymer structural alterations on the plastic surfaces (Wróbel et al., 2023).

3.5.5 Future Recommendations

Different pretreatments including UV radiations or addition of chemicals to facilitate the process of plastic breakdown are being applied broadly in several bioremediation studies. These pretreatments have been shown to aid biodegradation of synthetic plastics. For example, another strategy to increase the biodegradability of synthetic-based microplastics is to add pro-oxidant and biodegradable additives such as starch for LDPE and HDPE, PVC and PS polymers (Zadjelovic et al., 2020). The hydrophilicity of plastics is increased with the addition of these additives, which in turn increases the enzyme's efficiency for degrading microplastics. Cacciari et al. (1993) suggested that the presence of starch in turn increases the degradation of polypropylene and polyethylene via hydrolysis by microorganisms. Furthermore, the enzymatic action adds functional groups which leads to increased hydrophilicity of the polymer, rendering it more accessible for microbial degradation.

Moreover, it is recommended that hydrolysis of PET through enzyme action should proceed under an optimum temperature close to the PET glass transition temperature; hence, PET depolymerases with high thermostability are needed. Therefore, several methods have been analyzed to make PET hydrolases more thermostable in this regard. These methods include glycosylation of enzyme, genetically engineered enzymes, enzyme immobilization, use of surfactants and additives (Maurya et al., 2020). Moreover, in the degradation of PET microplastics by employing bacterial whole-cell, it is suggested that the combined effect of alkali hydrolysis and degradation through microbes accelerates the overall hydrolysis of PET.

By facilitating the interaction between polymers and microplastic-degrading microorganisms, enhanced biodegradation of microplastics could be achieved as reported by Tribedi and Alok (2013). The authors claimed that the biodegradation of PE was made more efficient with the addition of mineral oil which led to increased hydrophobicity and strengthened adhesion of polymer surface. It is a well-documented fact that biofilms play a significant role in the biodegradation of microplastics. Hence, the biotechnological approaches aimed at increasing the formation of biofilms could be applied for the acceleration of degradation of microplastics by microorganisms.

Moreover, the efficiency of enzymes involved in the process of biodegradation can be enhanced significantly by using Protein/enzyme engineering of microorganisms capable of conducting plastic degradation. For example, three mutant PETase (R61A, L88F and I179F) from *I. sakaiensis* obtained through protein engineering demonstrated 1.4, 2.1, and 2.5-fold higher enzymatic activity in contrast to wild type strain's enzyme efficiency (Ma et al., 2018). Strain engineering is a well-known biotechnological strategy applied on microorganisms to enhance their enzymatic efficiencies. Likewise, the production of PETase from *I. sakaiensis* was facilitated by a microbial cell factory from a photosynthetic microalga i.e., *Phaeodactylum tricornutum* (Moog et al., 2019). In addition, it is seen that microbiomes or complex microbial consortia when compared with single strains are more likely to survive in new environments owing to their enhanced adaptive capacity. In fact, they demonstrate greater degradative ability towards complex environmental pollutants as compared to single strains. Therefore, when selecting the most feasible method for biodegradation of microplastics, it is important to consider the efficiency as well as applicability of the method in real environments. Since, the enzyme efficiency to enhance the degradation process of microplastics can be improved using novel biotechnological approaches, the method of immobilized enzyme/microorganism complexes can be explored and manipulated for better outcomes in this regard.

Recent advances in the field of biotechnology have opened new avenues not only in designing new strategies for effective degradation of synthetic plastics, but also for the formation of novel biodegradable materials. The understanding of biochemical degradation mechanisms, help in the manufacturing of such polymers that may easily degrade when exposed to certain conditions, including varied temperature, salinity, and humidity. Such detailed understanding of the plastics degradation process may also help in developing materials that are easily convertible into productive materials as per requirement. For example, the conversion of non-degradable polymers into raw materials for sustainable supply chains gives promising prospects. Similarly, waste plastics can produce pure organic acids after a series of biochemical treatments is applied on them. Such degraded products can then be utilized as raw materials in the manufacturing of new degradable plastics.

Another important aspect in the biodegradation of microplastics is the need of finding more specific microbial strains and biocatalysts to get rid of plastic pollution. For this purpose, the researchers should consider the degradative ability of isolated microbial strains not only under specific laboratory conditions but also in natural environments to study their ecological flexibility as well. By doing so, the adaptive capacity of selected microbial strains to survive in new environments would get the required importance in the study of degradation. Therefore, the development of unique microbial degradation techniques are necessary to make the removal of microplastics pollutants from the environment possible.

The future studies on bioremediation of microplastics should consider the development of successful mathematical models for the depiction of microplastics in the surroundings. This would make the exploration of microplastics in the environment much more easy and accurate. In addition, a standard method for extraction of microplastics should be established by optimizing the several existing extraction procedures for microplastics. All the analytical techniques and instruments, including Micro-IR Spectroscopy, Gas Chromatography-Mass Spectrum, and others should be evaluated in this regard. Furthermore, there should be an accurate assessment of ecological risks posed by the presence of tiny plastic particles in the environment.

3.6 SUMMARY AND CONCLUSION

This chapter highlights the degradation of synthetic microplastics along with the comprehensive discussion on polymer biodegradation process, microbial communities as well as identified enzymes involved in the degradation process of microplastics. In order to provide better insights into plastic degradation, the standard analytical procedures generally used for biodegradation assessment have also been reviewed in this chapter. Existing substantial findings of polyethylene, polyvinylchloride, polystyrene, and polypropylene biodegradation studies under different environmental conditions with all the microorganisms involved are discussed thoroughly in this chapter. Moreover, some recommendations are made by keeping the literature available on plastic degradation in mind. To deal with microplastic pollution, the phenomenon of biodegradation is regarded as the most promising and least harmful process for the environment. Hence, the degradation of synthetic polymers through microorganisms has been considered an eco-friendly approach and more research studies need to be done to make enhanced progress in this respect.

In the end, it can be concluded that microplastics present a mixture of different polymers; therefore, the advanced methods of plastic degradation can be optimized for the removal of these tiny plastic particles from specific environments. Improving microbial strains involved in plastic degradation can also be used for enhancing polymer biodegradation process. Engineering of microorganisms, as well as optimization of different factors affecting rate of biodegradation such as pH, temperature and different culture conditions, can be exploited for obtaining best outcomes. Lastly, extensive research is needed for the manufacturing of environment-friendly biodegradable plastics from renewable sources and strict implementation of their use in several industries as an alternative of synthetic plastics is required.

REFERENCES

Ali, M. I., Ahmed, S., Robson, G., Javed, I., Ali, N., Atiq, N., & Hameed, A. (2014). Isolation and molecular characterization of polyvinyl chloride (PVC) plastic degrading fungal isolates. *Journal of Basic Microbiology*, *54*(1), 18–27.

Anwar, M., Negi, H., Zaidi, M., Haider, G., Gupta, S., & Goel, R. (2013). Biodeterioration studies of thermoplastics in nature using indigenous bacterial consortium. *Brazilian Archives of Biology and Technology*, *56*, 475–484.

Arkatkar, A., Juwarkar, A. A., Bhaduri, S., Uppara, P. V., & Doble, M. (2010). Growth of Pseudomonas and Bacillus biofilms on pretreated polypropylene surface. *International Biodeterioration & Biodegradation*, *64*(6), 530–536.

Asmita, K., Shubhamsingh, T., & Tejashree, S. (2015). Isolation of plastic degrading micro-organisms from soil samples collected at various locations in Mumbai, India. *International Research Journal of Environment Sciences*, *4*(3), 77–85.

Cacciari, I., Quatrini, P., Zirletta, G., Mincione, E., Vinciguerra, V., Lupattelli, P., & Giovannozzi Sermanni, G. (1993). Isotactic polypropylene biodegradation by a microbial community: physicochemical characterization of metabolites produced. *Applied and Environmental Microbiology*, *59*(11), 3695–3700.

Das, G., Bordoloi, N. K., Rai, S. K., Mukherjee, A. K., & Karak, N. (2012). Biodegradable and biocompatible epoxidized vegetable oil modified thermostable poly (vinyl chloride): Thermal and performance characteristics post biodegradation with Pseudomonas aeruginosa and Achromobacter sp. *Journal of Hazardous Materials*, *209*, 434–442.

De Oliveira, T. A., Barbosa, R., Mesquita, A. B., Ferreira, J. H., de Carvalho, L. H., & Alves, T. S. (2020). Fungal degradation of reprocessed PP/PBAT/thermoplastic starch blends. *Journal of Materials Research and Technology*, *9*(2), 2338–2349.

Devi, R. S., Ramya, R., Kannan, K., Antony, A. R., & Kannan, V. R. (2019). Investigation of biodegradation potentials of high density polyethylene degrading marine bacteria isolated from the coastal regions of Tamil Nadu, India. *Marine Pollution Bulletin*, *138*, 549–560.

Di Napoli, M., Silvestri, B., Castagliuolo, G., Carpentieri, A., Luciani, G., Di Maro, A., ... & Varcamonti, M. (2023). High density polyethylene (HDPE) biodegradation by the fungus Cladosporium halotolerans. *FEMS Microbiology Ecology*, *99*(2), fiac148.

Eashur, E. Y., & Jasim, M. A. (2020). Plastic waste biodegradation by local bacterial isolates in Ramadi City. *Journal of University of Anbar for Pure Science*, *14*(1), 1–5.

Esmaeili, A., Pourbabaee, A. A., Alikhani, H. A., Shabani, F., & Esmaeili, E. (2013). Biodegradation of low-density polyethylene (LDPE) by mixed culture of Lysinibacillus xylanilyticus and Aspergillus niger in soil. *Plos one*, *8*(9), e71720.

Fontanella, S., Bonhomme, S., Brusson, J. M., Pitteri, S., Samuel, G., Pichon, G., ... & Delort, A. M. (2013). Comparison of biodegradability of various polypropylene films containing pro-oxidant additives based on Mn, Mn/Fe or Co. *Polymer degradation and stability*, *98*(4), 875–884.

Fontanella, S., Bonhomme, S., Koutny, M., Husarova, L., Brusson, J. M., Courdavault, J. P., ... & Delort, A. M. (2010). Comparison of the biodegradability of various polyethylene films containing pro-oxidant additives. *Polymer degradation and stability*, *95*(6), 1011–1021.

Gajendiran, A., Krishnamoorthy, S., & Abraham, J. (2016). Microbial degradation of low-density polyethylene (LDPE) by Aspergillus clavatus strain JASK1 isolated from landfill soil. *3 Biotech*, *6*, 1–6.

Ghatge, S., Yang, Y., Ahn, J. H., & Hur, H. G. (2020). Biodegradation of polyethylene: a brief review. *Applied Biological Chemistry*, *63*, 1–14.

Giacomucci, L., Raddadi, N., Soccio, M., Lotti, N., & Fava, F. (2019). Polyvinyl chloride biodegradation by Pseudomonas citronellolis and Bacillus flexus. *New Biotechnology*, *52*, 35–41.

Gómez-Méndez, L. D., Moreno-Bayona, D. A., Poutou-Pinales, R. A., Salcedo-Reyes, J. C., Pedroza-Rodríguez, A. M., Vargas, A., & Bogoya, J. M. (2018). Biodeterioration of plasma pretreated LDPE sheets by Pleurotus ostreatus. *PloS one*, *13*(9), e0203786.

Gravouil, K., Ferru-Clément, R., Colas, S., Helye, R., Kadri, L., Bourdeau, L., ... & Ferreira, T. (2017). Transcriptomics and lipidomics of the environmental strain Rhodococcus ruber point out consumption pathways and potential metabolic bottlenecks for polyethylene degradation. *Environmental Science & Technology*, *51*(9), 5172–5181.

Gupta, K. K., & Devi, D. (2020). Characteristics investigation on biofilm formation and biodegradation activities of Pseudomonas aeruginosa strain ISJ14 colonizing low density polyethylene (LDPE) surface. *Heliyon*, *6*(7).

Horton, A. A., Svendsen, C., Williams, R. J., Spurgeon, D. J., & Lahive, E. (2017). Large microplastic particles in sediments of tributaries of the River Thames, UK–Abundance, sources and methods for effective quantification. *Marine Pollution Bulletin*, *114*(1), 218–226.

Hou, L., &Majumder, E. L. W. (2021). Potential for and distribution of enzymatic biodegradation of polystyrene by environmental microorganisms. *Materials*, *14*(3), 503.

Indumathi, A., Gayathri, T., Biotechnology, E., Nadu, T., & Nadu, T. (2016). Plastic Degrading ability of Aspergillus oryzae isolated from the garbage dumping sites of Thanjavur, India. *Int. J. Curr. Microbiol. Appl. Sci*, *8* (13), 2319–7706.

Jamil, S.U.u., S. Zada, I. Khan, W. Sajjad, M. Rafiq, A.A. Shah, and F. Hasan. (2015). Biodegradation of polyethylene by bacterial strains isolated from Kashmir Cave, Buner, Pakistan. *Journal of Cave and Karst Studies*. 79(1): 73–80.

Jeon, H. J., & Kim, M. N. (2016). Isolation of mesophilic bacterium for biodegradation of polypropylene. *International Biodeterioration & Biodegradation*, *115*, 244–249.

Jeon, J. M., Park, S. J., Choi, T. R., Park, J. H., Yang, Y. H., & Yoon, J. J. (2021). Biodegradation of polyethylene and polypropylene by Lysinibacillus species JJY0216 isolated from soil grove. *Polymer Degradation and Stability*, *191*, 109662.

Jeyakumar, D., Chirsteen, J., & Doble, M. (2013). Synergistic effects of pretreatment and blending on fungi mediated biodegradation of polypropylenes. *Bioresource Technology*, *148*, 78–85.

Johnston, B., Radecka, I., Hill, D., Chiellini, E., Ilieva, V. I., Sikorska, W., ... & Kowalczuk, M. (2018). The microbial production of polyhydroxyalkanoates from waste polystyrene fragments attained using oxidative degradation. *Polymers*, *10*(9), 957.

Kang, B. R., Kim, S. B., Song, H. A., & Lee, T. K. (2019). Accelerating the biodegradation of high-density polyethylene (HDPE) using Bjerkandera adusta TBB-03 and lignocellulose substrates. *Microorganisms*, *7*(9), 304.

Kaplan, D. L., Hartenstein, R., & Sutter, J. (1979). Biodegradation of polystyrene, poly (metnyl methacrylate), and phenol formaldehyde. *Applied and Environmental Microbiology*, *38*(3), 551–553.

Kawai, F., Kawabata, T., & Oda, M. (2019). Current knowledge on enzymatic PET degradation and its possible application to waste stream management and other fields. *Applied Microbiology and Biotechnology*, *103*, 4253–4268.

Liu, R., Zhao, S., Zhang, B., Li, G., Fu, X., Yan, P., & Shao, Z. (2023). Biodegradation of polystyrene (PS) by marine bacteria in mangrove ecosystem. *Journal of Hazardous Materials*, *442*, 130056.

Ma, Y., Yao, M., Li, B., Ding, M., He, B., Chen, S., ... & Yuan, Y. (2018). Enhanced poly (ethylene terephthalate) hydrolase activity by protein engineering. *Engineering*, *4*(6), 888–893.

Maurya, A., Bhattacharya, A., & Khare, S. K. (2020). Enzymatic remediation of polyethylene terephthalate (PET)–based polymers for effective management of plastic wastes: an overview. *Frontiers in Bioengineering and Biotechnology*, *8*, 602325.

Moog, D., Schmitt, J., Senger, J., Zarzycki, J., Rexer, K. H., Linne, U., ... & Maier, U. G. (2019). Using a marine microalga as a chassis for polyethylene terephthalate (PET) degradation. *Microbial Cell Factories*, *18*, 1–15.

Mor, R., & Sivan, A. (2008). Biofilm formation and partial biodegradation of polystyrene by the actinomycete Rhodococcus ruber: Biodegradation of polystyrene. *Biodegradation*, *19*, 851–858.

Muhonja, C. N., Makonde, H., Magoma, G., & Imbuga, M. (2018). Biodegradability of polyethylene by bacteria and fungi from Dandora dumpsite Nairobi-Kenya. *PloS one*, *13*(7), e0198446.

Nag, P., Mukhopadhyay, A., Debnath, A., Roy, A., & Chakrabarti, S. (2020). Degradable Plastic Composite Film—A Comparison Between Photocatalytic and Biodegradation. In *Advances in Bioprocess Engineering and Technology: Select Proceedings ICABET 2020* (pp. 249–256). Springer Singapore.

Nakamiya, K., Sakasita, G., Ooi, T., & Kinoshita, S. (1997). Enzymatic degradation of polystyrene by hydroquinone peroxidase of Azotobacter beijerinckii HM121. *Journal of Fermentation and Bioengineering*, *84*(5), 480–482.

Padmanabhan, L., Varghese, S., Patil, R. K., Rajath, H. M., Krishnasree, R. K., & Shareef, M. I. (2020). Ecofriendly degradation of polyethylene plastics using oil degrading microbes. *Recent Innovations in Chemical Engineering (Formerly Recent Patents on Chemical Engineering)*, *13*(1), 29–40.

Park, S. Y., & Kim, C. G. (2019). Biodegradation of micro-polyethylene particles by bacterial colonization of a mixed microbial consortium isolated from a landfill site. *Chemosphere*, *222*, 527–533.

Peixoto, J., Silva, L. P., & Krüger, R. H. (2017). Brazilian Cerrado soil reveals an untapped microbial potential for unpretreated polyethylene biodegradation. *Journal of Hazardous Materials*, *324*, 634–644.

Raddadi, N., & Fava, F. (2019). Biodegradation of oil-based plastics in the environment: Existing knowledge and needs of research and innovation. *Science of the Total Environment*, *679*, 148–158.

Restrepo-Flórez, J. M., Bassi, A., & Thompson, M. R. (2014). Microbial degradation and deterioration of polyethylene–A review. *International Biodeterioration & Biodegradation*, *88*, 83–90.

Roy, P. K., Hakkarainen, M., Varma, I. K., & Albertsson, A. C. (2011). Degradable polyethylene: fantasy or reality. *Environmental Science & Technology*, *45*(10), 4217–4227.
Russell, J. R., Huang, J., Anand, P., Kucera, K., Sandoval, A. G., Dantzler, K. W., ... & Strobel, S. A. (2011). Biodegradation of polyester polyurethane by endophytic fungi. *Applied and Environmental Microbiology*, *77*(17), 6076–6084.
Sangale, M. K., Shahnawaz, M., & Ade, A. B. (2019). Potential of fungi isolated from the dumping sites mangrove rhizosphere soil to degrade polythene. *Scientific Reports*, *9*(1), 5390.
Sarkhel, R., Sengupta, S., Das, P., & Bhowal, A. (2020). Comparative biodegradation study of polymer from plastic bottle waste using novel isolated bacteria and fungi from marine source. *Journal of Polymer Research*, *27*, 1–8.
Shabbir, S., Faheem, M., Ali. N., Kerr, P. G., Wang, L. F., Kuppusamy, S., & Li, Y. (2020). Periphytic biofilm: An innovative approach for biodegradation of microplastics. *Science of the Total Environment*, *717*, 137064.
Sheik, S., Chandrashekar, K. R., Swaroop, K., & Somashekarappa, H. M. (2015). Biodegradation of gamma irradiated low density polyethylene and polypropylene by endophytic fungi. *International Biodeterioration & Biodegradation*, *105*, 21–29.
Shimao, M., Tamogami, T., Kishida, S., & Harayama, S. (2000). The gene pvaB encodes oxidized polyvinyl alcohol hydrolase of Pseudomonas sp. strain VM15C and forms an operon with the polyvinyl alcohol dehydrogenase gene pvaA. *Microbiology*, *146*(3), 649–657.
Singh, G., Singh, A. K., & Bhatt, K. (2016). Biodegradation of polythenes by bacteria isolated from soil. *International Journal of Research and Development in Pharmacy & Life Sciences*, *5*(2), 2056–2062.
Singh, P., Mir, I. H., & Sharma, M. K. (2018). Biodegradation Of Plastics By Microorganisms As Plastic Degradation Agents. *World Journal of Pharmaceutical Research*, *7*(17), 490–501.
Sivan, A., Szanto, M., & Pavlov, V. (2006). Biofilm development of the polyethylene-degrading bacterium Rhodococcus ruber. *Applied Microbiology and Biotechnology*, *72*, 346–352.
Skariyachan, S., Manjunatha, V., Sultana, S., Jois, C., Bai, V., & Vasist, K. S. (2016). Novel bacterial consortia isolated from plastic garbage processing areas demonstrated enhanced degradation for low density polyethylene. *Environmental Science and Pollution Research*, *23*, 18307–18319.
Skariyachan, S., Patil, A. A., Shankar, A., Manjunath, M., Bachappanavar, N., & Kiran, S. (2018). Enhanced polymer degradation of polyethylene and polypropylene by novel thermophilic consortia of Brevibacillus sps. and Aneurinibacillus sp. screened from waste management landfills and sewage treatment plants. *Polymer Degradation and Stability*, *149*, 52–68.
Soleimani, Z., Gharavi, S., Soudi, M., & Moosavi-Nejad, Z. (2021). A survey of intact low-density polyethylene film biodegradation by terrestrial Actinobacterial species. *International Microbiology*, *24*, 65–73.
Sowmya, H. V., Ramalingappa, B., Nayanashree, G., Thippeswamy, B., & Krishnappa, M. (2015). Polyethylene degradation by fungal consortium. *International Journal of Environmental Research*, *9*(3), 823–830.
Tian, L., Kolvenbach, B., Corvini, N., Wang, S., Tavanaie, N., Wang, L., ... & Ji, R. (2017). Mineralisation of 14C-labelled polystyrene plastics by Penicillium variable after ozonation pre-treatment *New Biotechnology*, *38*, 101–105.
Tribedi, P., & Sil, A. K. (2013). Low-density polyethylene degradation by Pseudomonas sp. AKS2 biofilm. *Environmental Science and Pollution Research*, *20*, 4146–4153.
UNEP, UNEP Year Book 2014: Emerging issues in our global environment, UNEP, 2014.

Usha, R., Sangeetha, T., & Palaniswamy, M. (2011). Screening of polyethylene degrading microorganisms from garbage soil. *Libyan Agriculture Research Center Journal International*, *2*(4), 200–204.

Vimala, P. P., & Mathew, L. (2016). Biodegradation of polyethylene using Bacillus subtilis. *Procedia Technology*, *24*, 232–239.

Wang, P., Zhao, J., Ruan, Y., Cai, X., Li, J., Zhang, L., & Huang, H. (2022). Degradation of polypropylene by the Pseudomonas aeruginosa strains LICME WZH-4 and WGH-6. *Journal of Polymers and the Environment*, *30*(9), 3949–3958.

Ward, P. G., Goff, M., Donner, M., Kaminsky, W., & O'Connor, K. E. (2006). A two step chemo-biotechnological conversion of polystyrene to a biodegradable thermoplastic. *Environmental. Science & Technology*, *40*(7), 2433–2437.

Webb, J. S., Nixon, M., Eastwood, I. M., Greenhalgh, M., Robson, G. D., & Handley, P. S. (2000). Fungal colonization and biodeterioration of plasticized polyvinyl chloride. *Applied and Environmental Microbiology*, 66(8), 3194–3200

Webb, J. S., Nixon, M., Eastwood, I. M., Greenhalgh, M., Robson, G. D., & Handley, P. S. (2000). Fungal colonization and biodeterioration of plasticized polyvinyl chloride. *Applied and Environmental Microbiology*, *66*(8), 3194–3200.

Wilkes, R. A., & Aristilde, L. (2017). Degradation and metabolism of synthetic plastics and associated products by Pseudomonas sp.: capabilities and challenges. *Journal of Applied Microbiology*, *123*(3), 582–593.

Witt, U., Einig, T., Yamamoto, M., Kleeberg, I., Deckwer, W. D., & Müller, R. J. (2001). Biodegradation of aliphatic–aromatic copolyesters: evaluation of the final biodegradability and ecotoxicological impact of degradation intermediates. *Chemosphere*, *44*(2), 289–299.

Wróbel, M., Szymańska, S., Kowalkowski, T., & Hrynkiewicz, K. (2023). Selection of microorganisms capable of polyethylene (PE) and polypropylene (PP) degradation. *Microbiological Research*, *267*, 127251.

Wu, H., Liu, Q., Sun, W., Lu, Y., Qi, Y., & Zhang, H. (2023). Biodegradability of polyethylene mulch film by Bacillus paramycoides. *Chemosphere*, *311*, 136978.

Xu, Y., Xian, Z. N., Yue, W., Yin, C. F., & Zhou, N. Y. (2023). Degradation of polyvinyl chloride by a bacterial consortium enriched from the gut of Tenebrio molitor larvae. *Chemosphere*, *318*, 137944.

Yang, Y., Yang, J., Wu, W. M., Zhao, J., Song, Y., Gao, L., ... & Jiang, L. (2015). Biodegradation and mineralization of polystyrene by plastic-eating mealworms: part 2. Role of gut microorganisms. *Environmental Science & Technology*, *49*(20), 12087–12093.

Yoshida, S., Hiraga, K., Takehana, T., Taniguchi, I., Yamaji, H., Maeda, Y., ... & Oda, K. (2016). A bacterium that degrades and assimilates poly (ethylene terephthalate). *Science*, *351*(6278), 1196–1199.

Zadjelovic, V., Chhun, A., Quareshy, M., Silvano, E., Hernandez-Fernaud, J. R., Aguilo-Ferretjans, M. M., ... & Christie-Oleza, J. A. (2020). Beyond oil degradation: enzymatic potential of Alcanivorax to degrade natural and synthetic polyesters. *Environmental Microbiology*, *22*(4), 1356–1369.

4 Impact of Microplastics on the Environment and Human/Animal Health and Their Enzymatic Removal

Muhammad Uzair Javed and Hamid Mukhtar

4.1 INTRODUCTION

Plastic, one of the most imperative breakthroughs in the preceding 75 years, has permitted tremendous advancements in various industries because of its affordability, versatility, and flexibility. Though the applications of plastics have extended in everyday products, their waste management, and non-biodegradability could be a problem. In the environment, plastics are found to gradually disintegrate into small plastic particles in the influence of microbes, sunlight, and mechanical abrasion. There has been an increased amount of microplastic contamination due to the continuous disposal of plastic waste, which in turn poses a threat to the whole environment. Owing to the chemical and physical characteristics of MPs, they are difficult to degrade from the environment. Microplastic pollution has attracted much attention from scientists as it spreads around the globe (Yuan et al., 2020).

In 2019, the production of plastic around the globe reached 370 million tons, of which 58 million tons was only from Europe. According to a survey of Europe in 2020, packing and construction accounted for 20.4% and 39.6%, respectively. Although techniques have been established to recycle materials, only 32.5% of plastics are recycled, 42.6% of which is utilized to produce energy, and 24.9% of which is dumped in landfills. Although recycling has increased recently, it is crucial to remember that over 50% of plastics have applications that are single-use disposable, which causes the environment to get filled with plastics. Single-use plastics are mainly utilized in disposable consumer items, packaging, and agricultural films. Only 20 to 25% of plastics are utilized in long-lasting products such as structural materials, pipelines, and cable coatings, and the residual plastics are utilized for goods with intermediary life spans, like furniture, automobiles, and electronic items (Ferdous et al., 2021).

The chemical as well as physical nature of the particles determines the effects of MPs on living organisms. The chemical nature of the polymer and characteristics of

 DOI: 10.1201/9781003379386-4

the additives reflect the chemical properties of microplastics. Polystyrene (PS), polypropylene (PP), and polyethylene (PE) particles make up the majority of microplastic pollutants. The additives include plasticizers, lubricants, antioxidants, flame retardants, dyes, and UV stabilizers, as well as inert or reinforcing fillers that give the plastic certain properties. The physical properties of microplastics comprise size and shape, shear strength, elasticity, and surface charge of the particles. The extraordinary surface-to-volume ratio of MP makes them effective adsorbents of toxins and microorganisms (Campanale et al., 2020).

Over the previous 60 years, the world has created more than 6.3 billion tons of plastic, of which roughly 9% were recycled into secondary raw materials and around 12% were burned (Alabi et al., 2019). On the basis of its chemical makeup and environmental factors, a 1 mm thick plate may take decades to centuries to degrade. Plastic materials break down into tiny fragments as they age (Hirt & Body-Malapel, 2020). MPs are quickly distributed by wind and water because of the microscopic size of the particles. As a result, the particles may be found inside living organisms as well as in polar ice, soil, air, and water (Yee et al., 2021).

Both advantageous and detrimental effects on the properties of soil can be caused by microplastic contamination. Furthermore, research has shown that a number of contaminants, such as heavy metals, pesticides, persistent organic pollutants, and antibiotics, interact with microplastics (Haq et al., 2016a; 2016b; 2017, 2018, 2020; Haq and Kalamdhad, 2021; Haq et al., 2022a; 2022b; 2022c; Haq and Kalamdhad, 2022, 2023; Haq et al., 2024). They could therefore act as a vector for the spread of pollutants in terrestrial environments. Microplastics and the compounds they are associated with have the ability to move through food chains and accumulate at different trophic levels, possibly harming the health of people and other living things. Numerous findings have concentrated on the presence and effects of microplastic pollution in marine ecosystems, but less is known about the sources, implications, fate, and distribution of this pollution in terrestrial areas. It is also necessary to do more research on the transport, and effects of microplastics in connection to soil characteristics, polymer content and morphologies, and types of land use (Dissanayake et al., 2022).

The longevity of plastics is a concern since, although being a good material alternative and allowing for extended usage, research has revealed that plastic waste has a hard time decomposing. To mitigate the influence of plastics there is legislation in countries; nevertheless, the environment has been excruciating from the heap of plastic waste. For an improved insight of this nascent issue, it is important to recognize the plastic sources, the amount that is disseminated, and the treatment approaches for various environmental compartments, avoiding future dispersion, and understanding the many consequences that plastics could have on living organisms and human health (Wang et al., 2018).

4.2 MICROPLASTIC DISTRIBUTION AND SOURCES

The multitude of ways MPs may be made supports its ubiquitousness in the environment. The ultimate source of MPs, whether created by humans or fractured through time, is plastic itself. The first step in understanding how to reduce MP waste in the future is to comprehend how MPs are created and introduced into the environment. MPs can be divided into two groups on the basis of their size and source: primary and secondary microplastics (Auta et al., 2017).

Primary microplastics are small particles designed for industrial usage. They are produced industrially as microbeads of different sizes and are utilized in personal care products usually as exfoliants, in sand-blasting media or as the larger virgin plastic pellets which are utilized as raw materials for the purpose of fabrication of products. These pellets "leak" into the environment during production, shipping, or usage (Landeros et al., 2022).

Secondary microplastics are particles that result from the breakdown of larger plastic products. They include agricultural mulch film pieces that are left in the field after usage and textile fiber bits that are unavoidably discharged after washing synthetic garments. The most probable source of secondary MPs, however, is the weathering breakdown of plastic waste in the seashore environment. Secondary MPs exhibit extremely high geographical and temporal variability, despite their insurmountable quantities in the oceans (Andrady, 2017).

Regardless of the primary or secondary source, it is well acknowledged that 92% of the world's marine plastic waste is found in the ocean, and that seawater comprises 102,000 microplastic particles per cubic meter. The outcomes of this widespread exposure to microplastics are not well known, so the aim should be to comprehend these consequences, define a number of strategies that are effective for any quantity of microplastics present in the environment, and develop new strategies to prevent this accumulation in the environment (Prata et al., 2020).

4.3 INCIDENCE OF MICROPLASTICS IN THE ENVIRONMENT

MPs are exceptionally prevalent and can be found in soil sediments and aquatic environments such as rivers, oceans, lakes, and even aquatic commodities including salt and seafood (Andrady, 2017). Most detected polymer categories are polyethylene terephthalate (PET), polyethylene(PE), polystyrene (PS), polypropylene (PP), and polyvinylchloride (PVC).

PVC and PET are more prone to sink in water than PE, PP, and PS due to their higher densities. The dissemination of microplastics are also affected by other factors such as obstruction structure, flow rate, coastline shape, water quality, and biological uptake and digestion (Koelmans et al., 2019). Therefore, to better comprehend MPs, it is necessary to discuss the dispersion of both the aquatic system and the terrestrial environment.

4.3.1 Aquatic Environment

The incidence of microplastics in aquatic environment has become astonishingly evident in recent years. Microbeads from personal care items, tire wear fragments, decayed trash, and fibers from synthetic fabrics are only a few sources of microplastics. Paths to the aquatic habitats are also numerous and contain commercial, domestic and industrial wastewater (including plastic shavings and dust, synthetic fibers, and microbeads), wastewater treatment plant effluent and sludge (including synthetic fibers), stormwater runoff (including litter and tire wear particles), and atmospheric deposition (Wagner et al., 2018; Dris et al., 2017).

One of the most pressing issues with aquatic environments today is marine plastic waste pollution. Marine plastic waste has drawn interest from governments, the scientific community, the media, non-profit organizations, and the general public, on a global scale and has become a hot topic in contemporary marine ecology and environmental studies (Yuan et al., 2022).According to recent research, the improper disposal of plastic waste results in plastic litter, which makes about 80% of plastic isolated from marine waste. Microplastics utilized in the primary industries, air blasting and cosmetics, are broadly used worldwide. Plastics that are disposed of improperly have a significant likelihood of entering the ocean through rivers and wastewater systems (Miranda et al., 2020). Nevertheless, various sources directly contribute to the expansion of plastic waste in the oceans, for instance marine vessels, recreation and commercial fishing, coastal tourism, and marine industries. Unfortunately, the plastic emitted along coastal resorts and beaches is linked with tourism and recreation. Fishing represents 18% of the marine plastic waste in the oceans and is also described as an immediate provision to plastic pollution in marine habitat (Martinho et al., 2022).

The two key sources of microplastics in marine environments are marine and terrestrial activities. The terrestrial sources of microplastics comprises of industrial, agricultural, and domestic waste (Luo et al., 2021). On the other hand, ship transport, fishing, oil and gas exploration, and other marine activities that result in the release of microplastics directly into marine environments contribute to the marine sources of microplastics (Shao et al., 2020). Microplastics from the land get into marine habitats by means of tidal action or groundwater drainage. These particles either pile up in the surface sediment or get suspended in the water and then drift to the deep sea following the current (Wei et al., 2022). Undeniably, microplastics have been identified in various marine environments, including sediments, coastlines, seawater, and beaches, where they pose harm to the ecological systems. In particular, they stimulate detrimental effects on organisms such as false satiation brought on by ingesting microplastic, accumulation of toxic compounds (e.g., heavy metal pollutants and harmful additives), negatively affecting organism growth, survival, reproduction, and metabolism (Li et al., 2020).

The potential of microplastics to harm aquatic organisms is a primary concern. Microplastics are ingestible by all levels of the food chain, from plankton to mussels, fish, and birds, as they continue to degrade over time into even-smaller particles. Several studies have shown harmful impacts on individual species, but the consequences on populations as a whole are still unknown (Windsor et al., 2019).

4.3.2 Terrestrial Environment

Microplastics can invade terrestrial environment either directly as primary microplastics, or indirectly as secondary microplastics. Several studies have proven that the unconscious disposal of plastics has also affected the terrestrial systems.

Microplastics constitute a severe environmental issue because it is estimated that there may be 4 to 23 times as much of them in soil as there are in the oceans.

In terrestrial environment, there are three main types of sources for microplastics: inputs from agricultural practices, the stimulus of runoff, and the deprivation of large particles after improper disposal. Several practices used in agriculture, such as using sewage sludge as fertilizer, can contribute to the soil's disposal of microplastics. According to several studies, the bulk of microplastics are contained in the sludge's solid phase. The usage of microbeads in cosmetics and other industries, the release of fibers from washing synthetic clothing, tire debris, and fragments of plastic from urban runoff are a few causes that contribute to the prevalence of microplastics in sludges (Duis & Coors, 2016; Horton et al., 2017; Hurley & Nizzetto, 2018; Wang et al., 2019). Moreover, in agriculture, some methods used to enhance crop yields and decrease pests may also promote the presence of microplastics.

After use, plastic films are frequently discarded, and plastic mulch is frequently utilized, causing plastics build-up in soils, which will suffer transformation as a result of processes such as fragmentation and degradation, which produce plastics in a range of sizes, from micro to nano (Hurley & Nizzetto, 2018).

The incorporation of plastic particles into deep soil layers occurs by ploughing, or water infiltration brought on by digging animal activity including earthworm ingestion and egestion(Castelvetro et al., 2020). Various methods can be used to degrade microplastics present in the environment, namely UV radiation, thermal oxidation, physical abrasion, microbial action, or interface with soil colloids. Microplastics may age as an outcome of the collective influences of these methods; however, each method behaves in a different way depending on the time, the season, the location, and other circumstances. For instance, during the day, light irradiation is crucial, but at night, its effect may be minimal(P. Liu et al., 2021). Furthermore, a more noteworthy consequence of temperature on the aging of microplastics might be detected in desert areas in summer as compared to that in polar regions in winter. The physiochemical characteristics of microplastics such as crystallinity, color, chemical composition, and surface chemistry are altered during the process of aging. In addition, when microplastics age, several compounds, such as phthalates, pigments, stabilizers, oligomers, and oxygenated products (such as acetophenones, phenols, and carboxylated products), can be discharged into the soil(Ren et al., 2021).

4.3.3 MPs in Food Systems

Microplastic contamination of the ecosystem is a problem that is now receiving a lot of attention from both the scientific society and the public because of its increasing prevalence. Microplastic contamination can severely damage the biological ecosystems, affecting everything from producers like plants to the very top of the food chain (Sridhar et al., 2022).

Plastics waste breakdown into smaller fragments like nano-plastics or microplastics and then leak into the environment. Following the leakage, they get ingested by the aquatic and terrestrial biota, leading to the contamination of our food system. For instance, fine plastic particles having a diameter of nearly 0.2 μm, migrate from roots into the shoots of plants. MPs were discovered in plants (like lettuce) growing in hydroponic systems and in sand metric areas watered with wastewater that contained plastic components, indicating that they were contaminating food chains. Wheat grains raised on contaminated soil have also been found to contain phthalate esters (PAEs). The average daily MP intake was 80 g from edible plants (Li et al., 2020).

Due to their small size, aquatic creatures at the lower trophic levels, such as plankton and filter feeders, may unintentionally consume MPs. The biggest concentrations of MPs will be found in creatures at the higher trophic levels, including crustaceans and mollusks, where these compounds can consequently accumulate along the food chain. Highly dangerous seafood items that finally make it to human consumers may arise from such bioaccumulation of MPs, which may contain hazardous chemicals (Revel, 2018).

Additionally, an Italian study on fruits and vegetables revealed that due to a potential translocation chain from plant and soil pathways to the ecosystem, an assessed daily consumption of MPs via food systems ranged from 87,600 to 124,900 particles/g (Conti et al., 2020). Another study was made on the disposable cups and packaging material according to which the average content of microplastics ranged between 3 ± 1.13 and 38 ± 5.12 mg. Because of their small size, they can be consumed by any animal in the food chain, releasing toxic chemicals and additives into the environment (Fadare et al., 2020).

In recent times, microplastics have been found in a variety of food products, directly impacting human health. These complex chemicals and plasticizers, such as bisphenol A and phthalates, seep into the soil, causing the microorganisms to become endangered. Moreover, MPs have demonstrated an impact on nutrient cycle and toxicity in a specific biota through absorption. Additionally, MPs have demonstrated their presence in the air. In recent years, MPs have been analyzed in processed

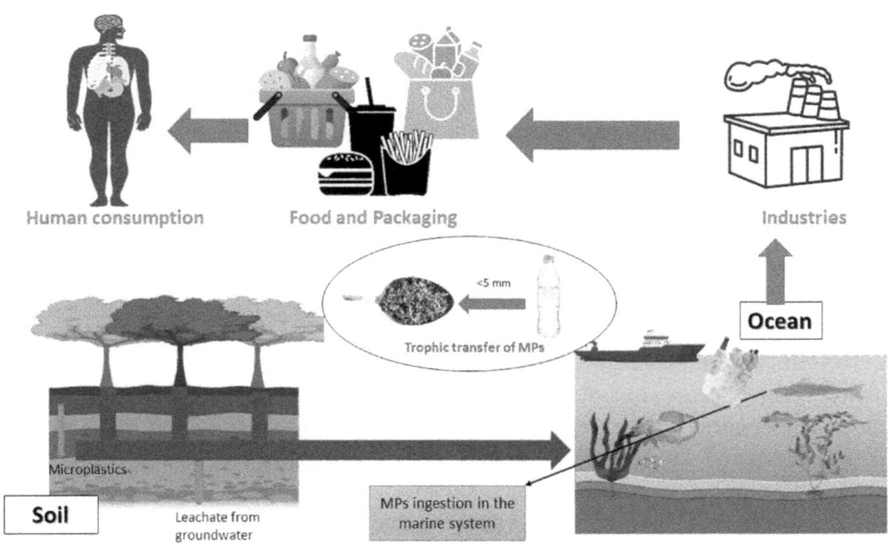

FIGURE 4.1 Transport of microplastics in the food system.

eatables such as honey, table salt, milk, water, and beer triggering intestinal disorders and influencing the overall health of humans (Khalid et al., 2020).

Most of these studies investigate the chemical components as well as the widespread distribution of MPs and their transfer to processed foods or animal parts that are ultimately consumed by people. Although there is less information on the physical impacts and toxicity of MPs on human health, preliminary research has shown that long-term exposure can harm marine species by inducing tissue inflammation, necrosis, problems with reproduction, and cellular growth. There is a big possibility that MPs will be exposed in the human diet since these marine organisms are concentrated with MPs. The paths of MPs exposure are also highlighted by numerous scientific studies, raising concerns about food safety on a global scale. Although there are disagreements over the exact size of these plastics, they have been divided into three categories: macroplastics (>200 mm), mesoplastics (5–200 mm), and microplastics (5 mm) (Sridhar et al., 2022). Nevertheless, for evaluating separation and detection technologies for MPs present in food systems, better assessments must be done.

4.4 IMPACT OF MICROPLASTICS

There have been several reports throughout the years from various parts of the world demonstrating the connection between MPs contamination in the soil, freshwater, and ocean environments, which had extremely hazardous consequences on both humans and animals. Given that more than 300 megatons of plastic are produced annually around the world, it is vital to take into account the pervasive presence of MPs in our environment as well as their potential consequences on both human and animal health. MPs may leak into the air, food, and water through the packaging of food, textiles, tires, and other plastic products. There is rising concern regarding the possible negative impacts of MPs on human health and the environment since leakage of plasticizer chemicals has been widely connected with endocrine-disrupting capabilities (Boucher & Friot, 2017).

4.4.1 Environmental Impacts of MPs

The widespread and ongoing usage of plastics in a variety of socioeconomic sectors, along with their durability and slow decomposition, has significantly contaminated the environment with plastic debris. It is challenging to evaluate the environmental effects of microplastics (MPs) due to changes in ecosystem processes, risky effects on biota, and different toxicities brought on by their composition. Additionally, it was suggested that there is a shortage of available data and a lack of understanding of the ecological impact of MPs (Rillig et al., 2019).

Nevertheless, it has been reported by recent research that within the composting environment, greater amounts of carbon dioxide are released by the un-weathered mulch that is enriched with Polybutylene adipate terephthalate (PBAT) as compared to the weathered biodegradable mulch. It is due to the fact that environmental weathering plays a crucial part in enhancing the degradation of biodegradable mulch. The constituents of a polymer also affect the rate of mulch degradation by microbes. For instance, the degradation of mulch by microbes is more persistent in bacterial communities as compared to the fungal ones (Wan et al., 2019).

Over the past decade, the scientific attention towards microplastics has been revived. Microplastics are considered one of the most potential threats to the normal functioning of ecosystem (J.-J. Guo et al., 2020). The most heavily polluted places are landfills, cities, and beaches, while agricultural ecosystems are also severely influenced. Concerns about the pollution of marine ecosystems by microplastic have grown recently. However, 4.8–12.7 metric tons of plastic are transported annually from land to marine ecosystems (Dissanayake et al., 2022).

The accumulation of microplastics in the terrestrial habitat can arise from both the primary and secondary sources. Microplastics that are produced in various industries as micrometer-sized particles constitute the primary sources. On the contrary, chemical, biological, and physical fragmentation of macroplastics falls under the category of secondary sources. Microplastics can contaminate the soil by various ways, such as the addition of compost, use of plastic mulching films, municipal solid waste, irrigation tools, the materials employed in greenhouses, and atmospheric inputs (de Souza Machado et al., 2018). According to estimates, sewage sludge application releases between 44,000 and 300,000 tons of microplastics yearly to farmlands in North America compared to between 63,000 and 430,000 tons annually to farmlands in Europe (Nizzetto et al., 2016). In 2017, more than 1.4 million tons of agricultural plastic mulching film were used in China alone (Gao et al., 2019). At this time, widespread microplastic contamination in all sorts of ecosystems is a significant global concern.

Apart from the short-term advantages of plastic materials, their long-term implications on soil quality and agricultural productivity are increasingly receiving more attention. Microplastics interact with and absorb organic pollutants like dioxins, polycyclic aromatic hydrocarbons (PAHs), dichlorodiphenyltrichloroethane (DDT), and polychlorinated biphenyls (PCBs), as well as inorganic pollutants like trace metals (Nizzetto et al., 2016).

Microplastics can operate as a vector for contaminants to be transmitted inside and across diverse ecological sections because of their surface interactions. Aside from the more apparent environmental issues caused by microplastic pollution, there is also a possibility that terrestrial plastic contamination will have an impact on human health and the food supply (Boots et al., 2019). There is limited research on the environmental effects of MP pollution, despite the fact that the ongoing presence of MPs in ecosystems offers a potential risk to all biota on the planet including soil, water, and food systems. Therefore, in order to assess the environmental severity of MP pollution and to mitigate its possible consequences, scientific studies on the environmental effects are required.

4.4.2 Health Effects of MPs

4.4.2.1 Impact of Microplastic on Human Health

Plastics have significantly improved packaging materials that enhanced their shelf life and decreased food waste, leading to the enhancement of food security. By preventing the transmission of infectious diseases using single-use medical equipment including facemasks, syringes, and medical tubing, the material also promotes public health. However, human exposure to microplastics will surely result from the extensive use of plastics in daily life and the subsequent environmental contamination caused by post-use plastics (Prata et al., 2018).

MPs can enter the human body in several ways because of their size, which is approximately 5 mm. One of the main ways by which people are exposed to MPs is through the oral intake of contaminated food or water. When consuming marine items, there is also a significant chance of ingesting MPs. Particles enter the body through the mouth and move through the gastrointestinal tract into the intestine. Intestinal cells will only get exposed to particles that exhibit the ability to penetrate through the intestinal mucus due to their microscopic size or the development of a surface "corona" of biocompatible coating. Here, microplastics can either be absorbed through perspiration, the paracellular transference of particles, which may lead to systemic circulation later, or by specialized M-cells that cover the Peyer's patches which is considered as a specialized lymphoid tissue. A significant impact of microplastics on cell viability, pro-inflammatory response as well as on gene expression has been observed through *in vitro* testing. It has also been reported by *in vivo* testing that the count of spermatogenic cells in mice was reduced after the ingestion of polystyrene microparticles, even in concentration as low as 0.01 mg of 5 μm. Moreover, the sperm count was also reduced after 42 days of ingestion (Xie et al., 2020).

The inhalation of airborne particles is another potential route by which humans can get exposed to microplastics. Our air is continuously being polluted by synthetic fibers and microplastics. According to an estimate, humans get exposed to nearly 272 microplastics per day by inhalation (Vianello et al., 2019). These particles are emitted from commonplace items and activities, and their concentrations and properties are greatly modified by them. The extent of depth to which the particles will reach in respiratory system is determined by their density and size. A dust overload response can be induced if the number of particles in the air is extremely high. Due to this response, chemotactic factors are intensely released that enhance the vascular permeability and inhibit the migration of macrophages, leading to chronic inflammation. In fact, microplastics in lungs are capable of causing inflammation and exhibit strong oxidative action. Chronic exposure to elevated amounts of airborne microplastics can cause interstitial lung illness and airway disease, which may eventually result in the growth of malignant lesions, as has been seen in industrial exposure (Prata, 2018).

Finally, as it is known that microplastics are so prevalent in the environment, cutaneous exposure cannot be completely ruled out. It has been proposed that plastics smaller than 100 nm might be able to pass through the epithelial barrier. Several skin-care products that are intended for direct application to the skin even include microplastics, thus making this subject worth researching. Human epithelial cells may experience cytotoxicity from oxidative stress if internalization does take place (Godoy et al., 2019). Although, uncertainty exists concerning the usage of *in vitro* and *in vivo* models, but by using the accessible toxicological information, it has been systematically interpreted by the Adverse Outcome Pathways that the production of Reactive Oxygen Species as well as the resulting inflammation and oxidative stress, are both the main toxicological hazards to animal and human health (Jeong & Choi, 2019).

Ecosystems on land and in the sea are both heavily contaminated with microplastics. By 2050, 12 billion metric tons of plastic garbage will be present in landfills or the environment, according to current forecasts, in contrast to the 4.9 billion metric tons (or 60% of all plastics ever created) discovered in 2015 (Geyer et al., 2017). An increasing body of research implies that animals consuming microplastics from the environment, pollution throughout manufacturing, and contamination by plastic packaging are all ways in which microplastics are getting into commonly used edible products. Microplastic particles (MPs) having a diameter less than 130 μm are capable of translocating into human tissues and triggering a restricted immunological action.

Moreover, they are also involved in releasing integral monomers, toxic substances supplemented throughout plastic making, and pollutants captivated from the atmosphere, containing heavy metals and obstinate organic pollutants such as DDT and PCBs. An enquiry into the increasing human contact with microplastics has not been shown, despite the mounting indication that microplastics pollute a wide range of foods and beverages, as well as outdoor and indoor environments, and also exhibits the potential for harmful impressions on human health after assimilation or by breath (Cox et al., 2019).

There is apparently no proof that microplastics have a direct harmful impact on human health. With the same constraints as those outlined for the evaluation of the impacts of microplastics on animal health, most of the observations now available are based on *in vitro* or *in vivo* research in model organisms. The Science Advice for Policy by European Academies came to a similar conclusion, highlighting the absence of proof for influence of microplastic on human health in current trace levels while also emphasizing the need for knowledge gaps to be filled (SAPEA, 2019).

4.4.2.2 Impact of Microplastic on Animal Health

Irrespective of their trophic level, almost all aquatic and terrestrial organisms encounter plastic garbage because it is so pervasive. Physical contact with microplastics, chemical exposure to plastic components (such as additives), sorbed toxins (manufactured or organic), or creatures attached to the surface (e.g., pathogens) are all potential causes of biological consequences. A range of different health or ecosystem impacts, including ingestion, entanglement, and mortality, can result from interactions. The interactions' nature in the context of microplastic debris is clear to see. Plankton and other miniature aquatic biota, as well as bulkier filter feeders like shellfish and whale sharks, might consume or encounter plastic particles.

The size, texture, and shape of the microplastics helps to investigate the presence of any negative impact. However, because of their microscopic size and complex behavior, it is not easily determined (Hale et al., 2020). Wright et al., (2013) reported the physical effects of microplastics on the biota living in marine ecosystems. Moreover, the consequential changes made in their behavior, metabolic activity and development have also been investigated after the exposure to microplastics. It has been found by Mattsson et al. (2017) that the survival rate of zooplankton was reduced and behavior (observation was made on their activity, time they take for feeding and distance they swam to find prey) of finfish was changed due to the presence of nanoplastic particles that usually have the diameter of 52–330 nm.

Despite the fact that only a limited number of investigations have been made on the impact of microplastics across different stages of life, a study was done by Luan et al. (2019) determining the toxicity of polystyrene microplastics. It was reported that polystyrene microplastics show greater toxicity to the growing clams while hatching as compared to the larval stages. Consequently, the rates of development are reduced due to the presence of microplastics. According to a study made on sea urchins, microplastics are also involved in the disruption of development, thus leading to abnormal physical features. Following the exposure to microplastics, oxidative stress has also been seen in several species, including the red tilapia and juvenile zebra fish.

Numerous studies have examined these consequences overall. However, it is challenging to make a direct comparison of results due to the variety of microplastic materials present and measures of impacts utilized, just as it is challenging to confidently anticipate toxicological outcomes due to the variety of microplastics in the natural environment (Hale et al., 2020; Prokić et al., 2019).

Furthermore, there is a vast variety of microplastics amid the environment, which exists as an intricate combination in a pool of additional potential stressors. As a result, linking effects to a specific plastic outside a carefully controlled laboratory setting can be quite challenging. To clarify the toxicological effects, more investigation on the transfer of microplastics within and among tissues is obviously required. Increasing our capacity to accurately trace small microplastics presents a significant barrier for this research. Table 4.1 provides a summary of different microplastics along with their applications, their potential contaminants, and their health effects.

4.4.3 One Health Perspective on Microplastics

Plastic is used everywhere in modern society. Since the 1950s, the need for plastic has increased dramatically due to its low cost, strength, durability, and light weight. As a consequence, plastic pollution is now widespread on Earth, including in soils, the atmosphere, open ocean, deep sea, and along coastlines. According to the latest estimates, there are at least 5.25 trillion plastic particles in the world's oceans, and that figure is expected to rise (Barrett et al., 2020). In fact, even if urgent action is done to reduce waste, it is anticipated that between 2016 and 2040, the volume of plastic pollution invading the aquatic and terrestrial ecosystem will increase by an extra 710 million metric tons (Lau et al., 2020).

In the environment, plastics breakdown into small chunks having distinct surface qualities by means of different biotic and abiotic activities including bacterial activity, UV radiation, temperature, and abrasion. The most common kind of solid trash is made up of these smaller plastics, which are referred to as microplastics (<5 mm) and nanoplastics (<1 m), particularly in the water habitats (González et al., 2019). Moreover, both varieties of minuscule plastics—known as micro- and nanoplastics—can be discovered in products used in commerce and industry. Plastics are so common in the ecosystem that interactions with both people and animals are now inevitable. Plastic pollution has an impact on a large variety of marine creatures. Mussels, clams, crabs, fish, and oysters intended for human food, along with seaweed, beer, table and

TABLE 4.1
Summary of different microplastics along with their applications, potential contaminants present and their health effects

Microplastic type	Applications	Contaminants	Health effects	Reference
Polyethylene	Packaging film Garbage bags Grocery bags Insulation for wires and cables Agricultural mulch Bottles Toys Houseware	Difenoconazole Diflubenzuron Carbendazim Malathion Dipterex Phenanthrene Tonalide Benzophenone Fipronil Ciprofloxacin	• Cancer development • Skin problems • Menstrual and pregnancy issues • Endocrine disruptor	(Kalogerakis et al., 2017;Fred-Ahmadu et al., 2020)
Polypropylene	Ropes Twines Bottle caps Netting	Imazamox Imazapic Imazethapyr Fipronil	• Cytotoxicity • Hypersensitivity • Unwanted immune response • Acute response like hemolysis.	(Hwang et al., 2019)
Polystyrene	Food containers Toys Foam cups Cosmetic containers	Hexaconazole Myclobutanil Triadimenol Fipronil	• Affects CNS (causing symptoms such as headaches, fatigue, malaise, dizziness etc.) • Thyroid problems • Breast cancer • Prostate cancer • Liver damage • Nerve tissue damage • Disrupts normal hormonal functions by mimicking estrogens in the body	(Fang et al., 2019)

(continued)

TABLE 4.1 (Continued)
Summary of different microplastics along with their applications, potential contaminants present and their health effects

Microplastic type	Applications	Contaminants	Health effects	Reference
Polyamide	Trap netting Electrical connectors Medical implants Gears	Sulfamethoxazole	• Aplastic anemia • Allergic reactions—skin rash, itching, hives, swelling of the face, lips, tongue, or throat.	(Hwang et al., 2019;Guo et al., 2019)
Polyester	Clothes Fibers Upholstered furniture	Ciprofloxacin Pentachlorobenzene Trifluralin Tonalide Phenanthrene Hexachlorobenzene Benzophenone Trichlorobenzenes (1,2,3-TeCB, 1,3,5-TeCB, 1,2,4-TeCB)	• Skin problems (rashes, itching, redness, eczema, dermatitis, blistering) • Lung problems • Miscarriages • Infertility	(F. Liu et al., 2022; Tubić et al., 2021)
Polyvinyl chloride	Plastic films Cups Window frames Stationary Automotive interiors Footwear Medical devices	17 β-Estradiol Fipronil 17 α-Ethynylestradiol	Genotoxic and mutagenic toxicant	(Lieshchova et al., 2019; Qiu et al., 2019)
Polyethylene terephthalate	Bottles Industrial strapping Construction materials Protective packaging	Sulfamethoxazole Antimony trioxide Toluene chlorobenzene phenol limonene benzophenone	No significant human health threat	(Rao, 2019)

sea salt, honey, tea, and bottled and tap water, all include microplastics. Microplastics have also been found in the human body, including in placentas, blood, feces, and lung tissues (Alang et al., 2021; Morrison et al., 2022; Zarus et al., 2021).

Apart from microplastics, there is no such anthropogenic pollutant that covers a broad range of direct disclosure extending to all organisms. Once ingested, plastics interfere with homeostasis at the level of the individual organism. By modifying habitats, causing harm to them, and upsetting the balance of species throughout ecosystems, plastic pollution can also interfere with the normal functioning of ecosystems. These modifications obviously have enigmatic repercussions on one's health. A strategy that goes beyond the usual species-level risk evaluations is necessary to mitigate the effects of plastic on the health of animals, people, and ecosystems. One Health is one such conceptual framework. One Health acknowledges the interdependence of plants, people, and animals, as well as how the health of each being is reliant on the wellbeing of the ecosystem they share. In order to address health concerns at the regional, national, and international levels, a multi-sectoral, multidisciplinary, and cooperative strategy is required (Morrison et al., 2022).

As was discussed in the earlier sections, plastic has an impact on every aspect of the One Health paradigm. One Health sees animal, human, and environmental health as three interrelated domains that are all directly and indirectly impacted by what happens in one. Every aspect of global health could be negatively impacted by plastic pollution, and novel approaches to deal with these effects at large scales are required because the majority of these effects are circumstantial based on dispersed literature.

4.5 PREVENTION COUNTERMEASURES FOR MICROPLASTICS IN PLANTS

During the production of plastics, several additives such as bisphenol A, polybrominated diphenyl ethers, and phthalates are used. Such chemicals can produce toxic changes in plants as well as humans when they ingest them through several routes. Moreover, due to the greater capacity of absorbing different materials, microplastics can transport several pollutants. Some of the efficient ways for removing microplastics from the environment are using Periphytic Biofilm, Hyperthermophilic Composting (hTC) Technology, and Whole-Cell Biocatalysis. Several microorganisms have also been investigated to determine their capacity of degrading MPs.

Microplastics offer a remarkable ecological niche for microorganisms that proposes them assistance for colonization as well as growth. Three steps that are required to complete a degradation process of microplastics are referred to as biodeterioration, biofragmentation, and assimilation. Numerous microorganisms that can break down MPs were isolated from various environmental sources in order to study the biodegradation of MPs. For instance algae (*Scenedesmus dimorphus*, *Anabaena spiroides*, and *Naviculapupula*, *Anabaena spiroides*), bacteria (*Paenibacillus sp., Rhodococcus sp., Bacillus sp*) and fungi (*Aspergillus sydowii*, *Zalerionmaritimum*, *Aspergillus terreus*, *A. tubingensis* VRKPT1 and *A. flavus* VRKPT2) (Campanale et al., 2022; Kumari et al., 2022).

4.5.1 Enzymatic Removal of MPs

Recently, it has become clear that depolymerizing the waste petro-plastics using microbial enzymes is a promising strategy for mineralizing or recycling plastics. Petro-polymers can be distributed into two classes from the viewpoint of enzymatic degradation: non-hydrolysable (Polyethylene, Polystyrene, Polypropylene, and Polyvinylchloride) and hydrolysable (Polyethylene Terephthalate and Polyurethane). These two groups employ biodegradation methods that are extremely distinct from one another (Wei & Zimmermann, 2017). Petro-plastic polymers have the same ester bonds that are present in other polyester polymers, and esterases are enzymes that cleave these ester bonds. Esterase appears in a number of forms, each with exclusive biological characteristics, substrate specificities, and protein forms. Esterase, lipase, hydrolase, laccase, cutinase, carboxylesterase, protease, and other frequently used microbial enzymes are used in the biodegradation of MPs (Mohanan et al., 2020).

Enzymatic processes such as oxidation, hydroxylation, and hydrolysis cause the polymer chains to break down into monomers or oligomers. It is assumed that in the process of degrading microplastics, extracellular enzymes released by microbes first break down large, high molecular weight polymers into smaller subunits (dimers or oligomers) that can be absorbed onto the microbial cells. When these breakdown products enter the cells, they are directed through the traditional catabolic pathways, which produce energy and reducing power for cell development. Petro-polymer degradation by microbes is considered to be a relatively slow phenomenon. These polymers' massive molecular weight as well as quite hydrophobic interfaces, which prevent these particles from moving across the cell wall, are mainly accountable for their great resistance. Various levels of degradability are believed to be dependent on the existence of strong C-C bonds, which are exceptionally resistant to being attacked by enzymes, the quantity of crystalline and amorphous forms, and other factors (Rübsam et al., 2018).

Cutinases are referred to as carboxylic ester hydrolases or α/β hydrolases and were first isolated from the plant pathogenic fungi, for instance, *Fusarium solanipisi*. Cutinases (EC 3.1.1.74) belong to the esterase class of enzymes and exhibit the ability to efficiently hydrolyze polyesters because of their high molar mass. Because of high phytopathogenicity, attention is being given to the enzyme as it can hydrolyze the suberin in bark as well as resolve the cuticular layer in leaves. The crystal structures of cutinases that hydrolyze PET have been examined to determine the substrate-binding domain and to induce mutations to create better enzymes and to speed up the pace of degradation. According to several studies made on cutinases, it has been reported that the fusion of enzyme with binding domains via protein engineering can increase the polymer breakdown (Chen et al., 2013).

Both enzymatic surface modification of polyester fibers and enzymatic hydrolysis of the polymer are possible methods for enzymatic hydrolysis of PET. PET surface-modifying enzymes or hydrolases, like lipases, carboxylesterases, cutinases, and proteases, carry out enzymatic surface modification. These enzymes destroy the PET polymer's surface rather than its chains, which they do not hydrolyze. However, significant breakdown of the PET chains is necessary for the treatment of PET waste.

Since the time when Müller (2005) initially discovered PET hydrolase, relatively few cutinases have been identified as PET hydrolases that can breakdown the inner block of PET (by at least 10%). For the surface hydrophilization of fibers, a number of hydrolases have been tested, including fungal cutinases from *Penicillium citrinum*, *Aspergillus oryzae*, and *Fusarium solani*, lipases from *Triticum aestivum*, *Candida antarctica*, *Thermomyceslanuginosus*, and *Burkholderiaspp*, and those from actinomycetes, *Saccharomonosporaviridis*, *T.fusca*, *Thermobifidacellulosilytica*, *Thermobifida alba*; carboxylesterases and esterases (*Thermobifidahalotolerans*) (Mohanan et al., 2020).

Enzymatic PET degradation is influenced by several factors, including reaction temperatures (>Tg, preferably 65–70°C in aqueous solution), the enzyme structure, and the surface topology, crystallinity, molecular size, and hydrophobicity of synthetic polymers. The amorphous portions of PET are hydrolyzed preferentially by polyesterases. Polymer chains are less available for enzymatic attack because of increased crystallinity, which restricts their capacity to move. None of the PET hydrolases that have been discovered so far can break down the crystalline polymers found in PET bottles, clothing, and biaxially stretched films. At higher temperatures, degradation of the polyesters was discovered to be encouraged. The disinfection of digested plastic garbage was also aided by higher temperatures (above 55°C) (Islam et al., 2019).

It has been demonstrated that blending the plastics with specific kinds of natural polymers, like starch, can accelerate their biodegradation. The quick enzyme-catalyzed hydrolysis of starch, which renders the polymer porous and prone to both biotic and abiotic degradations, has been identified as the mechanism for the accelerated biodegradation. The biodegradation of blended samples containing LDPE and starch by using an alpha-amylase aqueous solution was reported by Karimi & Biria (2019). Amylase was discovered to exhibit co-metabolic activity and hydrolyze both PE molecules and the major specified substrate (starch). Additionally, it was shown that adding hydrophilic starch to hydrophobic PE increased both polymers' biodegradability.

Therefore, many strategies for removing microplastics from the surroundings have been formed and used. Although each approach has its own set of benefits and drawbacks, there is an absence of enough information on these removal techniques, making it difficult to propose one method as the best one. Further research can be done on the different microorganisms and techniques that have been frequently employed to remove other toxins from environment and that can be coupled with one another to enhance the degradation process.

4.6 SUMMARY AND CONCLUSION

Microplastics, having the size of nearly 5 mm, are abundantly present around the globe and are considered as one of the main sources of environmental contamination. They can either have originated from intended production (primary microplastics) or by the fragmentation of large polymers (secondary microplastics). The health of humans, plants, and animals living in terrestrial or marine habitats, are affected by the presence of microplastics in these ecosystems. These MPs have the ability to absorb

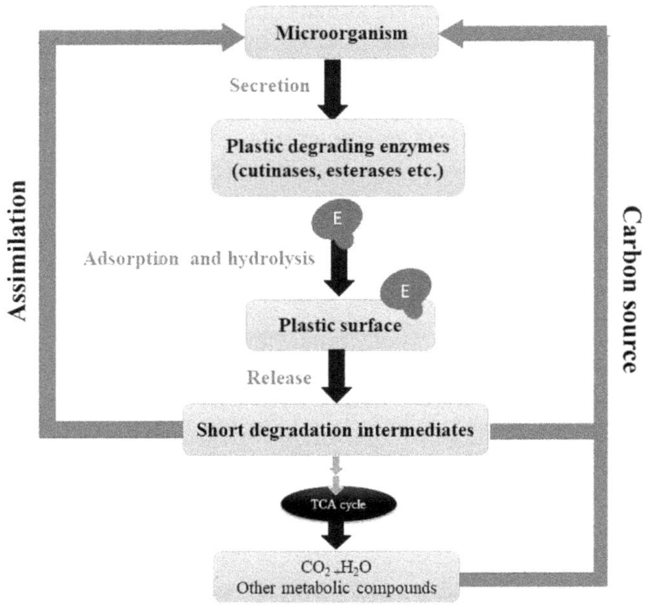

FIGURE 4.2 General mechanism for biological degradation of plastics.

toxic compounds, leading to negative impacts on both the soil quality and productivity. Though significant attempts are being made to replace synthetic plastic with biodegradable plastic to reduce plastic pollution in environment, extra attention must be paid to prevent the harmful impacts of biodegradable plastics on both humans and the environment. The issues related to microplastic waste may be resolved through enzymatic removal of microplastics. Enzymes especially linked to the depolymerization of ester-based PU and PET have been found and thoroughly investigated. To further understand the triggering molecular processes of biodegradation, extra research into the genes or gene products that hydrolyze the high molecular weight plastic polymers may be beneficial. Exclusively for persistent non-hydrolyzable polymers, study concentrating on the gut microorganisms or digestive enzymes of plastic-degrading invertebrates may also end in innovative methods for plastic degradation. On the basis of this knowledge, it may be preferable to use genetic engineering techniques to produce recombinant microbial strains or enzymes to accelerate the biodegradation of plastics. Furthermore, environmental health officers and other associated professions in the future will need a higher awareness and knowledge base of MPs in order to handle the growing public concerns about MPs and the progressive cultural shift towards sustainability. Additionally, it's important to prioritize and focus future research on the most hazardous MPs by using the proper risk assessment techniques.

Conclusively, this chapter highlights the pervasive threat posed by microplastic pollution to both environmental integrity and public health. Through an examination of its sources, distribution, and impacts on ecosystems and human well-being, we have gained insight into the urgency of addressing this pressing issue. Embracing a holistic approach, such as the One Health perspective, allows us to appreciate the interconnectedness of environmental, animal, and human health in the context of microplastic contamination. Moreover, the discussion on preventive measures, notably enzymatic degradation, offers promise in mitigating the adverse effects of microplastics. It is imperative for stakeholders across sectors to collaborate in developing and implementing effective strategies to combat microplastic pollution. By prioritizing interdisciplinary research, innovation, and proactive action, we can work towards a sustainable future where the harmful impacts of microplastics are mitigated, ensuring the well-being of both current and future generations.

REFERENCES

Alabi, O. A., Ologbonjaye, K. I., Awosolu, O., & Alalade, O. E. (2019). Public and environmental health effects of plastic wastes disposal: a review. *J Toxicol Risk Assess, 5*(021), 1–13.

Alang, T. A. I. T., Ibrahim, Y. S., Anuar, S. T., Azmi, A. A., Khalik, W. M. A. W. M., Lehata, S., Hamzah, S. R., Ismail, D., Ma, Z. F., & Dzulkarnaen, A. (2021). Detection of microplastics in human colectomy specimens: An observational study. *Journal of Gastroenterology and Hepatology, 36*, 58–59.

Andrady, A. L. (2017). The plastic in microplastics: A review. *Marine Pollution Bulletin, 119*(1), 12–22.

Auta, H. S., Emenike, C. U., & Fauziah, S. H. (2017). Distribution and importance of microplastics in the marine environment: a review of the sources, fate, effects, and potential solutions. *Environment International, 102*, 165–176.

Barrett, J., Chase, Z., Zhang, J., Holl, M. M. B., Willis, K., Williams, A., Hardesty, B. D., & Wilcox, C. (2020). Microplastic pollution in deep-sea sediments from the Great Australian Bight. *Frontiers in Marine Science*, 808.

Boots, B., Russell, C. W., & Green, D. S. (2019). Effects of microplastics in soil ecosystems: above and below ground. *Environmental Science & Technology, 53*(19), 11496–11506.

Boucher, J., & Friot, D. (2017). *Primary microplastics in the oceans: a global evaluation of sources* (Vol. 10). Iucn Gland, Switzerland.

Campanale, C., Galafassi, S., Savino, I., Massarelli, C., Ancona, V., Volta, P., & Uricchio, V. F. (2022). Microplastics pollution in the terrestrial environments: Poorly known diffuse sources and implications for plants. *Science of the Total Environment, 805*, 150431.

Campanale, C., Massarelli, C., Savino, I., Locaputo, V., & Uricchio, V. F. (2020). A detailed review study on potential effects of microplastics and additives of concern on human health. *International Journal of Environmental Research and Public Health, 17*(4), 1212.

Castelvetro, V., Corti, A., Bianchi, S., Ceccarini, A., Manariti, A., & Vinciguerra, V. (2020). Quantification of poly (ethylene terephthalate) micro-and nanoparticle contaminants in marine sediments and other environmental matrices. *Journal of Hazardous Materials, 385*, 121517.

Chen, S., Su, L., Chen, J., & Wu, J. (2013). Cutinase: characteristics, preparation, and application. *Biotechnology Advances, 31*(8), 1754–1767.

Conti, G. O., Ferrante, M., Banni, M., Favara, C., Nicolosi, I., Cristaldi, A., Fiore, M., & Zuccarello, P. (2020). Micro-and nano-plastics in edible fruit and vegetables. The first diet risks assessment for the general population. *Environmental Research, 187*, 109677.

Cox, K. D., Covernton, G. A., Davies, H. L., Dower, J. F., Juanes, F., & Dudas, S. E. (2019). Human Consumption of Microplastics. *Environmental Science and Technology*, *53*(12), 7068–7074. https://doi.org/10.1021/acs.est.9b01517

de Souza Machado, A. A., Kloas, W., Zarfl, C., Hempel, S., & Rillig, M. C. (2018). Microplastics as an emerging threat to terrestrial ecosystems. *Global Change Biology*, *24*(4), 1405–1416.

Dissanayake, P. D., Kim, S., Sarkar, B., Oleszczuk, P., Sang, M. K., Haque, M. N., Ahn, J. H., Bank, M. S., & Ok, Y. S. (2022). Effects of microplastics on the terrestrial environment: A critical review. *Environmental Research*, *209* (January), 112734. https://doi.org/10.1016/j.envres.2022.112734

Dris, R., Gasperi, J., Mirande, C., Mandin, C., Guerrouache, M., Langlois, V., & Tassin, B. (2017). A first overview of textile fibers, including microplastics, in indoor and outdoor environments. *Environmental Pollution*, *221*, 453–458.

Duis, K., & Coors, A. (2016). Microplastics in the aquatic and terrestrial environment: sources (with a specific focus on personal care products), fate and effects. *Environmental Sciences Europe*, *28*(1), 1–25.

Fadare, O. O., Wan, B., Guo, L.-H., & Zhao, L. (2020). Microplastics from consumer plastic food containers: Are we consuming it? *Chemosphere*, *253*, 126787.

Fang, S., Yu, W., Li, C., Liu, Y., Qiu, J., & Kong, F. (2019). Adsorption behavior of three triazole fungicides on polystyrene microplastics. *Science of the Total Environment*, *691*, 1119–1126.

Ferdous, W., Manalo, A., Siddique, R., Mendis, P., Zhuge, Y., Wong, H. S., Lokuge, W., Aravinthan, T., & Schubel, P. (2021). Recycling of landfill wastes (tyres, plastics and glass) in construction–A review on global waste generation, performance, application and future opportunities. *Resources, Conservation and Recycling*, *173*, 105745.

Fred-Ahmadu, O. H., Bhagwat, G., Oluyoye, I., Benson, N. U., Ayejuyo, O. O., & Palanisami, T. (2020). Interaction of chemical contaminants with microplastics: principles and perspectives. *Science of the Total Environment*, *706*, 135978.

Gao, H., Yan, C., Liu, Q., Ding, W., Chen, B., & Li, Z. (2019). Effects of plastic mulching and plastic residue on agricultural production: A meta-analysis. *Science of the Total Environment*, *651*, 484–492.

Geyer, R., Jambeck, J. R., & Law, K. L. (2017). Production, use, and fate of all plastics ever made. *Science Advances*, *3*(7), e1700782.

Godoy, V., Martín-Lara, M. A., Calero, M., & Blázquez, G. (2019). Physical-chemical characterization of microplastics present in some exfoliating products from Spain. *Marine Pollution Bulletin*, *139*, 91–99.

González, N., Cunha, S. C., Monteiro, C., Fernandes, J. O., Marques, M., Domingo, J. L., & Nadal, M. (2019). Quantification of eight bisphenol analogues in blood and urine samples of workers in a hazardous waste incinerator. *Environmental Research*, *176*, 108576.

Guo, J.-J., Huang, X.-P., Xiang, L., Wang, Y.-Z., Li, Y.-W., Li, H., Cai, Q.-Y., Mo, C.-H., & Wong, M.-H. (2020). Source, migration and toxicology of microplastics in soil. *Environment International*, *137*, 105263.

Guo, X., Chen, C., &Wang, J. (2019). Sorption of sulfamethoxazole onto six types of microplastics. *Chemosphere*, *228*, 300–308.

Hale, R. C., Seeley, M. E., La Guardia, M. J., Mai, L., & Zeng, E. Y. (2020). A Global Perspective on Microplastics. *Journal of Geophysical Research: Oceans*, *125*(1), 1–40. https://doi.org/10.1029/2018JC014719.

Haq, I., Kumar, S., Kumar, V., Singh, S. K., & Raj, A. (2016a). Evaluation of bioremediation potentiality of ligninolytic Serratia liquefaciens for detoxification of pulp and paper mill effluent. *J Hazard Mater* 305: 190–199.

Haq, I., Kumari, V., Kumar, S, Raj, A., Lohani, M., & Bhargava, R. N. (2016b). Evaluation of the phytotoxic and genotoxic potential of pulp and paper mill effluent using Vigna radiata and Allium cepa. *Adv Biol* 8065736.

Haq, I., Kumar, S., Raj, A., Lohani, M., & Satyanarayana, G. N. V. (2017). Genotoxicity assessment of pulp and paper mill effluent before and after bacterial degradation using Allium cepa test. *Chemosphere* 169: 642–650.

Haq, I., Raj, A., & Markandeya (2018). Biodegradation of Azure-B dye by Serratia liquefaciens and its validation by phytotoxicity, genotoxicity and cytotoxicity studies. *Chemosphere 196*, 58–68.

Haq, I., Mazumder, P., & Kalamdhad, A. S. (2020). Recent advances in removal of lignin from paper industry wastewater and its industrial applications–A review. *Bioresour. Technol* 312:123636.

Haq, I. & Kalamdhad, A. S. (2021). Phytotoxicity and cyto-genotoxicity evaluation of organic and inorganic pollutants containing petroleum refinery wastewater using plant bioassay. *Environ. Technol. Innov* 23:101651.

Haq, I., Kalamdhad, A. S., & Pandey, A. (2022a). Genotoxicity evaluation of paper industry wastewater prior and post-treatment with laccase producing Pseudomonas putida MTCC 7525. *J. Clean. Prod* 342:130981.

Haq, I., Singh, A., & Kalamdhad, A. S. (2022b). Integrated technologies for the remediation of paper industry waste in a circular bioeconomy, Editor(s): Sunita Varjani, Ashok Pandey, Mohammad J. Taherzadeh, Huu Hao Ngo, & R.D. Tyagi, *Biomass, Biofuels, Biochemicals*, Elsevier, pp. 351–363, ISBN 9780323885119.

Haq, I., Roy, S., & Kalamdhad, A. S. (2022c). Characterization of Pulp and Paper Mill Wastewater and Its Toxicity Analysis Using Vigna Radiata. In *Environmental Degradation: Monitoring, Assessment and Treatment Technologies*; Springer International Publishing: Cham, Switzerland, pp. 261–269. ISBN 9783030941475.

Haq, I. & Kalamdhad, A. S. (2022). Advanced and Ecofriendly Technologies for the Treatment of Industrial Wastewater to Constrain Environmental Pollution. In: Arora, S., Kumar, A., Ogita, S., Yau, Y. Y. (eds) *Innovations in Environmental Biotechnology*. Springer, Singapore. https://doi.org/10.1007/978-981-16-4445-0_22.

Haq, I. & Kalamdhad, A. S. (2023). Enhanced biodegradation of toxic pollutants from paper industry wastewater using Pseudomonas sp. immobilized in composite biocarriers and its toxicity evaluation. *Bioresour. Technol. Repor* 24:101674.

Haq, I., Kalamdhad, A. S., & Malik, A. (2024). Bioremediation of Petroleum Refinery Wastewater Using Bacillus subtilis IH-1 and Assessment of Its Toxicity. *Arch. Microbiol*, 206, 296.

Hirt, N., & Body-Malapel, M. (2020). Immunotoxicity and intestinal effects of nano-and microplastics: a review of the literature. *Particle and Fibre Toxicology*, *17*(1), 1–22.

Horton, A. A., Walton, A., Spurgeon, D. J., Lahive, E., & Svendsen, C. (2017). Microplastics in freshwater and terrestrial environments: evaluating the current understanding to identify the knowledge gaps and future research priorities. *Science of the Total Environment*, *586*, 127–141.

Hurley, R. R., & Nizzetto, L. (2018). Fate and occurrence of micro (nano) plastics in soils: Knowledge gaps and possible risks. *Current Opinion in Environmental Science & Health*, *1*, 6–11.

Hwang, J., Choi, D., Han, S., Choi, J., & Hong, J. (2019). An assessment of the toxicity of polypropylene microplastics in human derived cells. *The Science of the Total Environment*, *684*, 657–669. https://doi.org/10.1016/j.scitotenv.2019.05.071

Islam, S., Apitius, L., Jakob, F., & Schwaneberg, U. (2019). Targeting microplastic particles in the void of diluted suspensions. *Environment International*, *123*, 428–435.

Jeong, J., & Choi, J. (2019). Adverse outcome pathways potentially related to hazard identification of microplastics based on toxicity mechanisms. *Chemosphere*, *231*, 249–255.

Kalogerakis, N., Karkanorachaki, K., Kalogerakis, G. C., Triantafyllidi, E. I., Gotsis, A. D., Partsinevelos, P., & Fava, F. (2017). Microplastics generation: onset of fragmentation of polyethylene films in marine environment mesocosms. *Frontiers in Marine Science*, *4*, 84.

Karimi, M., & Biria, D. (2019). The promiscuous activity of alpha-amylase in biodegradation of low-density polyethylene in a polymer-starch blend. *Scientific Reports*, *9*(1), 1–10.

Khalid, N., Aqeel, M., & Noman, A. (2020). Microplastics could be a threat to plants in terrestrial systems directly or indirectly. *Environmental Pollution*, *267*, 115653.

Koelmans, A. A., Nor, N. H. M., Hermsen, E., Kooi, M., Mintenig, S. M., & De France, J. (2019). Microplastics in freshwaters and drinking water: Critical review and assessment of data quality. *Water Research*, *155*, 410–422.

Kumari, A., Rajput, V. D., Mandzhieva, S. S., Rajput, S., Minkina, T., Kaur, R., Sushkova, S., Kumari, P., Ranjan, A., Kalinitchenko, V. P., & Glinushkin, A. P. (2022). Microplastic Pollution: An Emerging Threat to Terrestrial Plants and Insights into Its Remediation Strategies. *Plants*, *11*(3), 1–14. https://doi.org/10.3390/plants11030340

Landeros, L., Williams, I., Shaw, P., Hudson, M., & Dominguez Cortinas, G. (2022). A review of the origins of microplastics arriving at wastewater treatment plants. *Detritus*, *20*, 41–55.

Lau, W. W. Y., Shiran, Y., Bailey, R. M., Cook, E., Stuchtey, M. R., Koskella, J., Velis, C. A., Godfrey, L., Boucher, J , & Murphy, M. B. (2020). Evaluating scenarios toward zero plastic pollution. *Science*, *369*(6510), 1455–1461.

Li, Z., Gao, C.-M., Yang, J.-L., Wu, L.-Z., Zhang, S., Liu, Y.-H., & Jin, D. D. (2020). Distribution characteristics of microplastics in surface water and sediments of Haizhou Bay, Lianyungang. *Huan Jing Ke Xue= Huanjing Kexue*, *41*(7), 3212–3221.

Lieshchova, M. A., Brygadyrenko, V. V, Tishkina, N. M., Gavrilin, P. M., & Bohomaz, A. A. (2019). Impact of polyvinyl chloride, polystyrene, and polyethylene on the organism of mice. *Regulatory Mechanisms in Biosystems*, *10*(1), 50–55.

Liu, F., Gao, X., Shi, R., Edmund, C. M., & Chen, Y. (2022). A general electrochemical strategy for upcycling polyester plastics into added-value chemicals by a CuCo 2 O 4 catalyst. *Green Chemistry*, *24*(17), 6571–6577.

Liu, P., Shi, Y., Wu, X., Wang, H., Huang, H., Guo, X., & Gao, S. (2021). Review of the artificially-accelerated aging technology and ecological risk of microplastics. *Science of the Total Environment*, *768*, 144969.

Luan, L., Wang, X., Zheng, H., Liu, L., Luo, X., & Li, F. (2019). Differential toxicity of functionalized polystyrene microplastics to clams (Meretrix meretrix) at three key development stages of life history. *Marine Pollution Bulletin*, *139*, 346–354.

Luo, Y. M., Shi, H. H., & Tu, C. (2021). Research progresses and prospects of microplastics in the environment. *Chin Sci Bull*, *66*, 1547–1562.

Martinho, S. D., Fernandes, V. C., Figueiredo, S. A., & Delerue-Matos, C. (2022). Microplastic Pollution Focused on Sources, Distribution, Contaminant Interactions, Analytical Methods, and Wastewater Removal Strategies: A Review. *International Journal of Environmental Research and Public Health*, *19*(9). https://doi.org/10.3390/ijerph19095610

Mattsson, K., Johnson, E. V, Malmendal, A., Linse, S., Hansson, L.-A., & Cedervall, T. (2017). Brain damage and behavioural disorders in fish induced by plastic nanoparticles delivered through the food chain. *Scientific Reports*, *7*(1), 1–7.

Miranda, M. N., Silva, A. M. T., & Pereira, M. F. R. (2020). Microplastics in the environment: A DPSIR analysis with focus on the responses. *Science of the Total Environment*, *718*, 134968.

Mohanan, N., Montazer, Z., Sharma, P. K., & Levin, D. B. (2020). Microbial and enzymatic degradation of synthetic plastics. *Frontiers in Microbiology, 11*, 580709.

Morrison, M., Trevisan, R., Ranasinghe, P., Merrill, G. B., Santos, J., Hong, A., Edward, W. C., Jayasundara, N., & Somarelli, J. A. (2022). A growing crisis for One Health: Impacts of plastic pollution across layers of biological function. *Frontiers in Marine Science, 9*(November), 1–14. https://doi.org/10.3389/fmars.2022.980705

Müller, R.-J. S. H. (2005). Profe J. Dresler K. & Deckwer, W.-D. *Macromol. Rapid Commun, 26*, 1400–1405.

Nizzetto, L., Futter, M., & Langaas, S. (2016). *Are agricultural soils dumps for microplastics of urban origin?* ACS Publications.

Prata, J. C. (2018). Airborne microplastics: consequences to human health? *Environmental Pollution, 234*, 115–126.

Prata, J. C., da Costa, J. P., Lopes, I., Duarte, A. C., & Rocha-Santos, T. (2020). Environmental exposure to microplastics: An overview on possible human health effects. *Science of the Total Environment, 702*, 134455.

Prata, J. C., Lavorante, B. R. B. O., Maria da Conceição, B. S. M., & Guilhermino, L. (2018). Influence of microplastics on the toxicity of the pharmaceuticals procainamide and doxycycline on the marine microalgae Tetraselmis chuii. *Aquatic Toxicology, 197*, 143–152.

Prokić, M. D., Radovanović, T. B., Gavrić, J. P., & Faggio, C. (2019). Ecotoxicological effects of microplastics: Examination of biomarkers, current state and future perspectives. *TrAC Trends in Analytical Chemistry, 111*, 37–46.

Qiu, Y., Zheng, M., Wang, L., Zhao, Q., Lou, Y., Shi, L., & Qu, L. (2019). Sorption of polyhalogenated carbazoles (PHCs) to microplastics. *Marine Pollution Bulletin, 146*, 718–728.

Rao, M. (2019). Review Microplastics in the aquatic environment: implications for post-harvest fish quality. *Indian Journal of Fisheries, 66*, 142–152. https://doi.org/10.21077/ijf.2019.66.1.83125-20

Ren, Z., Gui, X., Xu, X., Zhao, L., Qiu, H., & Cao, X. (2021). Microplastics in the soil-groundwater environment: aging, migration, and co-transport of contaminants–a critical review. *Journal of Hazardous Materials, 419*, 126455.

Revel, M. (2018). Micro (nano) plastics: A threat to human health?, Cur. Opin. *Environ. Sci. Health, 1*, 17.https://doi.org/10.1016/j.coesh.2017.10.003

Rillig, M. C., Lehmann, A., de Souza Machado, A. A., & Yang, G. (2019). Microplastic effects on plants. *New Phytologist, 223*(3), 1066–1070.

Rübsam, K., Weber, L., Jakob, F., & Schwaneberg, U. (2018). Directed evolution of polypropylene and polystyrene binding peptides. *Biotechnology and Bioengineering, 115*(2), 321–330.

SAPEA. (2019). Science Advice for Policy by European Academies. In *A Scientific Perspective on Microplastics in Nature and Society*. SAPEA Berlin, Germany.

Shao, Y., Zhang, F., & Liang, Q. (2020). Research on Microplastic Pollution in Terrestrial-marine Ecosystems. *Ecology and Environmental Sciences, 29*(10), 2118–2129.

Sridhar, A., Kannan, D., Kapoor, A., & Prabhakar, S. (2022). Extraction and detection methods of microplastics in food and marine systems: A critical review. *Chemosphere, 286*(P1), 131653. https://doi.org/10.1016/j.chemosphere.2021.131653

Tubić, A., Lončarski, M., Apostolović, T., Kragulj Isakovski, M., Tričković, J., Molnar Jazić, J., & Agbaba, J. (2021). Adsorption mechanisms of chlorobenzenes and trifluralin on primary polyethylene microplastics in the aquatic environment. *Environmental Science and Pollution Research, 28*(42), 59416–59429.

Vianello, A., Jensen, R. L., Liu, L., & Vollertsen, J. (2019). Simulating human exposure to indoor airborne microplastics using a Breathing Thermal Manikin. *Scientific Reports, 9*(1), 1–11.

Wagner, S., Hüffer, T., Klöckner, P., Wehrhahn, M., Hofmann, T., & Reemtsma, T. (2018). Tire wear particles in the aquatic environment-a review on generation, analysis, occurrence, fate and effects. *Water Research*, *139*, 83–100.

Wan, Y., Wu, C., Xue, Q., & Hui, X. (2019). Effects of plastic contamination on water evaporation and desiccation cracking in soil. *Science of the Total Environment*, *654*, 576–582.

Wang, F., Wong, C. S., Chen, D., Lu, X., Wang, F., & Zeng, E. Y. (2018). Interaction of toxic chemicals with microplastics: a critical review. *Water Research*, *139*, 208–219.

Wang, J., Liu, X., Li, Y., Powell, T., Wang, X., Wang, G., & Zhang, P. (2019). Microplastics as contaminants in the soil environment: A mini-review. *Science of the Total Environment*, *691*, 848–857.

Wei, R., & Zimmermann, W. (2017). Microbial enzymes for the recycling of recalcitrant petroleum-based plastics: how far are we? *Microbial Biotechnology*, *10*(6), 1308–1322.

Wei, Y., Ma, W., Xu, Q., Sun, C., Wang, X., & Gao, F. (2022). Microplastic Distribution and Influence Factor Analysis of Seawater and Surface Sediments in a Typical Bay With Diverse Functional Areas: A Case Study in Xincun Lagoon, China. *Frontiers in Environmental Science*, 60.

Windsor, F. M., Tilley, R. M., Tyler, C. R., & Ormerod, S. J. (2019). Microplastic ingestion by riverine macroinvertebrates. *Science of the Total Environment*, *646*, 68–74.

Wright, S. L., Thompson, R. C., & Galloway, T. S. (2013). The physical impacts of microplastics on marine organisms: a review. *Environmental Pollution*, *178*, 483–492.

Xie, X., Deng, T., Duan, J., Xie, J., Yuan, J., & Chen, M. (2020). Exposure to polystyrene microplastics causes reproductive toxicity through oxidative stress and activation of the p38 MAPK signaling pathway. *Ecotoxicology and Environmental Safety*, *190*, 110133.

Yee, M. S.-L., Hii, L.-W., Looi, C. K., Lim, W.-M., Wong, S.-F., Kok, Y.-Y., Tan, B.-K., Wong, C.-Y., & Leong, C.-O. (2021). Impact of microplastics and nanoplastics on human health. *Nanomaterials*, *11*(2), 496.

Yuan, J., Ma, J., Sun, Y., Zhou, T., Zhao, Y., & Yu, F. (2020). Microbial degradation and other environmental aspects of microplastics/plastics. *Science of the Total Environment*, *715*, 136968.

Yuan, Z., Nag, R., & Cummins, E. (2022). Human health concerns regarding microplastics in the aquatic environment – From marine to food systems. *Science of the Total Environment*, *823*, 153730.https://doi.org/10.1016/j.scitotenv.2022.153730

Zarus, G. M., Muianga, C., Hunter, C. M., & Pappas, R. S. (2021). A review of data for quantifying human exposures to micro and nanoplastics and potential health risks. *Science of the Total Environment*, *755*, 144010.

5 Biotechnological Advances in Treating Micro-plastics

A Sustainable Approach towards the Environment

Ranjana Das, Aishwarya Das, Annesha Basu, and Chiranjib Bhattacharjee

5.1 INTRODUCTION

One of the most threatening pollutants recently been identified severely causing damage of ecosystem is "Micro-plastic" (MP). Though, the effect of plastic pollution is already a critical issue to address globally moreover, its micro form has created more grim issue to deal with. Plastic is categorized as organic polymer derived from petroleum resources with wide usage involving household products, building and construction material, packaging, vehicle parts, electronics and agricultural products. As per present statistical reports, over 400 million tons of plastic are produced every year, half of which is used to create single-use plastics creating harm to the environment and damaging biodiversity. About 14 million tons of plastic ends up in the ocean every year, while it makes up 80% of all marine debris found from surface waters to deep-sea sediments, as mentioned in a report by the International Union of Conservation of Nature (2021). As per this report, the plastic debris perceived in the ocean are all land-based, from urban and storm water runoff, sewer overflows, littering, inadequate waste disposal and management, industrial activities, tire abrasion, construction, and illegal dumping. Under the influence of solar UV radiation, wind, currents and other natural factors, plastic breaks down into small particles termed as microplastics (particles <5 mm) or nanoplastics (particles < 100 nm). The small size makes them easy for marine life to ingest accidentally, which may causes severe injuries and death to aquatic life. Micro and nanoplastics have also threatened food safety and quality, human health, coastal environment, and climate changes (Haward, 2018). Recently, during the COVID-19 pandemic, the worldwide production and disposal of face masks, sanitizer bottles, plastic laboratory and medical materials have drastically amplified, contributing significant plastic and micro-plastic litter in the environment (Fadare & Okoffo, 2020). Furthermore, the UNEP has predicted, as per current tread of demand and supply, the total plastic production is about to reach 33

DOI: 10.1201/9781003379386-5

billion tons by 2050 (Watson et al., 2015; Barnes et al., 2009). Several studies related to plastic size distribution status have evidenced continuous occurring of micro plastic and nanoplastics (< 25 μm), in the ocean (Tanaka and Takada, 2016). The presence of micro and nanoplastic in water is alarming as the conventional filter may confuse it for plankton and end up consuming plastic litters (Barboza et al., 2018; Alfaro-Núñez et al., 2021). Microplastics assimilation has been reported for marine organisms from different trophic levels. The scientific research has evidenced that marine organisms of human consumption consume microplastics directly from the seawater or from lower trophic levels ultimately manages its trail to human body. However not only the micro plastic litters itself but it also acts as carrier of some hydrophobic contaminants (persistent, bioaccumulative, and toxic substances) to wildlife. Infiltration of the microplastic-based chemicals into cells may react with biomolecules and cause endocrine disruption. Microplastics not only transport contaminants, but it also increases its environmental persistence (Alfaro-Núñez et al., 2021).

The intentional addition of microplastic to daily-use personal care products have been reported as substantial contributing to microplastic pollution in remote aquatic and terrestrial environments causing detrimental effects to wide taxa of microorganisms, plant, protist, and animals. The effects of MPs have been reported for fish, amphibians, reptiles' birds and mammals, which was found to vary with nature of species, types of microplastics, level of concentrations, trails duration of exposure (Malafaia and Barceló, 2023). The ethical constraints and strict biosecurity measures to handle human samples are the main constraints for analyzing the effect of microplastics on human health. Hence the challenge of dealing with microplastic-related issues are primarily, current state of art for isolation, detection, and characterization of microplastic in biological samples and to emphasize current trends of beating microplastic, associated limitations, and to explore more advantageous techniques and methodologies for the future. A study published in 2023 has reported the trail of MP in the human food chain through plant-derived food. Their studies have presented the pathway of MP from contaminated soil in several vegetables such as lettuce, radish, wheat, and corn, at very early stage of growth (Liang et al., 2023). Hence, it is challenging to assess the actual level of MP uptake by human. Their study has explored the occurrence of MNPs, the probable impact of MP on human health, and key issues to be addressed in future. Figure 5.1 represents the human exposure routs of micro-nano plastics. Scientists have presented a comprehensive review on present state of art for research trends on micro- and nano-plastics as an emerging threat to the global environment with emphasis on its potential impacts on the feasibility, lifecycle, societal development, and productiveness of living organisms via several potential mechanisms, like oxidative tissue damage, impaired fatty acid metabolism, change in gut microbial communities' dynamics, and molecular damage (Kumar et al., 2021).

Numerous organizations worldwide are trying to address the issue of microplastic waste by approach of sustainable waste management. Various strategies and actions plans have been taken in both developing and developed countries worldwide to discuss the issue of plastic pollution and plausibility to deal with this emerging environmental issue. This chapter aims to address the need of advancements in biodegradation processes in treating microplastic pollutants from waste streams. This chapter intends

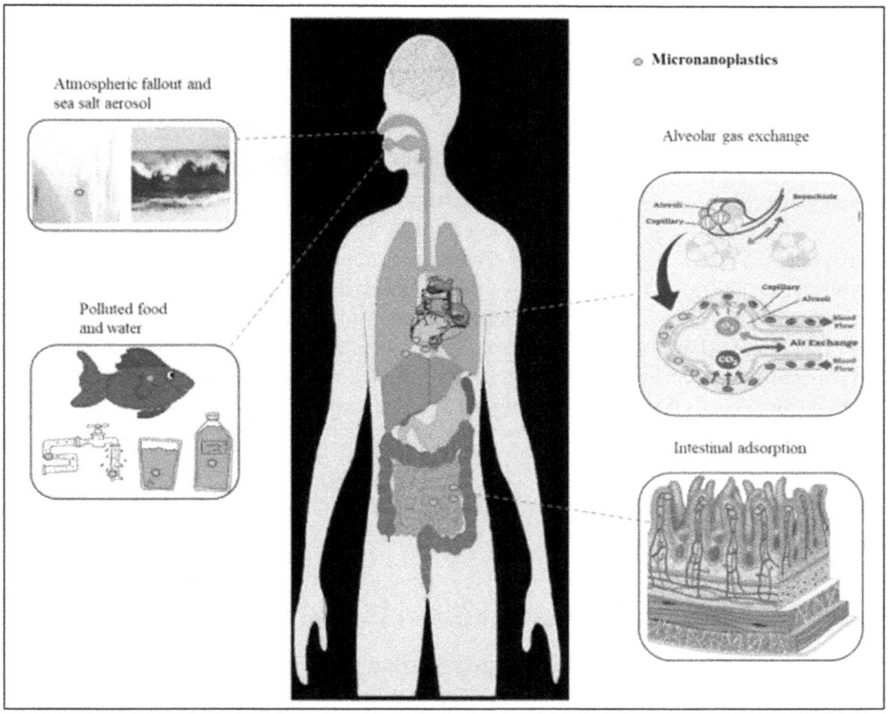

FIGURE 5.1 The trail of micro nano plastic to human body.

to present a comprehensive review on existing chemical and biological technologies to treat microplastics and biotechnological advancements in treating microplastics in an eco-friendly way, focusing on the potential of sustainable biotechnological approaches of processing microplastic waste.

5.2 OCCURRENCE OF MICROPLASTICS IN THE ENVIRONMENT

Plastics are basically synthetic polymers, which are often mixed with other chemicals to improve their performance. Plastic polymers are often broken down into smaller fragments known as macro (>25 mm), meso (25-5 mm), micro (5mm – 0.1 μm), and nano (<0.1 μm) plastics (Crawford and Quinn, 2017). In Table 5.1, Polymer fragments within this micro range have been found in the food chain are represented with their sources. Among them, microplastics are most abundantly found in food chain. Depending on their production process they can be of two types, namely primary microplastics and secondary microplastics. Primary microplastics are made intentionally by humans for commercial purpose (microbeads in toothpaste or in face wash etc.). On the other hand, secondary microplastics are the result of photooxidative, chemical, biological, or mechanical changes in original plastic polymers. In the form of foams, fragments, fibers, and flakes, microplastics are found in nature. Nowadays, microplastics have polluted seas, lakes, soils, and air due to the improper plastic

TABLE 5.1
Common plastic polymer fragments found in microplastics and their uses

Polymer fragments causing microplastics pollution	Common uses of these polymers
Polyethylene (PE)	Garbage bags, packaging film, insulation for wires and cables, toys
Polystyrene (PS)	Food containers, rubber tires, surfboards
Polypropylene (PP)	Toys, car bumpers, drain pipes, CD & DVD boxes
Polyamide (PA)	Synthetic fabric, fishing net, carpets
Polyvinylchloride (PVC)	Water pipes, cables & wire insulator
Polyethylene terephthalate (PET)	Water bottle, food containers, jars, beverages bottle
Polyvinyl alcohol (PVA)	Surgical threads, textile glue

waste management. The main sources of microplastic contamination are landfill and leachate, sludge, and food waste.

5.2.1 Microplastics from Landfills and Leachate

Almost 21 to 42% generated plastics globally find their home in some landfills, which are abundant sources of microplastics. Most of the microplastics present in the landfills are of irregular shape, which proves they are secondary microplastics. Because of the high and fluctuating temperature, pH range 4–9, presence of microorganisms and various chemical reactions, plastic polymers are easily broken down into microplastic form inside the landfills. Due to improper waste management these microplastics get their way out through the leachate and contaminate nearby aquatic environment and are deposited in agricultural soil. This is a main route through which they end up in local food chains (Golwala et al., 2021).

5.2.2 Microplastics in Sludge

Wastewater treatment plants are the main source of microplastics in sludge. Increased use of personal care products, industries, and modern agricultural practice in developed and developing countries is causing this pollution. Current wastewater treatment processes are efficient to a great extent. Primary and secondary treatment alone can remove 97.8% of microplastics from the wastewater. After tertiary treatment, more than 99% of microplastics can be removed from wastewater. The remaining microplastics in the sludge end up polluting the surrounding environment depending on their handling (Ziajahromi et al., 2017).

Microplastic concentration in wastewater greatly depends upon the country or geographical area. More populated places with larger industrial areas show higher concentrations of microplastic in wastewater. For example, microplastic concentration in Finland was reported as 301.4 items/kg sludge (Lares et al., 2018), whereas in China the concentration was reported as 271,700 items/kgL (Dong et al., 2019).

5.2.3 Microplastics in Food Waste

Excessive use of plastic in food processing and packaging is one of the reasons of microplastic pollution. During tea brewing, any crack or tear in fiber-made tea bags can liberate near about 11.6 billion of microplastic particles inside the tea cup. Human consumption of that tea will cause direct ingestion of plastic unknowingly. Even maximum paper tea cups used nowadays have plastic coating inside and when this comes into contact with hot water for tea brewing it causes 102 million microplastic to be oozed out inside the tea. From bottled water to plastic packaged meat, fish and other sea foods, even vegetables, are reported to have microplastic traces inside of them. 18.7 items/kg microplastic have been found in processed chicken sold with plastic packaging (Kurniawan, 2021). The amount of microplastic items per species of marine fishes, mussels or oysters has been reported as 15, 2.7-3.7, 1.4-7 microplastic items respectively. Natural salt can be a very concentrated source of microplastics depending on the microplastic concentration of the sea. So, natural salts produced from different geographic location may have 7-681 items of microplastic/Kg. In commercial salt microplastic concentration is relatively low, that is 1–10 items/Kg. Freshly packed fruits are reported to contain more microplastic than green vegetables (Golwala et al., 2021).

5.2.4 Source of Microplastic in Air

When plastics are exposed to high temperature, they fragment easily and produce microplastic. Natural phenomena like wildfire, or human actions such as burning plastic debris, release microplastic into the surrounding air (Jung et al., 2022). Currently, marine environment all over the world is most polluted by microplastic. Microplastic in the range between 7 to 33 μm is reported to be aerosolized through bubble burst or discharge in sea spray. In air, microplastics can be transported with it. In terrestrial environment, microplastics degrade more easily than in sea with continuous UV exposure and temperature fluctuation. Degraded microplastics from surface level can be suspended in air at a rate of 0.08 to 1.48 $mg/m^{2}/min$ easily through wind erosion or dust storm. Indoor activity of synthetic textile like blanket drying by using mechanical dryers cause emission of microplastic in the air at a rate of more than 1 microplastic fiber/m^3 in just 20 minutes. Activities in nail salon also cause emission of microplastic in the surrounding air. Inside the nail salon ~18 microplastics/m^3 have been found in the air. Extreme use of polyurethane, acrylic and rubber in the nail treatments cause this pollution (O'Brien et al., 2023).

5.3 EFFECT OF MICROPLASTICS OCCURRENCE ON AQUATIC AND TERRESTRIAL ECOSYSTEMS

Microplastic pollution is capable of deteriorating water and soil quality and through food sources from sea, air, and soil microplastics have found their way into the human body. This section involves focus on effects of microplastics on marine, soil environment, and on human health.

5.3.1 Effect of Microplastic Pollution on the Marine Environment

Microplastic pollution is becoming a great threat to our surrounding environment. These plastic debris are now accumulated in our marine environment and soil, thus infecting our food chain and finally ending up in our bodies. Microplastic pollution is nothing new; studies conducted in 1977 showed the presence of microplastic in sandy beaches of Canada, New Zealand, Lebanon, and Bermuda. Microplastic samples have also been found in phytoplankton. As a result, fish eating those planktons are getting the microplastics in their systems (Ivar do Sul and Costa, 2014). Studies showed the existence of microplastic inside teleost fish also (Carpenter et al., 1972). Their study reported that those fish ingested the microplastics in early phases of their life. Lantern fish (Myctophid), Catfish, Mojarra fish all over the world are contaminated with microplastics. Blue threads of nylon are also seen inside of them as source of microplastic. Fishing nets are the source of these blue nylon threads. 35% of planktivorous fish contains microplastics in their gastrointestinal area (Ivar do Sul and Costa, 2014). Not just in fish, microplastic debris are also being found inside of invertebrates. Molluscs, such as snails, slugs, mussels, octopuses; Crustaceans such as crabs, lobster, shrimp etc.; Annelids and Echinoderms ingest microplastics from the surrounding environment. Mytilusedulis or blue mussels can intake and accumulate microplastics in their system very quickly and in huge quantities (Browne et al., 2008). 83% lobsters in Clyde Sea have microplastics in their stomach. Dosidicusgigas

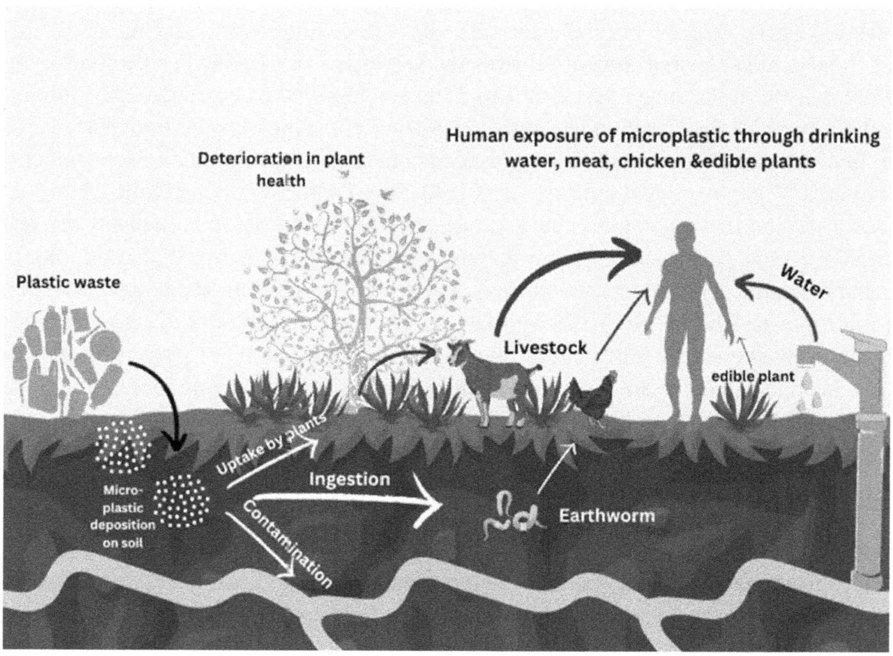

FIGURE 5.2 Microplastic pollution and its ingestion by humans and other animals from soil environment.

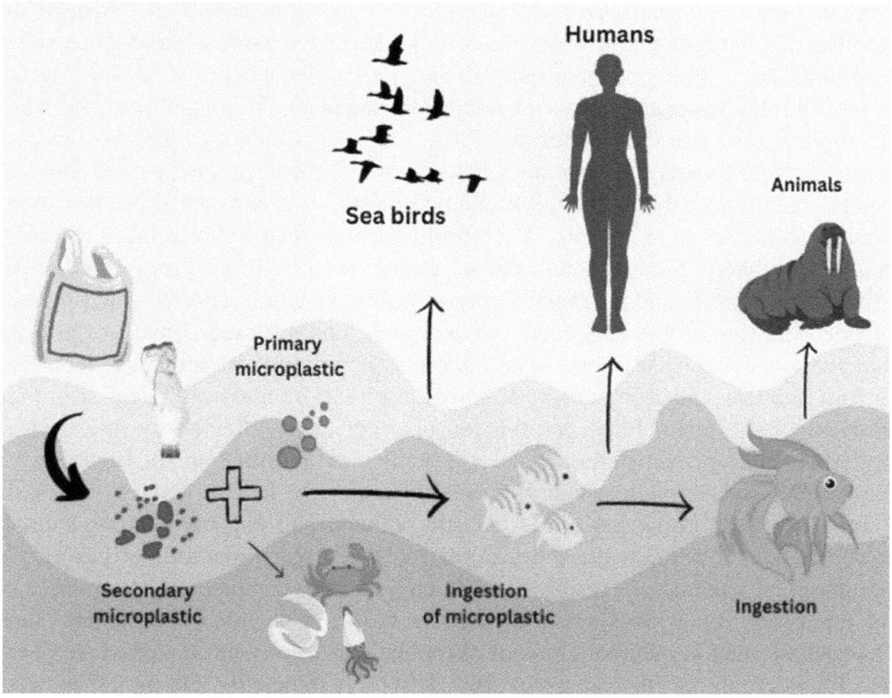

FIGURE 5.3 Pathway of microplastic pollution and its exposure to humans through the marine environment.

or Humboldt squids, Holothuria, Talitrussaltator also show the trend of ingesting and accumulating microplastics in their body. PE & PP microplastics are most seen inside of these living beings. The vastness of the microplastic pollution can be understood by the presence of microplastics inside fish or other invertebrates from sea shores which are not exactly hotspots of plastic pollution. Seabirds, namely Fulmarusglacialis, Cory shearwaters, Calonectrisleucomelas and Puffinustenuirrostris, ingest plastics in great quantities (Colabuono et al., 2010). Plastic fragments are commonly found in 80% of the Fulmar birds. More microplastics are found in the guts of young birds than in adults. Animals that depend upon fish as food sources such as fin whales (Balaenopteraphysalus), fur seals (Fossi et al., 2012) (Arctocephalustropicalis and Arctocephalusgazella) are contaminated by microplastic pollution. Microplastics from the marine environment end up in humans, sea birds, and in other animals living near the sea through sea foods, as represented in Figure 5.3.

5.3.2 Effect of Microplastic Pollution on Soil Environment

Leachates from landfills, agricultural plastics, and industrial processing are the main source of microplastic pollution in soil environment. Accumulation of these microplastics cause changes in physical properties of soil. Increased accumulation of PES, PET, PP, PS, PMMA microplastics decrease soil bulk density which in turn increases the risk of soil erosion. Microplastic deposition in soil also disturbs the stability of soil aggregation. Pore size of soil determines its water-holding capacity. Soil with larger pore size cause easy drainage of water whereas soil with smaller pores holds the water and makes it available for the plants for longer time. But when microplastics stay in the soil for more than one year it increases the existing pore sizes. Increased water drainage does not only affect plantation but also it damages natural microbial habitat. Pollution of microplastic also enhance soil water evaporation (Mbachu et al., 2021). Microplastic contaminated soil affects the plant health very badly. Studies conducted on spring onion (Allium fistulosum), wheat (Triticumaestivum), and perennial ryegrass (Loliumperenne) reported that, presence of microplastics on the soil decreases biomass, length, tissue density, total area, and average diameter of roots. Also, microplastic pollution of soil cause reduction in fruit biomass, nitrogen content of leaves, root-leaf biomass ratio, leaf size, total biomass of the plant. Plant germination also get badly affected by microplastic pollution, microplastic pollution decreases rates of germination. In Figure 5.2 a classic pathway, through which microplastics enter the human body by ingesting plant-based food grown on a microplastic polluted soil and herbivorous animals is depicted. It is not just plant health microplastic contamination that has effects on microbial habitat, soil respiration and enzyme activity fluctuate in presence of microplastics. But it does not always cause reduction in microbial habitat. Sure, changes are seen in metabolic rate of microbial community but the presence of PE and PA microplastic increases microbial activity significantly. On the other hand, PES and PMMA reduces microbial activity. Also changes in microbial activity are different in case of different soil. For example, clay soil, when polluted by microplastics heavily alter both the fungal and bacterial communities but at the same time cinnamon soil, paddy field soil even after being polluted by microplastics do not cause any changes in microbial community. So, soil type and texture are most important to understand the effect of microplastic pollution as each soil type responds in different ways (Mbachu et al., 2021).

5.3.3 Effect of Microplastic Pollution on Human Health

Microplastic ingestion is caused mainly by eating seafood, in case of humans, but not only seafoods, commercial salts made from sea or saline lakes or saline rocks are also a potential source of microplastic in the human body. As we have already discussed, the sea is one of the main areas infected by microplastics and these contaminants are often transported via the water waves from contaminated places to non-contaminated places. Though many people all around the world are vegetarian, everyone has salts, to some extent, almost with every food. Microplastics can enter the living cells and lymphatic systems of humans and finally they can circulate all over the body by getting inside the circulatory systems. Not only in humans but this same trend follows in case of rabbits, mice, and dogs. 0.2 to 150 μm particle-sized microplastics have been found in the human body (Waring et al., 2018). Possibly via Peyer's patch in the intestine, microplastics enter the cell and lymphatic system. But only 10% of the microplastics from food get absorbed in the body, the rest is excreted. Microplastics are biopersistent and bioaccumulate easily but the threat does not end there. Chemical pollutants' adsorption onto the microplastics surface is quite evident. Studies conducted in Japanese and Portuguese coast reported that microplastics have adsorbed polychlorinated biphenyls (PCB), organochlorines (DDE), nonylphenols (NP), polyacrylic aromatic hydrocarbons (PAH), DDT from the sea. The effects of these chemical pollutants are also common with microplastics ingestion (Peixoto et al., 2019). Lung and liver damage in humans caused by prolonged ingestion of microplastic through food sources has been reported (Prata et al., 2020). Consumption of these microplastics with contaminants like pesticides and other chemical pollutants causes inflammation, tissue necrosis, cell apoptosis, genotoxicity, increased oxidative stress plus localized tissue and cell damage, fibrosis. Some of these pollutants are potential carcinogens which increases the chance of carcinogenesis. Nano plastics are capable of passing the blood brain barrier. Longer exposure to these particles can alter human chromosomes and cause infertility, cancer, or obesity in the long run (Sharma and Chatterjee, 2017).

These microplastics are often the vector of harmful bacteria, through ingestion of these microplastics body get exposed to these pathogens and they affect immunity and development of the body adversely. Sometimes harmful chemicals are also used in the production of plastic bottles, for example antimony is used in making plastic bottles (PET), and it acts as a catalyst. When the plastic bottle faces higher temperature, preferably in between 60–85° C antimony is released. Sudden exposure to such chemicals or heavy metals cause nausea, vomiting, diarrhea in humans. PET microplastics have been found in salt which is produced by high temperature conditions; even when used in cooking foods, sea salts face high temperature condition again (Revel, Châtel, and Mouneyrac, 2018).

5.4 EXISTING CHEMICAL AND BIOLOGICAL TECHNOLOGIES TO TREAT MICRO-PLASTICS

The ubiquitous and inevitable application of plastics in our daily lives and industries leads to a huge amount of plastic waste in drinking waters, stormwaters, wastewaters, sludge, and receiving waters. The micro-nano plastics (MNP) is one of the most threatening challenges to humans because of their large specific surface area for adsorption of hazardous pollutants like persistent organic chemicals, geogenic contaminants and their potential of easy transmission, capacity to permeate biological membranes and accumulation in organisms' body (Kumar et al., 2021). MNP have been spreading and polluting all environmental media, including soil ecosphere, sea ecosphere, fresh water, and the frigid-zone. With the MNP being non-biodegradable in nature, its removal from contaminated water is highly challenging and has driven the researchers to explore the efficacy of existing water treatment processes and hybrid processes for treatment of MNP. Bioplastics being non-biodegradable and of microscale, there are difficulties in getting rid of MNP in water. To comprehend the prospect of MNP removal, studies have been conducted in various directions involving, physical separation techniques, chemical processes, biological approaches and hybrid technologies of any combination of the mentioned processes. This section emphasizes on some effective recent methodologies practiced and developed to remove MNP from aquatic resources.

Conventional water treatment plant and existing water treatment technologies are effective in removal of the MNP from water rendering the tiny size and hydrophobicity of the MNP which facilitate accumulation in adsorptive removal (Keerthana Devi et al., 2022). Some authors have also reported emerging technologies for their potential for MNP removal, such as advanced oxidation process, membrane technology, biodegradation, and electrocoagulation but with a concern of proper strategies and management of the remediation technology to remove the MNP from the environment (Haq et al., 2016a, b; 2017, 2018, 2020; Haq and Kalamdhad, 2021; Haq et al., 2022a, b, c; Haq and Kalamdhad 2022, 2023; Haq et al., 2024). In the context of MNP removal from drinking water, removal efficiency is low for coagulation, sedimentation, and filtration. In the coagulant treatment process removal efficiency reported was 16.5 %with the aid of coagulant, and 2% without the aid of coagulant for size range within 45–53 μm while in filtration process, removal efficiency achieved was 86.5 to 99.9% (Rout et al., 2022). The advanced filtration process, membrane filtration (ultrafiltration) has been explored in many dimensions for MNP removal. The properties of nano-ultrafiltration membranes are more effective in adsorptive removal of MNP. The membrane filtration has the removal efficiency of 32 ±12 % in a 3 μm filter; whereas, a filter pore size of 0.22 μm removed 92 ± 3 %. Figure 5.4 presents a brief account of the advanced technologies for its removal efficiency against MNP. Several other techniques suitable for the large volume of MNP separation from water samples being cloud point extraction, pressurized fluid extraction. The study presented the details on comparative efficiency of all the processes (Keerthana Devi et al., 2022).

In the present context of microplastics separation and treatment, agglomeration and coagulation have been explored and practiced with functioning treatment plants owing to its unique features of high capacity to get adsorbed on any species. Agglomeration and coagulation, processes have also been used to form larger particles to facilitate the removal of MPs from water with fine tuning of the process depending on the size of MPs.

In the category of the filtration process, advanced filtration techniques like membrane filtration and ultrafiltration have mostly emphasized separating micro and

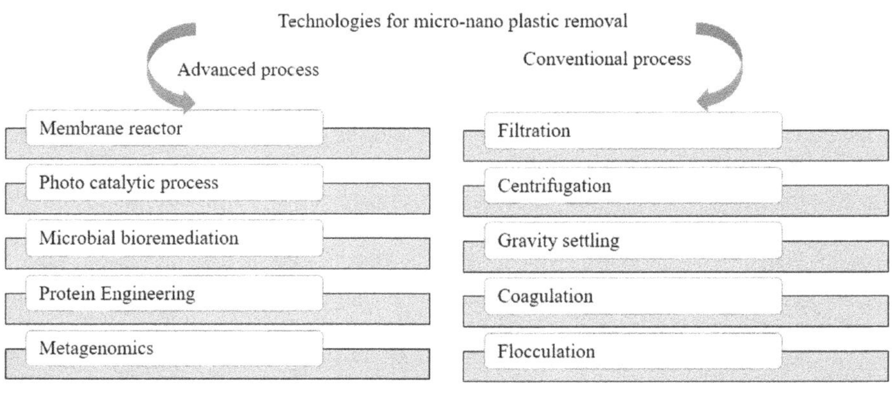

FIGURE 5.4 Conventional and emerging technologies for micro nano plastic remediation.

nano plastic from aquatic resources due to the flexibility in choice of varying pore size with targeted particle size of the plastic material (Gupta et al., 2021; Kundu et al., 2021; Mohana et al., 2021). Membrane filtration has also been modified with adsorptive removal of MNP using advanced membranes and charged membrane (Ali et al., 2021). The limiting issue associated with the membrane filtration process is the disposal pattern of the stream rich in MNP which may create a huge impact on living beings if discarded directly to the environment. Hence, filtration may not be a standalone technique for MNP separation but coupled with other processes like coagulation and adsorption serves the purpose in better way. Hence, there is an immense scope for efficient removal of MNP from water by the combination of two or more processes. A study has focused on the removal behavior of polyethylene (PE) in drinking water by ultrafiltration and coagulation processes using a Fe-based coagulant (Ma, 2019). A detailed study on the prospect of membrane filtration process for MNP serration has also been presented previously (Poerio et al., 2019). Dynamic Membranes Technology is also an emerging technology with the prospect of surface water treatment, wastewater treatment, oily water treatment, industrial wastewater treatment and sludge treatment, and studies are also underway to utilize the principle of dynamic membrane technology for microplastics separation. Reverse osmosis process has also been practiced in several fields to separate microplastics from water but limited to streams having high load of microplastic (Poerio et al., 2019).

A recent process intensification of the membrane filtration process is the design of Membrane Bioreactors (MBR), which may offer more competitiveness, better process control, and environmental friendliness (Poerio et al., 2019). MBR techniques have shown a significant MP removal (99%), good quality of final effluent, decreased process stages, and have replaced the conventional secondary clarifiers. MBR based on "plastic preserving enzyme maceration" may efficiently produce a purified MP solution. It has been reported that in WWTP, the use of MBR has shown 70% higher removal of Di(2-ethylhexyl) phthalate (DEHP) of conventional treatment (3%) and has escalated to about 83%while associated with adsorption processes. The process efficiency of MBR is associated with physicochemical properties of the phthalate esters and operative conditions such as hydraulic retention time (HRT), initial feed concentration, and microbial consortium responsible for degradation of microplastics. Hence, the MBR technology has now been focused not only on process intensification but also on isolation of a novel bacterium (*Idonellasakaiensis*) able to use polyethylene terephthalate (PET) to convert it to less harmful monomers (terephthalic acid and ethylene glycol).Since MBR technique involves enzyme systems, it is probably possible to degrade the MP in the enzymatic membrane reactor. A complete removal of MP through MBR pathway was associated with a preliminary anaerobic treatment and followed by an RO filtration (Poerio et al., 2019).

A promising technology emerging for remediation of microplastic is being catalytic conversion of plastic to nanocarbon and hydrocarbons, while both the products have unique application potential. Nanocarbon is well recognized for excellent adsorbent material while the hydrocarbon provides the platform of fuel synthesis. Catalytic processes can convert plastics into nanocarbons or hydrocarbons. Several studies have already been reported on photocatalytic degradation of the plastic to value-added products (Ge et al., 2022; Du et al., 2022). Existing literatures with a comprehensive review on photo-catalytic degradation of microplastics using various photocatalyst system in bare and composite systems have been demonstrated (Ge et al., 2022). Photocatalytic degradation is a complex mechanism and differs for both bare and composite photocatalysts system. In a class of bare photocatalyst, titanium dioxide was reported as an excellent photocatalyst system which causes weight loss rate of the nanocomposite film (LDPE) about 68% under solar irradiation for 200 h, while that of the bare film was only reported at 1.02%. Another report showed that LDPE-2% TiO_2 composite film losses weight at rate of 40.99% than the 4.97% pure LDPE film (Malafaia et al., 2023) with increased brittleness and generation of micron-sized pores specifying structural demolition of the nanocomposite film. Photocatalytic mineralization has also been achieved for PS and PE plastics in the presence of TiO_2 nanoparticle and the degradation has been confirmed from the generation of oxygen-containing functional groups in the processed products. ZnOnanorod photocatalysts have been explored for degradation of LDPE MPs deposits in the presence of sensitizing dye (Eosin Y). In the presence of merely dye-sensitizer weight loss rate of 5% was reported over a period of 2 hours while the addition of ZnOhaS raised the degradation rate to 20% under similar process conditions, other process parameters being catalyst loading capacity, dye sensitizer, UV irradiation intensity, and time of exposure. Several other potential photocatalyst material, as reported in published literature, are: unicryptomelane-type manganese oxide (OMS-2) photocatalyst, and ultrathinhydroxyl-rich BiOCl, polypyrrole/TiO_2nanocomposites,Goethite and Fe/Ag-doped TiO_2(14.34% weight loss for PE film, 300 h UV exposure). Based on the available data, a comparative performance study of novel photocatalyst materials, like tungstophosphoric acid (HPW), CdS and H_3BO_3on PVC for 250 h, in the presence of visible light irradiation has shown the weight loss rate of HPW-PVC film being 69.6%, for CdS-PVC film 47.2% and lowest weight loss rates for H_3BO_3-PVC film were 14.9% (Ge et al., 2022). DopingofTiO_2 with V, Mo, or W have shown increased degradation rate while a decreased rate was observed for Cr or MndoppedTiO_2catalyst. 0.7 wt% copper phthalocyanine (CuPc)-modified TiO_2 was reported as most promising dopant regarding PS and PE degradation. Platinum decorated on ZnOnanorods (Pt/ZnO) has also been reported having about 13% higher photocatalytic activity than bare ZnOnanorod under visible light induced photocatalytic degradation of LDPE-MPs (Du et al., 2022). A timeline for the development of the photocatalyst intended for photocatalytic degradation of microplastic is shown in Figure 5.5 (Du et al., 2022). P-type semiconductor nanomaterials like, Cu_2O and CuO, NiO NPs, CdS, CdS/CdOx, Nb_2O_5 have also been reported for significant photocatalytic activity. Another class of nanomaterial based on carbon, namely, graphene, grapheme oxide, carbon nanotubes, have also been reported for photocatalytic degradation of MPs/NPs. The limiting issue of the photocatalytic degradation of the microplastic has been identified by the researchers as being the incomplete mineralization which has driven the researches to develop novel composite catalyst system/surface-modified catalyst/metal and non-metal doped catalyst to realize complete mineralization of the microplastic. In photocatalytic process, the selective site of attack is critical to specify which may lead to formation of complex products with significant toxicity for living species.

Another approach for removal of microplastic is the application of magnetic nanoparticles (Shi et al., 2022). The efficacy of magnetic nanoparticle for extraction

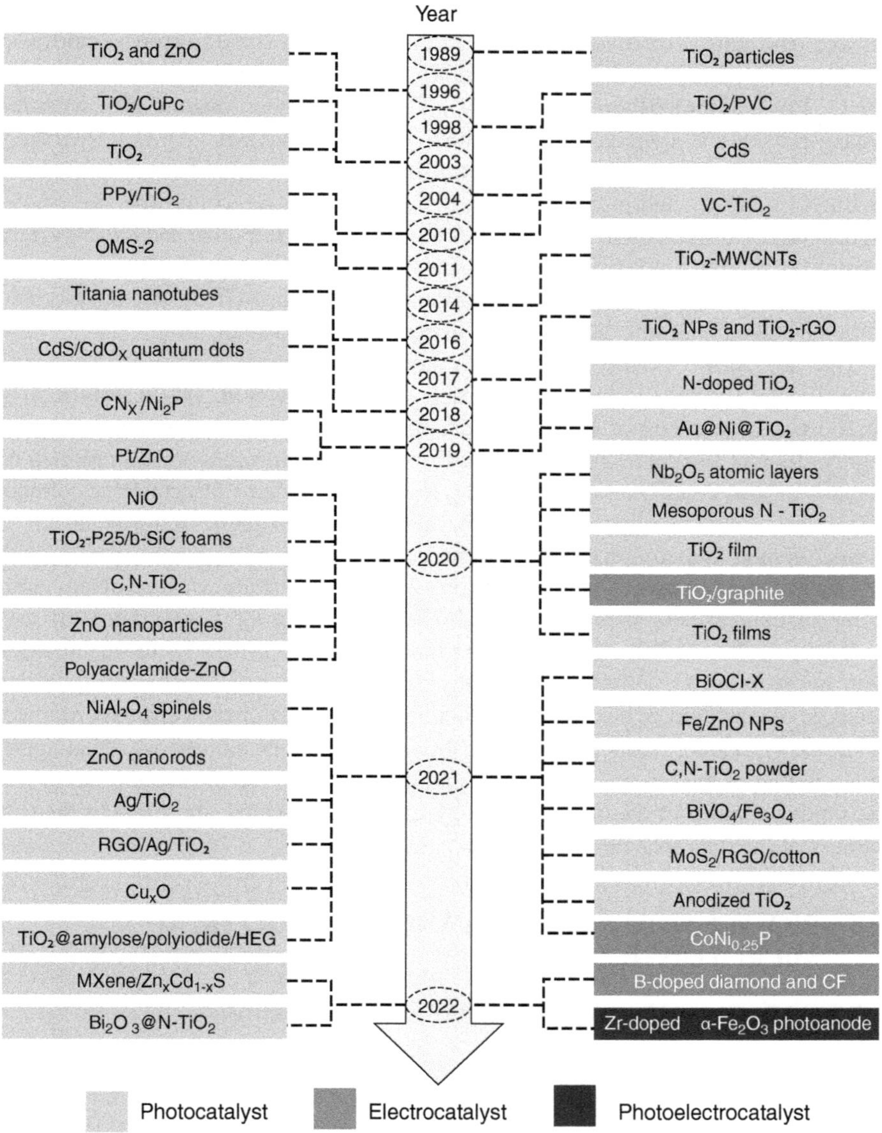

FIGURE 5.5 Timeline of the photocatalyst intended for photocatalytic degradation of microplastic.

of microplastics from environmental samples was demonstrated in earlier studies (Grbic et al., 2019). The study presented MPs removal by magnetic nano-Fe_3O_4 with a brief that, with 1.3 $g{\cdot}L^{-1}$ nano-Fe_3O_4 for 150 min treatment time, the average removal rate polyethylene, polypropylene, polystyrene and polyethylene terephthalate of size of approximately 200–900 μm was 86.87 ± 6.92%, 85.05 ± 4.70%, 86.11 ± 6.21%, and 62.83 ± 8.34%, respectively. The study also concluded about

80% removal of MP from environmental water bodies like, river water, domestic sewage, and natural seawater. A study in 2022 reported the adsorptive removal of various microplastic residues from portable and sea water using nanotricomposites (Fe_3O_4-PWA/amine) (Bhore et al., 2022). The hydrophobic magnetically separable organo-polyoxometalatenanotricomposites (Fe_3O_4-PWA/nBut, Fe_3O_4-PWA/nHex, and Fe_3O_4-PWA/nOct) were synthesized using Fe_3O_4, PWA, and respective amines (n-Butyl, n-hexyl, and n-octyl amines). Fe_3O_4-PWA/nOct composite has shown highest removal capacity of microplastics (83%) for polysulphone and 99% for polystyrene. Fe_3O_4-PWA/nOctnanotricomposite synthesized using n-octylamine can perform almost complete extraction of various microplastics (PET, PS, polystyrene) from the spiked microplastics solutions in potable and seawater.

The adsorptive exclusion of micro polystyrene using magnetic iron oxide nanoparticles has been reported in an article in 2022 (Heo et al., 2022) and the study demonstrated potential of iron oxide nanoparticle for removing microplastics via highly efficient and environmental-friendly procedure. Some authors has reported the potency of bare Fe_3O_4–nanozymenano aggregates for efficient adsorption of common microplastics via hydrogen bonding, permitting full magnetic removal at 1% of plastic weight, and complete microplastic degradation due to its nanozyme activity at elevated temperature (Zandieh & Liu, 2022). This study was involved modifying the nanoparticle with induced hydrophilicity unlike the more commonly used hydrophobically modified Fe_3O_4 nanoparticles, resulting a bare Fe_3O_4 nanoparticle with peroxidase-like activity of its exposed surface which is capable of additional catalytic degradation of microplastics with nearly 100% efficiency and easy recovery of the Fe_3O_4. Hence, from the pertinent literature it can be summarized that appropriate tuning of the chemical processes and integration of the chemical process with the unique feature of the nanotechnology can be a potential approach to beat the harness of the microplastics contamination in aquifers.

5.5 BIOTECHNOLOGICAL ADVANCEMENTS TO TREAT MICROPLASTICS

5.5.1 Microbial Valorization of MPs: Bioremediation

Chemical and physical mechanisms are used to operate traditional cleanup techniques. The transfer of contaminants into another medium, such as air or water, results in elevated levels and eventual destruction or decomposition in a smaller and benign form. Chemical processes, however, can have low efficiency and are expensive, and chemical compounds used in breakdown reactions may produce hazardous intermediates that must also be required to be removed. There are several bio-based technologies to treat microplastics in the environment. Bioremediation is being counted among one of the promising technologies used to degrade MPs. Bioremediation is an eco-friendly technique to valorize contaminants in the environment. Bioremediation is an umbrella term that brings under several processes, such as microbial bioremediation, plant-mediated bioremediation etc. Microbial bioremediation is an eco-friendly and economical process to treat MPs. Recently, microbial valorization has been devised to get rid of nearly un-degradable MP wastes and plastic debris (Ru et al., 2020). Bacteria-mediated remediation is widely studied and executed in removal of microplastics from soil, aquatic life, and also waste water treatment plants. Microplastics have been documented to be degraded by a wide variety of bacterial taxa, including *Vibrio, Campylobacter, Arcobacter, Acetobacteriodes, Saccharomonospora, Thermobifida, Alteromonadaceae, Burkholderiales, and Pseudomonas*. In microbial bioremediation, the microplastic is used as a source to energy that is consumed by the microbes and utilizes it to run their metabolic reactions. Compared to other cleanup techniques, bioremediation has significant benefits. Bioremediation produces little byproducts because it only relies on the regular metabolic functions of living systems, minimizing the harm to the environment. Microbes being the smallest factory or unit, it works as an overall functional system to carry out the entire process. This substantially minimizes the requirement of any specialized equipment that would consume additional energy. Bioremediation, although a slow process and there are several factors dependency like the nature of the polymer, types of organism used and the reaction conditions, it is still considered as one of the economical and simple technique. Despite substantial progress in the biotech of microplastics depletion, it is still unclear how it might be used. Aquatic and soil flora and fauna are immensely affected with microplastics, which majorly get incorporated by anthropogenic activities. There are several steps involved in the microbial biodegradation of MPs, including: 1) initial degradation of polymers in smaller particles from large polymeric structures; 2) deterioration of polymers to their oligomer, dimer, and monomer; and 3) mineralization of MPs by microbial biomass (Blair Espinoza et al., 2019). Microbial approach of bioremediation mainly follows the above-mentioned mechanism. The microplastics in aquatic and soil environment are exposed to the microbes to adhere on the surface forming a biofilm and changing the nature and morphology of the MPs. This is the initiation step of the degradation of MPs by microbes. The biofilm formation is followed by the secretion of some specific enzymes, which are responsible for breaking down the respective polymer into their oligomers, dimers, or monomers. As a result, microbes can utilize/absorb these simpler components as their energy source through simple of facilitated diffusion for biomass formation. In the penultimate phase, daughter metabolites such CO_2, H_2O, and CH_4 are produced. Although the pathway is similar for all the microbes that are capable of degrading plastics, the cellular metabolites that they secrete vary respectively. The extracellular enzymes like esterase, laccase, manganese peroxides etc. secreted by microbes tend to increase the hydrophilicity of the polymers, thus making it a step easier for the next biodegradation pathways. These enzymes may attach certain sensitive bonds to the side chains of plastics or chemical groups to facilitate chain cleavage during biodegradation. However, due to their size, they are unable to penetrate into the polymer structure; as a result, the degradation may take place on the surface, leading to fissures. MPs are typically defined as synthetic materials that are insoluble in water and do not degrade naturally. Possessing the proper enzymatic and metabolic pathways is one of the favorable conditions for effective biodegradation. In addition, favorable environmental factors like pH, salinity, temperature, and moisture content are necessary to promote biodegradation.

In order to study the biotechnological advancements (such as biostimulation, bioaugmentation, and enzymatic biodegradation) that can be applied for MPs removal by biodegradation, a broad spectrum of MPs are taken into consideration that are found in the environment. The biotechnological advancements in bioremediation has also been assisted by several other parameters. The synthetic polymers are generally slow degrading in nature. Studies have formulated a microbial consortium to treat these synthetic polymers. With improved PE and PP biodegradation, new consortia like *Enterobacter* and *Pseudomonas* developed from cow dung showed up to 15% weight loss after 120 days (Ru et al., 2020). Most of the synthetic polymer biodegradation employed hybrid technology that is pre-treatment followed by biogenic processing. Given that PE, PP, and PVC's high molecular weight significantly impedes their biodegradation, physicochemical pretreatments like chemical oxidizing chemicals, uv-irradiation, and thermo-oxiodation etc. can enhance the accessibility of MPs for biodegradation. Literature survey has also provided with the fact that microbes utilize this pre-treated MPs as a carbon source in better way. In order to break down these complicated polymers, bacteria and fungi like *Phaeosphaeriaspartinicola*, P. halima, and *Mycosphaerellasp* assemblages may form biofilms on the outermost layers of synthetic polymers and consume microplastic particles (Kawai et al., 2019).

Relatively low molecular weight plastics like PET is widely used in daily life for packaging materials, personal care etc. Even though PET is thought to be non-toxic, its tiny granules are pervasive, resilient in aquatic and terrestrial environments, carry harmful substances, and build up in living things. Employing microbial treatment for degrading PET is one alternative to the conventional chemical processes. According to Gong et al. (2018), who investigated the biodegradation of MPs made from PET using bacterial whole-cell, the interaction between biodegradation and alkali hydrolysis improved the hydrolysis of PET [49]. Yoshida et al. (2016) found a bacterial strain (Ideonellasakaiensis) that can degrade PET and demonstrated that this strain attaches to PET and produces two hydrolytic enzymes (PETase, MHETase) that can degrade PET [50]. PETase is an extracellular enzyme that hydrolyzes PET to create the main intermediates terephthalic acid (TPA) and mono (2-hydroxyethyl) terephthalic acid (MHET). The enzyme MHETase breaks down MHET into TPA and ethylene glycol (EG). They both come from crude petroleum and are PET monomers. The TPA transporters typically carry TPA to the cytoplasm, where oxidative and ring breakage enzymes catabolize it. Biotechnological advancements in the field of bioremediation has accompanied with hybrid pathways as the presence of MPs in ecosystem is in broad spectrum. Low-density polyethylene 30- and 40-mm sheets were exposed to isolated pure bacterial cultures from a dumpsite Muhonja et al. (2018). Bacteria from the genera of *Bacillus*, *Brevibacillus*, *Cellulosimicrobium*, *Lysinibacillus*, *Ochrobactrum*, and *Pseudomonas* were included in the study. The study revealed that after 16 weeks of incubation, *Bacillus cereus* and *Brevibacillusborstelensis* showed the highest biodegradations, of 35.7 and 20.4%, respectively (Muhonja et al., 2018). *Bacillus cereus* and *Bacillus gottheilii* were used (Auta et al., 2017) in the biodegradation of polyethylene microparticles. Bacterial consortium made up of *Enterobacter* and *Pseudomonas* degraded polypropylene. Even bacterial species *Rhodococcus* also contributed to polypropylene degradation (Skariyachan et al., 2021).

There is disagreement over whether bacteria or fungus are better at dissolving microplastics. According to some (Auta et al., 2017), bacteria are more effective because they develop more quickly and are better able to adapt to a variety of harmful environments though the majority of people believe that fungi are more effective (Muhonja et al., 2018). Fungi do really offer some advantages. Fungi can grow on a variety of substrates, to start with. Additionally, because of their absorptive nourishment, they are ideally suited to live in the "plastisphere." The fungus is growing apically. Fungi can also create biofilm. It is possible for fungi to utilize polymers, such as those found in microplastics, as a source of carbon and energy by secreting hydrophobic proteins from their cells. The biodegradation of complex-structured polymers, such as aromatic ones, is most effectively accomplished by fungi. Numerous non-specific enzymes that can catalyze a variety of processes are present in fungi. They are hence ideal for the biodegradation of microplastics. Their ability to adapt is significantly influenced by the fungal intracellular enzymatic system, which serves as an internal detoxifying mechanism. Polyethylene is the polymer that has been the subject of the most research when it comes to microorganisms degrading microplastics. *Aspergillusflavus* and *Pleurotusostreatus* are two fungi that are capable of biodegrading polyethylene microparticles because they both release the laccase enzyme (Zhang et al., 2021). Low-density polyethylene is more efficiently degraded by *Aspergillus* fungi than by some *Bacillus*, *Brevibacillus*, *Cellulosimicrobium*, *Lysinibacillus*, *Ochrobactrum*, or *Pseudomonas* bacteria (Muhonja et al., 2018) in one of their studies. Polyethylene (PE), polypropylene (PP), and polystyrene (PS) are the most often observed types of microplastics. Polyvinyl chloride (PVC) and polylactic acid (PLA) are two other forms of microplastics that have been reported. Biodegradation or bio-treatment of microplastics can be enhanced and improved by implanting certain techniques, crucial tactics like pretreatment, utilizing produced biosurfactant, and utilizing technological advancements like biostimulation (e.g., using oxidative shocks or providing more nutrition for the enrichment of microbial consortia.

WWTPs are frequently seen as sources of microplastics, because they concentrate microplastics in streams of waste water and then release them back into the environment with little mass loss. The waste water sludge contains microplastic pools that are removed from the environment. Treatment of this sludge is critical. One of the processes is anaerobic treatment, which explores microorganisms. Bioaugmentation has recently gained importance as an advanced bioremediation technique to treat microplastics containing sludge. Anaerobic bacterium species that can break down microplastics have been detected in far lower numbers than aerobic ones. Thus, currently, microbial consortia from sludge and water treatment plants have been used for bioaugmentation more frequently than pure or mixed microbial cultures. A study showed, the use of inoculum made up of microbial consortia from sludge or WWTPs is a relatively popular bioaugmentation technique that has been shown to cause greater ageing and erosion of the surface of polylactic acid microplastics over incubation periods of 43 days and 30 days, as well as up to 52.3% weight loss in polyhydroxyalkanoates (PHA) (Ma et al., 2022). *Brevibacillus sp.* and *Aneurinibacillus sp.* working together to biodegrade three different forms of microplastics more efficiently than single cultures (Skariyachan et al., 2021). Utilizing bio-surfactants is another option to increase the effectiveness of biodegradation techniques. The impact of biosurfactants on the dissemination of microorganisms on plastic films has been the subject of extensive research. Different kinds of biosurfactants can be used to regulate the surface hydrophobicity and breakdown of plastic films. Studies reveal that this active surface chemical changes the surface tension of both plastic and microorganisms and strengthens their functional relationship. During the biodegradation of plastic, some strains (such as *Bacillus subtilis and Pseudomonas aeruginosa*, respectively) are capable of producing biosurfactants such as surfactin and rhamnolipid (Taghavi et al., 2022). The major application of biostimulation is the infusion of nutrients to promote biodegradation by native microorganisms. The slow rate of degradation of certain plastics stimulates their breaking down into smaller pieces and also escalates the rate of persistence in the ecosystem. On the other side, using inducers and stimulants may aid in boosting enzyme activity to speed up the breakdown of MPs.

5.5.2 Enzymatic Degradation of MPs

Enzymatic surface moderation of the microplastics has already been the focus of the study previously. With promising advancements in biotechnological field of remediating microplastics, exploring enzymes as one of the stimulants has been a significant research interest. In a nutshell, surface modification makes polymers more wettable, dyeable, and anti-pilling, which eventually makes plastics more hydrophilic (Kawai et al., 2019). Additionally, the surface-to-volume ratio is crucial to the degradation process; the greater the ratio or smaller the polymer size, the quicker fibers or films will disintegrate. Over the past ten years, an enormous quantity of research has been done in an effort to discover a novel enzyme that can break down microplastics. It has been suggested that several types of enzymes can break down polymers into their subunit forms. The degradation of polymers is facilitated by several classes of enzymes, including oxidases, amidases, laccases, hydrolases, and peroxidases. The most prevalent plastic material throughout the range of sectors is PE. Consequently, PE-caused microplastic contamination is widespread in terrestrial habitats. The biodegradation of PE microplastics has already been research interest, but the core point that stands with the degradation ability of the microbial sources is the enzymes they produce. Staphylococcus epidermis supernatant subjected to LDPE for three months was analyzed using SDS-PAGE to show that LDPE is broken down by enzymes (Chatterjee et al., 2010). The mechanism was further described in a review of an LDPE-degrading enzyme published in 2015. According to this review, enzyme-based disintegration occurs in two stages. Enzymes that originate outside the cell play a crucial role in the process of extracellular de-polymerization of the plastic, which occurs first. Mineralization refers to the subsequent stage of the procedure. The more brief LDPE chain is mineralized to the final product at this stage, which includes CO_2, H_2O, and CH_4. These final products will serve as an alternative source of carbon for all metabolisms in microorganisms. Alkane hydrolase is one such enzyme that has gained importance due to its novel functional properties (Gajendiran et al., 2016). The most frequent microbes identified to breakdown this microplastic are fungi and bacteria, with bacterium phyla split into three distinct categories. *Actinobacteria, Firmicutes*, and *Proteobacteria* are the three common phyla that have been linked to the breakdown of microplastics. The enzyme most frequently linked to HDPE breakdown is laccase. The enzyme laccase, which belongs to the oxidase group, is reported to depolymerize polymers by oxidatively cleaving the amorphous area of HDPE. This creates a readily accessible carbonyl region inside the polymer chain. Manganese peroxidase from the fungi *Phanerochaetechrysosporium* was reported to be able to reduce and decrease the tensile strength and total molecular weight of PE (Ghatge et al., 2020). PET degrading enzymes are already discussed in the bioremediation section. The PET-degrading enzyme cutinase's structure was compared to that of PETase, which revealed that they are noticeably different. Cutinase has an isoelectric point (pI) value of 9.6 and cutinase 9.3, while PETase has a polarized surface. In other studies, the structure and mechanism of the enzyme were discussed further for better understanding. The effectiveness of PETase is enhanced by an improved comprehension of its molecular makeup and operation. To enhance PETase effectiveness, multiple analyses including mutation and overexpression were carried out.

5.5.3 Engineered Microbes in Valorization of MPs

Despite the remarkable natural capacities of microbes, there has recently been an increase in the need for created synthetic microbial scavengers with higher tolerability to harmful pollutants. However, the isolated natural microorganisms evolved to thrive in the contaminated site rather than efficiently converting toxins. Due to their low enzymatic activity, lack of substrate specificity, or susceptibility to the poisonous chemicals, they frequently exhibit poor metabolic efficiency in breaking down toxic substances. The term "synthetic microbial scavengers" refers to microorganisms that have been rationally created via the use of cutting-edge genetic engineering techniques in the domains of synthetic biology and biotechnology, as well as computational methodologies in the field of systems biology (Yang et al., 2020). Present-day status in advancements in technical grounds, microbial technology driven by synthetic biology and molecular engineering have gained importance to integrate synthetic technology with living organisms. This has helped in creating synthetic microbial scavengers for recalcitrant remediation. Recent developments in bioinformatic techniques and sequencing technologies have made it possible to analyze genomic sequences and identify intrinsic enzymes that regulate metabolism and routes for bioremediation. Synthetic biology has made ways to engineered degradation pathways, augmented the enzymatic activity. Gene expression techniques are being used to escalate the synthetically driven pathways. The bacterial species may be used to degrade PET microplastics, or its hydrolyzing enzymes may be added to those of other bacteria to create stronger microbial scavengers. In this regard, *Phaeodactylumtricornutum*, a photosynthetic microalga, was developed as a novel framework to create and secrete PET-hydrolyzing enzymes into the environment (Moog et al., 2019). *P. aeruginosa* was created with the explicit purpose of capturing and accumulating microparticles within its adhesive exopolymeric compounds in its biofilm. The deletion of the wspF gene improved P. aeruginosa capture efficiency, because WspF controls WspR, which produces c-di-GMP, a secondary signaling molecule essential for biofilm development. By adding sugar residues in key locations, a yeast strain was genetically modified to manufacture bacterial Cutinase. This keeps the enzyme folded at high temperatures and prevents the enzyme from clustering by forming physical barriers. The artificially produced enzyme might break down PET at the right temperature and concentration for the environment (Shirke et al., 2018). From the plant compost metagenome, Yan et al. (2020) identified a thermophilic Cutinase that can break down PET at temperatures as high as 70 C. The Cutinase was cloned in Clostridium thermocellum, and in order to achieve high levels of gene expression and extracellular secretion of the enzyme, a strong promoter of a continuously expressed gene was utilized (Yan et al., 2020). As mentioned previously, a variety of bacteria, fungi, and actinomycetes can be found in contaminated areas, and ecosystems can season themselves to plastic through employing it as a source of carbon and nitrogen. The drawback remains here that each microbe is not capable of exploiting different components of plastics as their source of energy. Synthetically engineered consortia serve as an alternative to boost the process. The compatibility of these species of bacteria/fungi can be evaluated in silico by filtering them using simulation. The in silico tool known as FLYCOP, which stands for FLexiblesYnthetic Consortium OPtimisation, can be deployed to construct and improve communities of microbes. Thus, by utilizing genetic engineering techniques, it is possible to significantly improve the thermal stability along with the capacity of enzymes to break down plastics, as well as the conversion of plastic waste into chemicals with added value, thus promoting circular economy.

5.5.4 Omics Approach in MPs Remediation

Conventional bioremediation efficiency may differ from site to site depending on various factors. Additionally, microbe-based systems that effectively eliminate contaminants in the lab might not perform equally well in the field. Poor understanding of the mechanisms governing the microbial development in contaminated areas restricts the use of bioremediation. Therefore, it is crucial to take a variety of elements into account while constructing the bioremediation process in order to understand and predict the effects of toxins in the environment. Researchers are utilizing metabolomics, genomes, transcriptomics, proteomics, and fluxomics tools in the bioremediation process to better comprehend the fundamental mechanism and devise novel strategies. Researchers can examine biomolecules including proteins, RNA, DNA, and metabolites from specific species as well as the entire community at once using omics techniques, a branch of molecular biology (Gutierrez et al., 2018). Microbial technology has gained importance in the field of bio-based treatment of recalcitrant. Metagenomics approach is one such cutting-edge technique frequently used in environmental microbiology. By studying uncultured organisms, the branch of study known as "metagenomics" seeks to recognize microbial diversity, their activities, relationships, collaboration, and growth in a range of situations. Metagenomics has filled in the gaps left by culture-based methods by providing cognizance into the in-situ microbial structure. Consequently, bioremediation methods are getting better. In a number of microbiological species, transcriptomic methods were employed to explore genome-wide transcriptional processes, develop regulons and stimulations, delineate operon structures, locate DNA-binding sites, and carry out comparative genotyping. Investigations into the metabolic activities of the bacteria can be done using the results from the various omic techniques utilized for exploring the metabolic functioning of the microorganism. In recent years, finding strategies to control these contaminants in an environmentally friendly way has grown increasingly dependent on advanced technologies, including genomics, proteomics, transcriptomics, metabolomics, and fluxomic.

5.6 SUMMARY AND CONCLUSION

Single-use non-biodegradable plastics in all its forms are emerging as one of the most critical pollutant, the severity of which on living species has not yet been fully explored. The partial exploration of its effect on the ecosystem is alarming and needs to be fully controlled with utmost importance. The researches the on adverse effects of micro-plastic pollution on bionetwork, its identification protocol and sensing device, its elimination methodologies and the treatment technologies are still in its infancy level. As per data statistics, to date no specific treatment technologies have been practiced in any Waste Water Treatment Plant (WWTP) all over the world for the remediation of microplastics. Present state-of-the-art on microplastics treatment and remediation is still in the inception stage and only practiced at the lab-scale. The occurrence of very small plastic fragments in the open ocean was first reported in the 1970s and the term "microplastic" coined in early 2004 and it has been categorized as pollutant on the same time period. Till its onset, researchers have continuously tried to grip the microplastic pollution at the level of source, occurrence, detection, and remediation. Several studies have been done to cultivate robust, feasible sensing techniques for the investigation of microplastics but, an available solution is still unrealizable. Though several approaches have been explored involving physical separation method, chemical treatment methods, biotechnological approaches and hybrid technologies within this category, with significant prominence, the domain is yet not fully pierceable.

In this chapter is a brief portrayal of the microplastic associated problems and the potential environmental benign methodologies to beat microplastic pollution with specific emphasis on bacterial degradation, sunlight-driven photocatalyst, conversion to fuels, and biodegradable plastics. These could be game-changers in future research on plastic pollution control. However, the recently explored technologies are not sufficient and able to control widespread contamination of soil, aquatic environment and its occurrence in living species at this tenderfoot stage. The research on the effect of microplastic on human brains to blood and digestive systems are also scanty, which has put a significant challenge on proper regulation of environmental microplastics and the most reasonable way to restrain plastic pollution. Hence, for a sustainable future, advanced strategies on rapid identification techniques, effective treatment, and remediation are needed to fill a void of knowledge in the field of plastic mitigation and to stop a microplastic tsunami.

REFERENCES

Alfaro-Núñez, A., Astorga, D., Cáceres-Farías, L., Bastidas, L., Soto Villegas, C., Macay, K., & Christensen, J. H. (2021). Microplastic pollution in seawater and marine organisms across the Tropical Eastern Pacific and Galápagos. In *Scientific Reports* (Vol. 11, Issue 1). Springer Science and Business Media LLC. https://doi.org/10.1038/s41598-021-85939-3

Ali, I., Cheng, Q., Ding, T., Yiguang, Q., Yuechao, Z., Sun, H., Peng, C., Naz, I., Li, J., & Liu, J. (2021). Micro- and nanoplastics in the environment: Occurrence, detection, characterization and toxicity – A critical review. In *Journal of Cleaner Production* (Vol. 313, p. 127863). Elsevier BV. https://doi.org/10.1016/j.jclepro.2021.127863

Auta, H. S., Emenike, C. U., & Fauziah, S. H. (2017). Screening of Bacillus strains isolated from mangrove ecosystems in Peninsular Malaysia for microplastic degradation. In *Environmental Pollution* (Vol. 231, pp. 1552–1559). Elsevier BV. https://doi.org/10.1016/j.envpol.2017.09.043

Barboza, L. G. A., Dick Vethaak, A., Lavorante, B. R. B. O., Lundebye, A.-K., & Guilhermino, L. (2018). Marine microplastic debris: An emerging issue for food security, food safety and human health. In *Marine Pollution Bulletin* (Vol. 133, pp. 336–348). Elsevier BV. https://doi.org/10.1016/j.marpolbul.2018.05.047

Barnes, D. K. A., Galgani, F., Thompson, R. C., & Barlaz, M. (2009). Accumulation and fragmentation of plastic debris in global environments. In *Philosophical Transactions of the Royal Society B: Biological Sciences* (Vol. 364, Issue 1526, pp. 1985–1998). The Royal Society. https://doi.org/10.1098/rstb.2008.0205

Bhore, R. K., & Kamble, S. B. (2022). Nano adsorptive extraction of diverse microplastics from the potable and seawater using organo-polyoxometalate magnetic nanotricomposites. In *Journal of Environmental Chemical Engineering* (Vol. 10, Issue 6, p. 108720). Elsevier BV. https://doi.org/10.1016/j.jece.2022.108720

Blair Espinoza, R. M. (2019). Microplastics in wastewater treatment systems and receiving waters. University of Glasgow. https://doi.org/10.5525/GLA.THESIS.76781

Browne, M. A., Dissanayake, A., Galloway, T. S., Lowe, D. M., & Thompson, R. C. (2008). Ingested Microscopic Plastic Translocates to the Circulatory System of the Mussel, Mytilusedulis (L.). In *Environmental Science & Technology* (Vol. 42, Issue 13, pp. 5026–5031). American Chemical Society (ACS). https://doi.org/10.1021/es800249a

Carpenter, E. J., Anderson, S. J., Harvey, G. R., Miklas, H. P., & Peck, B. B. (1972). Polystyrene Spherules in Coastal Waters. In *Science* (Vol. 178, Issue 4062, pp. 749–750). American Association for the Advancement of Science (AAAS). https://doi.org/10.1126/science.178.4062.749

Chatterjee, S., Roy, B., Roy, D., & Banerjee, R. (2010). Enzyme-mediated biodegradation of heat treated commercial polyethylene by Staphylococcal species. In *Polymer Degradation and Stability* (Vol. 95, Issue 2, pp. 195–200). Elsevier BV. https://doi.org/10.1016/j.polymdegradstab.2009.11.025

Colabuono, F. I., Taniguchi, S., & Montone, R. C. (2010). Polychlorinated biphenyls and organochlorine pesticides in plastics ingested by seabirds. In *Marine Pollution Bulletin* (Vol. 60, Issue 4, pp. 630–634). Elsevier BV. https://doi.org/10.1016/j.marpolbul.2010.01.018

Crawford, C. B., & Quinn, B. (2017). Plastic production, waste and legislation. In *Microplastic Pollutants* (pp. 39–56). Elsevier. https://doi.org/10.1016/b978-0-12-809406-8.00003-7

Du, H., Wang, Q., Chen, G., & Wang, J. (2022). Photo/electro-catalytic degradation of micro- and nano-plastics by nanomaterials and corresponding degradation mechanism. In *TrAC Trends in Analytical Chemistry* (Vol. 157, p. 116815). Elsevier BV. https://doi.org/10.1016/j.trac.2022.116815

Fadare, O. O., & Okoffo, E. D. (2020). Covid-19 face masks: A potential source of microplastic fibers in the environment. In *Science of The Total Environment* (Vol. 737, p. 140279). Elsevier BV. https://doi.org/10.1016/j.scitotenv.2020.140279

Fossi, M. C., Panti, C., Guerranti, C., Coppola, D., Giannetti, M., Marsili, L., & Minutoli, R. (2012). Are baleen whales exposed to the threat of microplastics? A case study of the

Mediterranean fin whale (Balaenopteraphysalus). In *Marine Pollution Bulletin* (Vol. 64, Issue 11, pp. 2374–2379). Elsevier BV. https://doi.org/10.1016/j.marpolbul.2012.08.013

Gajendiran, A., Krishnamoorthy, S., & Abraham, J. (2016). Microbial degradation of low-density polyethylene (LDPE) by Aspergillusclavatus strain JASK1 isolated from landfill soil. In *3 Biotech* (Vol. 6, Issue 1). Springer Science and Business Media LLC. https://doi.org/10.1007/s13205-016-0394-x

Ge, J., Zhang, Z., Ouyang, Z., Shang, M., Liu, P., Li, H., & Guo, X. (2022). Photocatalytic degradation of (micro)plastics using TiO2-based and other catalysts: Properties, influencing factor, and mechanism. In *Environmental Research* (Vol. 209, p. 112729). Elsevier BV. https://doi.org/10.1016/j.envres.2022.112729

Ghatge, S., Yang, Y., Ahn, J.-H., & Hur, H.-G. (2020). Biodegradation of polyethylene: a brief review. In *Applied Biological Chemistry* (Vol. 63, Issue 1). Springer Science and Business Media LLC. https://doi.org/10.1186/s13765-020-00511-3

Golwala, H., Zhang, X., Iskander, S. M., & Smith, A. L. (2021). Solid waste: An overlooked source of microplastics to the environment. In *Science of The Total Environment* (Vol. 769, p. 144581). Elsevier BV. https://doi.org/10.1016/j.scitotenv.2020.144581

Gong, J., Kong, T., Li, Y., Li, Q., Li, Z., & Zhang, J. (2018). Biodegradation of Microplastic Derived from Poly(ethylene terephthalate) with Bacterial Whole-Cell Biocatalysts. In *Polymers* (Vol. 10, Issue 12, p. 1326). MDPI AG. https://doi.org/10.3390/polym10121326

Grbic, J., Nguyen, B., Guo, E., You, J. B., Sinton, D., & Rochman, C. M. (2019). Magnetic Extraction of Microplastics from Environmental Samples. In *Environmental Science & Technology Letters* (Vol. 6, Issue 2, pp. 68–72). American Chemical Society (ACS). https://doi.org/10.1021/acs.estlett.8b00671

Gupta, B., Ambekar, R. S., Tromer, R. M., Ghosal, P. S., Sinha, R., Majumder, A., Kumbhakar, P., Ajayan, P. M., Galvao, D. S., Gupta, A. K., & Tiwary, C. S. (2021). Development of a schwarzite-based moving bed 3D printed water treatment system for nanoplastic remediation. In *RSC Advances* (Vol. 11, Issue 32, pp. 19788–19796). Royal Society of Chemistry (RSC). https://doi.org/10.1039/d1ra03097c

Gutierrez, D. B., Gant-Branum, R. L., Romer, C. E., Farrow, M. A., Allen, J. L., Dahal, N., Nei, Y.-W., Codreanu, S. G., Jordan, A. T., Palmer, L. D., Sherrod, S. D., McLean, J. A., Skaar, E. P., Norris, J. L., & Caprioli, R. M. (2018). An Integrated, High-Throughput Strategy for Multiomic Systems Level Analysis. In *Journal of Proteome Research* (Vol. 17, Issue 10, pp. 3396–3408). American Chemical Society (ACS). https://doi.org/10.1021/acs.jproteome.8b00302

Haward, M. (2018). Plastic pollution of the world's seas and oceans as a contemporary challenge in ocean governance. In *Nature Communications* (Vol. 9, Issue 1). Springer Science and Business Media LLC. https://doi.org/10.1038/s41467-018-03104-3

Haq, I., Kumar, S., Kumar, V., Singh, S.K., Raj, A. (2016a). Evaluation of bioremediation potentiality of ligninolyticSerratialiquefaciens for detoxification of pulp and paper mill effluent. *J Hazard Mater* 305: 190–199.

Haq, I., Kumari, V., Kumar, S., Raj, A., Lohani, M., & Bhargava, R.N. (2016b). Evaluation of the phytotoxic and genotoxic potential of pulp and paper mill effluent using Vignaradiata and Allium cepa. *Adv Biol* 8065736.

Haq, I., Kumar, S., Raj, A., Lohani, M., & Satyanarayana, G.N.V. (2017). Genotoxicity assessment of pulp and paper mill effluent before and after bacterial degradation using Allium cepa test. *Chemosphere* 169: 642–650.

Haq, I., Raj, A., & Markandeya (2018). Biodegradation of Azure-B dye by Serratialiquefaciens and its validation by phytotoxicity, genotoxicity and cytotoxicity studies. *Chemosphere* 196, 58–68.

Haq, I., Mazumder, P., & Kalamdhad, A.S. (2020). Recent advances in removal of lignin from paper industry wastewater and its industrial applications–A review. *Bioresour. Technol* *312*:123636.

Haq, I., & Kalamdhad, A.S. (2021). Phytotoxicity and cyto-genotoxicity evaluation of organic and inorganic pollutants containing petroleum refinery wastewater using plant bioassay. *Environ. Technol. Innov* 23:101651.

Haq, I., Kalamdhad, A.S., & Pandey, A. (2022a). Genotoxicity evaluation of paper industry wastewater prior and post-treatment with laccase producing Pseudomonas putida MTCC 7525. *J. Clean. Prod* 342:130981.

Haq, I., Singh, A., & Kalamdhad, A.S. (2022b). Integrated technologies for the remediation of paper industry waste in a circular bioeconomy, Editor(s): Sunita Varjani, Ashok Pandey, Mohammad J. Taherzadeh, HuuHao Ngo, & R.D. Tyagi, *Biomass, Biofuels, Biochemicals*, Elsevier, Pages 351–363, ISBN 9780323885119.

Haq, I., Roy, S., & Kalamdhad, A.S. (2022c). Characterization of Pulp and Paper Mill Wastewater and Its Toxicity Analysis Using Vigna Radiata. In *Environmental Degradation: Monitoring, Assessment and Treatment Technologies*; Springer International Publishing: Cham, Switzerland, pp. 261–269. ISBN 9783030941475.

Haq, I. & Kalamdhad, A.S. (2022) Advanced and Ecofriendly Technologies for the Treatment of Industrial Wastewater to Constrain Environmental Pollution. In: Arora, S., Kumar, A., Ogita, S., & Yau, Y.Y. (eds) *Innovations in Environmental Biotechnology*. Springer, Singapore. https://doi.org/10.1007/978-981-16-4445-0_22.

Haq I & Kalamdhad, A.S. (2023). Enhanced biodegradation of toxic pollutants from paper industry wastewater using Pseudomonas sp. immobilized in composite biocarriers and its toxicity evaluation. *Bioresour. Technol. Repor* 24:101674.

Haq, I., Kalamdhad, A.S. & Malik, A. (2024). Bioremediation of Petroleum Refinery Wastewater Using Bacillus subtilis IH-1 and Assessment of Its Toxicity. *Arch. Microbiol*, 206, 296.

Heo, Y., Lee, E.-H., & Lee, S.-W. (2022). Adsorptive removal of micron-sized polystyrene particles using magnetic iron oxide nanoparticles. In *Chemosphere* (Vol. 307, p. 135672). Elsevier BV. https://doi.org/10.1016/j.chemosphere.2022.135672

International Union of Conservation of Nature, (2021). www.iucn.org

Ivar do Sul, J. A., & Costa, M. F. (2014). The present and future of microplastic pollution in the marine environment. In *Environmental Pollution* (Vol. 185, pp. 352–364). Elsevier BV. https://doi.org/10.1016/j.envpol.2013.10.036

Jung, Y. S., Sampath, V., Prunicki, M., Aguilera, J., Allen, H., LaBeaud, D., Veidis, E., Barry, M., Erny, B., Patel, L., Akdis, C., Akdis, M., & Nadeau, K. (2022). Characterization and regulation of microplastic pollution for protecting planetary and human health. In *Environmental Pollution* (Vol. 315, p. 120442). Elsevier BV. https://doi.org/10.1016/j.envpol.2022.120442

Kawai, F., Kawabata, T., & Oda, M. (2019). Current knowledge on enzymatic PET degradation and its possible application to waste stream management and other fields. In *Applied Microbiology and Biotechnology* (Vol. 103, Issue 11, pp. 4253–4268). Springer Science and Business Media LLC. https://doi.org/10.1007/s00253-019-09717-y

Keerthana Devi, M., Karmegam, N., Manikandan, S., Subbaiya, R., Song, H., Kwon, E. E., Sarkar, B., Bolan, N., Kim, W., Rinklebe, J., & Govarthanan, M. (2022). Removal of nanoplastics in water treatment processes: A review. In Science of The Total Environment (Vol. 845, p. 157168). Elsevier BV. https://doi.org/10.1016/j.scitotenv.2022.157168

Kumar, M., Chen, H., Sarsaiya, S., Qin, S., Liu, H., Awasthi, M. K., Kumar, S., Singh, L., Zhang, Z., Bolan, N. S , Pandey, A., Varjani, S., & Taherzadeh, M. J. (2021). Current research trends on micro- and nano-plastics as an emerging threat to global environment: A review. In *Journal of Hazardous Materials* (Vol. 409, p. 124967). Elsevier BV. https://doi.org/10.1016/j.jhazmat.2020.124967

Kundu, A., Shetti, N. P., Basu, S., Raghava Reddy, K., Nadagouda, M. N., & Aminabhavi, T. M. (2021). Identification and removal of micro- and nano-plastics: Efficient and cost-effective methods. In *Chemical Engineering Journal* (Vol. 421, p. 129816). Elsevier BV. https://doi.org/10.1016/j.cej.2021.129816

Kurniawan, S. B., Said, N. S. M., Imron, M. F., & Abdullah, S. R. S. (2021). Microplastic pollution in the environment: Insights into emerging sources and potential threats. In Environmental Technology & Innovation (Vol. 23, p. 101790). Elsevier BV. https://doi.org/10.1016/j.eti.2021.101790

Lares, M., Ncibi, M. C., Sillanpää, M., & Sillanpää, M. (2018). Occurrence, identification and removal of microplastic particles and fibers in conventional activated sludge process and advanced MBR technology. In Water Research (Vol. 133, pp. 236–246). Elsevier BV. https://doi.org/10.1016/j.watres.2018.01.049

Liang, Y., Cao, X., Mo, A., Jiang, J., Zhang, Y., Gao, W., & He, D. (2023). Micro(nano)plastics in plant-derived food: Source, contamination pathways and human exposure risks. In *TrAC Trends in Analytical Chemistry* (Vol. 165, p. 117138). Elsevier BV. https://doi.org/10.1016/j.trac.2023.117138

Lv, X., Dong, Q., Zuo, Z., Liu, Y., Huang, X., & Wu, W.-M. (2019). Microplastics in a municipal wastewater treatment plant: Fate, dynamic distribution, removal efficiencies, and control strategies. In *Journal of Cleaner Production* (Vol. 225, pp. 579–586). Elsevier BV. https://doi.org/10.1016/j.jclepro.2019.03.321

Ma, B., Xue, W., Hu, C., Liu, H., Qu, J., & Li, L. (2019). Characteristics of microplastic removal via coagulation and ultrafiltration during drinking water treatment. In *Chemical Engineering Journal* (Vol. 359, pp. 159–167). Elsevier BV. https://doi.org/10.1016/j.cej.2018.11.155

Ma, J., Gong, Z., Wang, Z., Liu, H., Chen, G., & Guo, G. (2022). Elucidating degradation properties, microbial community, and mechanism of microplastics in sewage sludge under different terminal electron acceptors conditions. In *Bioresource Technology* (Vol. 346, p. 126624). Elsevier BV. https://doi.org/10.1016/j.biortech.2021.126624

Malafaia, G., & Barceló, D. (2023). Microplastics in human samples: Recent advances, hotspots, and analytical challenges. In *TrAC Trends in Analytical Chemistry* (Vol. 161, p. 117016). Elsevier BV. https://doi.org/10.1016/j.trac.2023.117016

Mbachu, O., Jenkins, G., Kaparaju, P., & Pratt, C. (2021). The rise of artificial soil carbon inputs: Reviewing microplastic pollution effects in the soil environment. In *Science of The Total Environment* (Vol. 780, p. 146569). Elsevier BV. https://doi.org/10.1016/j.scitotenv.2021.146569

Mohana, A. A., Farhad, S. M., Haque, N., & Pramanik, B. K. (2021). Understanding the fate of nano-plastics in wastewater treatment plants and their removal using membrane processes. In *Chemosphere* (Vol. 284, p. 131430). Elsevier BV. https://doi.org/10.1016/j.chemosphere.2021.131430

Moog, D., Schmitt, J., Senger, J., Zarzycki, J., Rexer, K.-H., Linne, U., Erb, T., & Maier, U. G. (2019). Using a marine microalga as a chassis for polyethylene terephthalate (PET) degradation. In *Microbial Cell Factories* (Vol. 18, Issue 1). Springer Science and Business Media LLC. https://doi.org/10.1186/s12934-019-1220-z

Muhonja, C. N., Makonde, H., Magoma, G., & Imbuga, M. (2018). Biodegradability of polyethylene by bacteria and fungi from Dandora dumpsite Nairobi-Kenya. In P. K. Arora (Ed.), PLOS ONE (Vol. 13, Issue 7, p. e0198446). Public Library of Science (PLoS). https://doi.org/10.1371/journal.pone.0198446

O'Brien, S., Rauert, C., Ribeiro, F., Okoffo, E. D., Burrows, S. D., O'Brien, J. W., Wang, X., Wright, S. L., & Thomas, K. V. (2023). There's something in the air: A review of sources, prevalence and behaviour of microplastics in the atmosphere. In *Science of The Total Environment* (Vol. 874, p. 162193). Elsevier BV. https://doi.org/10.1016/j.scitotenv.2023.162193

Peixoto, D., Pinheiro, C., Amorim, J., Oliva-Teles, L., Guilhermino, L., & Vieira, M. N. (2019). Microplastic pollution in commercial salt for human consumption: A review. In *Estuarine, Coastal and Shelf Science* (Vol. 219, pp. 161–168). Elsevier BV. https://doi.org/10.1016/j.ecss.2019.02.018

Poerio, Piacentini, & Mazzei. (2019). Membrane Processes for Microplastic Removal. In *Molecules* (Vol. 24, Issue 22, p. 4148). MDPI AG. https://doi.org/10.3390/molecules24224148

Prata, J. C., da Costa, J. P., Lopes, I., Duarte, A. C., & Rocha-Santos, T. (2020). Environmental exposure to microplastics: An overview on possible human health effects. In *Science of The Total Environment* (Vol. 702, p. 134455). Elsevier BV. https://doi.org/10.1016/j.scitotenv.2019.134455

Revel, M., Châtel, A., & Mouneyrac, C. (2018). Micro(nano)plastics: A threat to human health? In *Current Opinion in Environmental Science & Health* (Vol. 1, pp. 17–23). Elsevier BV. https://doi.org/10.1016/j.coesh.2017.10.003

Rout, P. R., Mohanty, A., Aastha, Sharma, A., Miglani, M., Liu, D., & Varjani, S. (2022). Micro- and nanoplastics removal mechanisms in wastewater treatment plants: A review. In *Journal of Hazardous Materials Advances* (Vol. 6, p. 100070). Elsevier BV. https://doi.org/10.1016/j.hazadv.2022.100070

Ru, J., Huo, Y., & Yang, Y. (2020). Microbial Degradation and Valorization of Plastic Wastes. In *Frontiers in Microbiology* (Vol. 11). Frontiers Media SA. https://doi.org/10.3389/fmicb.2020.00442

Sharma, S., & Chatterjee, S. (2017). Microplastic pollution, a threat to marine ecosystem and human health: a short review. In *Environmental Science and Pollution Research* (Vol. 24, Issue 27, pp. 21530–21547). Springer Science and Business Media LLC. https://doi.org/10.1007/s11356-017-9910-8

Shirke, A. N., White, C., Englaender, J. A., Zwarycz, A., Butterfoss, G. L., Linhardt, R. J., & Gross, R. A. (2018). Stabilizing Leaf and Branch Compost Cutinase (LCC) with Glycosylation: Mechanism and Effect on PET Hydrolysis. In *Biochemistry* (Vol. 57, Issue 7, pp. 1190–1200). American Chemical Society (ACS). https://doi.org/10.1021/acs.biochem.7b01189

Shi, X., Zhang, X., Gao, W., Zhang, Y., & He, D. (2022). Removal of microplastics from water by magnetic nano-Fe3O4. In *Science of The Total Environment* (Vol. 802, p. 149838). Elsevier BV. https://doi.org/10.1016/j.scitotenv.2021.149838

Skariyachan, S., Taskeen, N., Kishore, A. P., Krishna, B. V., & Naidu, G. (2021). Novel consortia of Enterobacter and Pseudomonas formulated from cow dung exhibited enhanced biodegradation of polyethylene and polypropylene. In *Journal of Environmental Management* (Vol. 284, p. 112030). Elsevier BV. https://doi.org/10.1016/j.jenvman.2021.112030

Taghavi, N., Zhuang, W.-Q., & Baroutian, S. (2022). Effect of rhamnolipidbiosurfactant on biodegradation of untreated and UV-pretreated non-degradable thermoplastics: Part 2. In *Journal of Environmental Chemical Engineering* (Vol. 10, Issue 1, p. 107033). Elsevier BV. https://doi.org/10.1016/j.jece.2021.107033

Tanaka, K., & Takada, H. (2016). Microplastic fragments and microbeads in digestive tracts of planktivorous fish from urban coastal waters. In *Scientific Reports* (Vol. 6, Issue 1). Springer Science and Business Media LLC. https://doi.org/10.1038/srep34351

Waring, R. H., Harris, R. M., & Mitchell, S. C. (2018). Plastic contamination of the food chain: A threat to human health? In *Maturitas* (Vol. 115, pp. 64–68). Elsevier BV. https://doi.org/10.1016/j.maturitas.2018.06.010

Watson, R. A., Nowara, G. B., Hartmann, K., Green, B. S., Tracey, S. R., & Carter, C. G. (2015). Marine foods sourced from farther as their use of global ocean primary production increases. In *Nature Communications* (Vol. 6, Issue 1). Springer Science and Business Media LLC. https://doi.org/10.1038/ncomms8365

Yan, F., Wei, R., Cui, Q., Bornscheuer, U. T., & Liu, Y. (2020). Thermophilic whole-cell degradation of polyethylene terephthalate using engineered Clostridium thermocellum. In *Microbial Biotechnology* (Vol. 14, Issue 2, pp. 374–385). Wiley. https://doi.org/10.1111/1751-7915.13580

Yang, F., Huang, F., Feng, H., Wei, J., Massey, I. Y., Liang, G., Zhang, F., Yin, L., Kacew, S., Zhang, X., & Pu, Y. (2020). A complete route for biodegradation of potentially carcinogenic cyanotoxinmicrocystin-LR in a novel indigenous bacterium. In *Water Research* (Vol. 174, p. 115638). Elsevier BV. https://doi.org/10.1016/j.watres.2020.115638

Yoshida, S., Hiraga, K., Takehana, T., Taniguchi, I., Yamaji, H., Maeda, Y., Toyohara, K., Miyamoto, K., Kimura, Y., & Oda, K. (2016). A bacterium that degrades and assimilates poly(ethylene terephthalate). In *Science* (Vol. 351, Issue 6278, pp. 1196–1199). American Association for the Advancement of Science (AAAS). https://doi.org/10.1126/science.aad635

Zandieh, M., & Liu, J. (2022). Removal and Degradation of Microplastics Using the Magnetic and Nanozyme Activities of Bare Iron Oxide Nanoaggregates. In *AngewandteChemie International Edition* (Vol. 61, Issue 47).Wiley. https://doi.org/10.1002/anie.202212013

Zhang, Y., Jiang, H., Bian, K., Wang, H., & Wang, C. (2021). A critical review of control and removal strategies for microplastics from aquatic environments. In *Journal of Environmental Chemical Engineering* (Vol. 9, Issue 4, p. 105463). Elsevier BV. https://doi.org/10.1016/j.jece.2021.105463

Ziajahromi, S., Neale, P. A., Rintoul, L., & Leusch, F. D. L. (2017). Wastewater treatment plants as a pathway for microplastics: Development of a new approach to sample wastewater-based microplastics. In *Water Research* (Vol. 112, pp. 93–99). Elsevier BV. https://doi.org/10.1016/j.watres.2017.01.042

6 Removal of Microplastics from Wastewater by Biological Process

Challenge in Health and Sustainable Development

Hayfa Rajhi and Anouar Bardi

6.1 INTRODUCTION

Plastic is a durable, lightweight, and inexpensive synthetic polymer. These characteristics make it a material of choice in several sectors of activity. Serving as a raw material in the manufacture of several objects used daily, the production of plastic was estimated at nearly 350 million tons in 2017 (Plastics Europe, 2018). The inadequate management of plastic waste and its low recycling rate result in serious drawbacks to the environment. Carried by the wind, rainwater or discharged directly into nature (e.g. by accidents, industrial discharges or via discharges of treated or untreated wastewater). Microplastics found in the environment come from various origins and therefore have different shapes, sizes, densities, and chemical compositions (APEL, 2019).

In recent years, the number of publications on the subject has steadily increased. However, although some of the sampling and laboratory analysis methods are more frequently used, no standardized international protocol has been established concerning the analytical steps: sampling, processing, quantification, and identification. Therefore, the comparison of data and results between each of these studies becomes particularly problematic. A standardized international protocol would therefore make it possible to better compare these data. In general, microplastics can be classified according to their size, small (1µm to 1 mm) and longest (1 mm to 5 mm). Different classes are used from one study to another, such as 1 mm to 5 mm (Baztan et al., 2014), 250 µm to 4 mm or 0.7 µm to 1 mm (Vianello et al., 2013). Most studies refer to two classes, namely from 500 µm to 5 mm and from 1–500 µm (Hidalgo-Ruz et al., 2012). Microplastics are defined as plastic particles smaller than 5 mm. In addition, we can speak of nanoplastics when their size is less than 100 nm. The plastic microparticles can be in the form of beads, fragments, or fibers. They can form as a result of the slow decay of larger objects and fragments or be directly produced and emitted in this form. In the environment, they come mainly from roads, packaging,

DOI: 10.1201/9781003379386-6

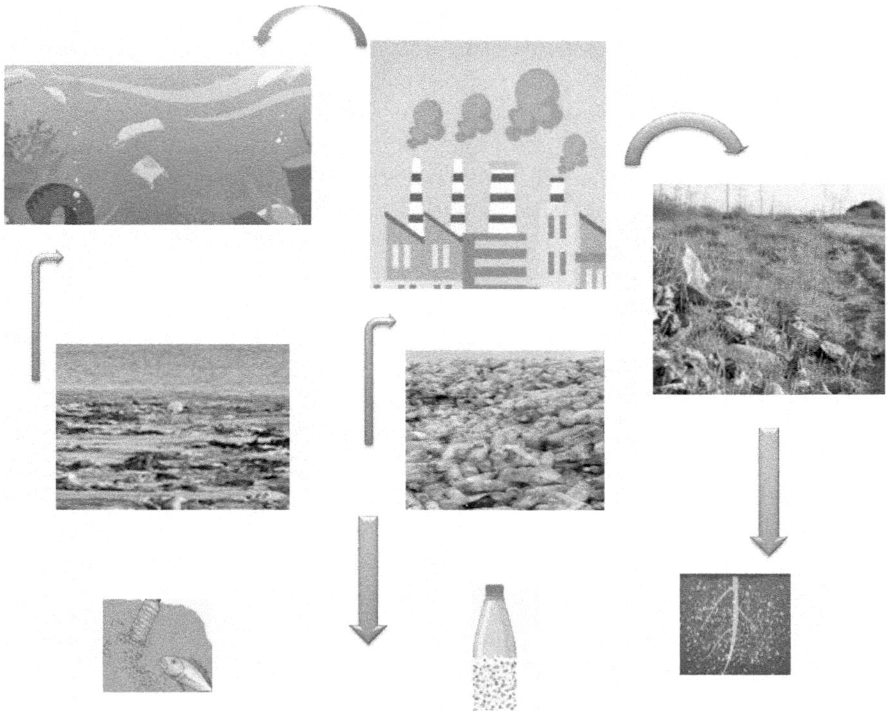

FIGURE 6.1 Plastics and microplastics in ecosystems.

sports fields, and building sites. More moderate quantities of microplastics from synthetic clothing, cleaning products and cosmetics are discharged into the aquatic environment by transiting through treatment plants. For its part, the plastic emitted by agriculture ends up directly in the soil, where their degradation into nature is very slow. Despite the large number of works recently devoted to microplastics, studies are limited to the identification and optimization of the degree and type of contamination without giving model biological processes to completely eliminate this threat (Figure 6.1). In this chapter, we try to give a study on the possibilities provided and hypotheses that can be considered to reduce biologically in this case by bioremediation on the total digestion of plastic waste for the assurance of sustainable development (APEL, 2019).

6.2 MICROPLASTICS IN WASTEWATER

Microplastics found in the environment come from different sources and therefore have different shapes, sizes, densities, and chemical compositions. Primary microplastics such as microbeads (from cosmetic, medical, or cleaning products) or microfibers (Vermaire et al., 2017) are manufactured to have a millimetric size. Secondary microplastics result from the erosion (physical, chemical, or biological) and fragmentation of larger plastic particles. The forms frequently encountered are microfibers, microbeads, fragments, and plastic foams. Color can also be used as a category to describe microplastics. The main plastic particles generally found in water and sediment samples are namely: Expanded polystyrene foam, Polypropylene, High density polyethylene, Polyvinyl chloride, Polyurethane, Polystyrene, Polyesters and polyethylene terephthalate, Polycarbonate, Polymethyl methacrylate (acrylic), polytetrafluoroethylene (Teflon). Their main origins are: Cups, food packaging, bottles, textiles and building or automobile components (Rodriguez-Seijo and Pereira 2017).

The most used plastics and microplastics are very resistant to degradation, which makes them a source of long-term pollution when they reach the environment. Several studies thus estimate that apart from the incinerated part, most of the plastics that have been produced in history are still present in the environment in one form or another (Rodriguez-Seijo and Pereira, 2017). This is particularly due to the fact that widely used polymers do not depolymerize over time, but, through the interplay of biotic and abiotic factors, simply break apart, forming smaller fragments (and then releasing the additives). The temporal variability (daily or seasonal) seems less than the spatial variability (Mintenig et al., 2020), even if, on the Seine, it could not be correlated either with the flow rate or with the presence of suspended solids (Dris et al., 2015). The long-term accumulation depends on the hydrodynamic conditions of the water masses, but also on the kinetics of microplastic degradation (Dris et al., 2015).

6.3 MICROPLASTICS WASTEWATER TREATMENT BY CONVENTIONAL TECHNIQUES

There is little information on the degradation process of plastic materials in the environment. Therefore, knowledge is focused on laboratory studies, which analyze certain processes such as biodegradation. Few experimental environments bring together all the necessary environmental conditions, bringing together the interaction of several degradation processes as in reality. These studies focus on weight loss, changes in tensile strength, or on the reasons for using this or that polymer in relation to the breakdown of the molecular structure and the identification of certain determined microbial strains (Lambert et al., 2014). The rate and extent of biodegradation of polymers can be affected by the type of organisms present in the water as well as the abundance of biomass (Lambert et al., 2014),as well as their sensitivity and adaptability to environmental parameters (Lambert et al., 2013).

6.3.1 Removal of Microplastics from Municipal Water Treatment Stations

Currently, plastic materials are mainly eliminated by physico-mechanical processes (settling, sedimentation, screening, flotation, etc.), physic-chemical (coagulation-flocculation, aeration and clarification, membrane separation), biological (biofilm/activated sludge, etc.), and chemicals (oxidation, chlorination, disinfection, etc.). The processes occurring in wastewater treatment plants can therefore be classified into two categories: separation processes (coagulation-flocculation, membrane separation, etc.) and degradation processes (photo-oxidation, biological treatment, etc.) (Collard et al., 2015).

In fact, they are considered preliminary, primary, secondary, and tertiary processing steps. Nevertheless, the smallest plastic debris between 20 and 100 nm can bypass all these processing steps, and are released into the environment (Cole et al., 2011). To date, no large-scale studies have been performed on the release of microplastics to surface waters, mainly because they are difficult to detect and quantify (Da Costa et al., 2016).

Numerous studies suggest that despite their upgrading, wastewater treatment plants still represent an important route for the daily release of micro- and nanoplastics into the environment (Collard et al., 2015). On the other hand, we can find the operation of conventional wastewater treatment processes (slowdown of rapid sand filtration, clogging of membrane filters, etc.). Moreover, the main disadvantage of most conventional processes is that they are non-destructive and only transfer this pollution to other sites (on filters or adsorbents). For example, coagulation/flocculation or electrocoagulation processes produce large volumes of sludge containing high concentrations of microplastics (Dalberg Advisors, 2019). The development of environmentally friendly degradation and mineralization processes is therefore necessary to eliminate the microplastics as well as the nanoplastics, which can be present in drinking water or in wastewater.

In this context, advanced oxidation processes (AOP) can represent an interesting solution for the degradation of microplastics and nanoplastics (Dalberg Advisors, 2019).

The general principle of AOP is based on the in-situ generation of powerful oxidizing agents, such as hydroxyl radicals (HO), at a sufficient concentration to effectively decontaminate water. Over the past 30 years, several studies have compared the efficiency of the main advanced oxidation processes. However, most of them have focused on the degradation of organic molecules (pesticides, dyes, drugs, etc.) and their by-products. There have only been a few attempts on microplastics.

A recent study showed that an electro-Fenton type technology based on a TiO2/graphite cathode was able to degrade polyvinyl chloride (PVC).However, all studies on the catalytic, photocatalytic, and electrochemical degradation of microplastics and nanoplastics have been performed at the laboratory scale, and further research is needed to assess the challenges associated with scaling up.

Recently, a research group in Strasbourg was one of the first to demonstrate the feasibility of the photocatalytic degradation of calibrated polymethyl methacrylate (PMMA) and polystyrene (PS) nanoparticles with TiO2 - P25/β- SiC foams under light. UV-A (Dekiff et al., 2014). Thus, during the photocatalytic degradation of suspensions of polystyrene nanobeads, they showed a mineralization of nearly 80% after 96 hours of treatment. The microscope image after only 48 hours of irradiation shows the presence of polymer debris with a wide variety of sizes and shapes (Dalberg Advisors, 2019).

6.4 MICROPLASTICS WASTEWATER TREATMENT BY BIOLOGICAL PROCESS

Ecosystems are set up on these pieces of plastic floating in the oceans, sea, and rivers. The term plastisphere was proposed in 2013 to designate this new ecosystem. Many micro-organisms are found on the surface of these plastic (microplastics): bacteria, micro-algae, molds, etc. Often the bacteria, found associated with this debris, develop on the plastic by forming biofilms (Delacuvellerie et al., 2019).

Plastic is composed mainly of carbon, which is a basic element for the chemistry of living organisms. Bacteria capable of feeding on plastic and using it as a carbon source are found on these plastic fragments. These bacteria therefore allow the biodegradation of plastic. Nevertheless, the rate of biodegradation is very long. For example, an Indian research team had immersed pieces of plastic (between 20 and 35 grams) in the Bay of Bengal (Indian Ocean) to study their biodegradation. These pieces of plastic lost a maximum of 1.9% of their mass in 12 months (Delacuvellerie et al., 2019).

The carbon atoms present in the plastic can be used by bacteria to produce various molecules (sugars, lipids, proteins, DNA, etc.). To recover these nutrients, the bacteria produce enzymes to "fragment" the chains of plastic into small fragments which they can then assimilate. After using these molecules, cellular "waste" is released into the environment by the bacteria. This "waste" can then be used as nutrients by other micro-organisms, thus forming a food chain.

Microorganisms not involved in the biodegradation of plastic are also present in this ecosystem (Haq and Kalamdhad, 2021; Haq and Kalamdhad, 2022, 2023; Haq et al., 2024). We find, for example, on this plastic debris, micro-algae, and phototrophic bacteria (e.g. cyanobacteria) capable of using sunlight in the context of photosynthesis. Uni- or multi-cellular micro-organisms feeding on bacteria are also present in this ecosystem. These play a role of predator vis-à-vis the bacteria(Delacuvellerie et al., 2019).

However, the micro-organisms present in this ecosystem can also be harmful to health or to human activities. Bacteria of the genus *Vibrio* were found on these pieces of plastic. Among this bacterial genus, we find in particular *Vibrio cholerae*, a pathogenic bacterium, responsible for cholera as well as other strains that cause disease in fish. Species of micro-algae that are toxic or involved in algal blooms can also be found in these ecosystems.

The bacteria present on this debris were identified using a sequence of their DNA: 16S rDNA. This DNA sequence is considered a biomarker because it would be specific to each bacterial species. It can be considered as a "bar code." By comparing these sequences with those present in databases, it is possible to identify the bacteria present in the plastisphere.

This census made it possible to highlight the presence of certain bacteria only in the plastic debris and their absence in the surrounding sediments. Some bacteria identified are specific to floating or sedimented plastic debris. This study made it possible for the first time to show differences in the bacterial community between plastic debris from the same geographical area but from different depths (Delacuvellerie et al., 2019).

Moreover, the bacterial strain *Alcanivoraxborkumensis*, a bacterium that degrades plastic, was identified for its ability to feed on hydrocarbon compounds derived from petroleum. It is also able to feed only with these compounds. *A. borkumensis* has been found in small quantities in all the oceans of the world. On the other hand, during maritime pollution by hydrocarbons, it is found in large quantities (Delacuvellerie et al., 2019).

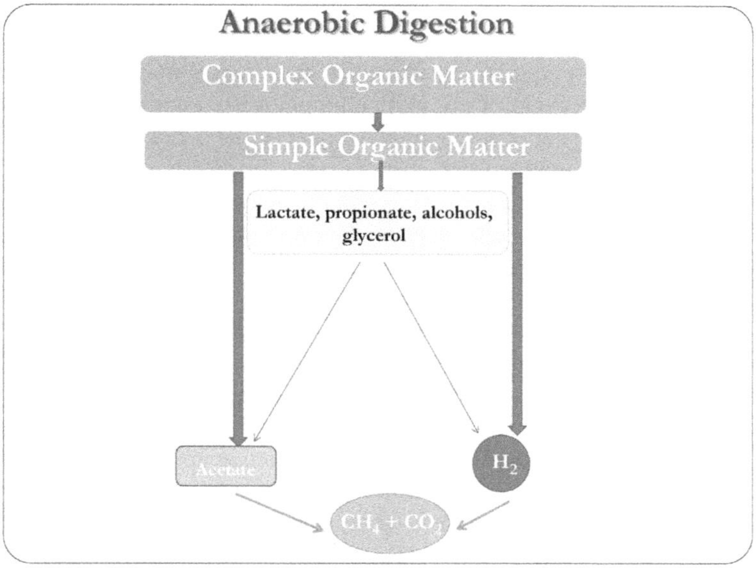

FIGURE 6.2 Anaerobic digestion a promote proposal process for microplastics removal.

Based on the following result: (I) Plastic is composed mainly of carbon which is a basic element for the chemistry of living organisms, (II) The carbon atoms present in the plastic can be used by bacteria to produce various molecules (sugars, lipids, proteins, DNA, etc.). To recover these nutrients, the bacteria produce enzymes to "fragment" the chains of plastic into small fragments which they can then assimilate. We propose the approach of anerobic digestion as bioprocess efficiency to degrade the polymers including microplastics and produce biogas (Rajhi, 2012).

During anaerobic digestion (hydrolytic and fermentative microorganisms), the complex macromolecules are transformed into monomers by simple hydrolysis; afterwards, these monomers are transformed during acidogenic into fatty acids and organic acids (except acetate), thereafter the latter are transformed into acetate and H_2 (this acetogenesis)

At this level, either the acetoclasticmethanogenic arches, which use acetate to produce biogas, or the hydrogenetrophicmethanogenic arks, which consume hydrogen to produce biogas, intervene (Rajhi, 2012) (Figure 6.2).

6.4.1 Removal of Microplastics from Wastewater by Phytoremediation

Wastewater treatment plants contribute to the discharge of micropollutants of domestic or industrial origin into surface water. The quality of their effluents is therefore checked before they are discharged in order to detect problematic compounds.

In general, these checks are carried out on the basis of time-limited samples which are subject, in the laboratory, to chemical analyses (and sometimes biotests) and data analysis which requires a certain amount of time. The composition of the effluents can, however, vary very abruptly, in particular when it receives industrial discharges.

Online biomonitoring systems for the early detection of malfunctions have long been recognized. In the event of unusual pollution, it is essential to be able to react quickly to guarantee the maintenance of water quality and avoid accidents. Currently, online biomonitoring is mainly used for monitoring the quality of drinking water and surface water. There is still little experience in the field of purification.

Online biomonitoring systems consist of three elements: i) an organism with which the test is carried out, which reacts to the environment under study by changes in its photosynthetic activity or its behavior, for example, ii) an automatic detection system, which monitors the reactions of the exposed organism, and iii) an alarm system, which triggers a signal as soon as a reaction threshold is exceeded. Various organisms can be chosen to serve as "detectors": bacteria, algae, micro-crustaceans, fish, etc. Their role is to report changes in water quality on behalf of ecosystem organisms. The parameters measured can be, for example, photoluminescence, fluorescence, swimming behavior or else respiration.

All these parameters can be influenced by water pollution. To be suitable for the online sewage effluent biomonitoring system, a detecting organism must meet several conditions: it must react sensitively to the pollutants to be detected, be relatively tolerant of other components of the wastewater. These contain compounds far more complex than drinking water or river water, which requires particularly effective detection methods.

Since not all organisms react in the same way to potentially present micropollutants, there is no single biomonitor that is suitable for all compounds. The ideal is therefore to implement a battery of tests allowing additional detections.

For this, the scientists selected three tests covering different trophic levels: a test with the unicellular green algae *Chlorella vulgaris* based on the measurement of photosynthesis and two tests on freshwater crustaceans, namely the water flea *Daphnia magna* and the *Gammaruspulex*, in which different behavioral parameters are considered.

Tests in a pilot treatment plant, firstly carried out tests to ensure that the organisms survived well in the effluent and that the tests were sensitive enough to react to the presence of the pollutants.

The unicellular green alga *Chlorella vulgaris* reacts to pollutants by modifying its photosynthesis activity. The latter is followed by a device that measures it through fluorescence. Algae are grown in an integrated fermenter and then automatically extracted for measurement. The device compares the effect of the water sample on the algae with the effect of reference water.

The tests will have to specify, among the substances identified, those which are detected by the biomonitoring system and the sensitivity with which they are detected. It is also planned to use the tests in other wastewater treatment plants. In fact, solid basis was created for making online biomonitoring an additional step in the quality control of wastewater treatment effluents (Guasch et al., 2022).

In addition, a study conducted by scientists (BhujuRajita, 2020) (in the USA, proved that slimy plants can filter microplastic particles from our wastewater).

then excreted. There are currently no data on the toxicity of plastic microparticles for humans, only the results of tests on animals and cell cultures. The latter attest to possible cytotoxicity, inflammatory reactions, and effects on growth. This is, however, at concentrations well above those found in the environment. The human digestive system is unable to break down microplastics once they have been swallowed. It therefore seems that plastic microparticles are unlikely to be absorbed into cells and therefore should not be particularly toxic. However, the data currently available are still insufficient to give a definitive opinion.

The risk posed by plastic nano and microparticles to the health of living organisms is assessed by comparing environmental concentrations with the maximum concentrations that still have no effect on organisms. The scientists therefore compared the concentrations measured with the concentrations having had a proven ecotoxicological effect. This comparison showed that the areas of concentration hardly overlapped. A risk to the environment therefore seems, for the moment, unlikely. Similarly, the risk to human health is now considered non-critical. These results must, however, be considered with caution because the particles present in the environment differ from those used for the tests.

Plastic is all around us in everyday life. We cannot do without. However, some plastics are used once and then thrown away and pollute the environment for years.

The benefit of recycling is that it reduces the amount of new plastic produced. The European Commission is planning a transition towards a more circular economy. The aim is to recycle and reuse plastic waste in a stricter way, also making it possible to bring significant economic benefit to companies. The European Commission quotes: "Thanks to this new strategy, all plastic packaging on the EU market will be recyclable by 2030, the consumption of single-use plastics will be reduced, and the intentional use of microplastics will be limited" (European Commission, 2017). However, recycling can present some challenges. For example, products made from these recycled plastics are of lower quality and therefore may fragment more easily (UNEP and GRID-Arendal, 2016). Recycling is therefore not the only solution to reducing plastic pollution (UNEP and GRID-Arendal, 2016; European Commission, 2017).

For the proponents of the sustainability of development, it was a question of transmitting to future generations the same global capital stock composed of four types of substitutable capital (natural, human, physical, and institutional). It is above all about transmitting or preserving to future generations the same stock of natural capital. As a matter of fact, natural capital can be substituted for other forms of capital (human, social, technical), which incorporates new technology. In this case, it is up to the State (institutional capital) to promote substitution between the different types of capital, by supporting, for example, technological changes that save nature as well as sensitizing individuals about the benefits of sustainable development (Rajhi et al. 2023).

6.7 SUMMARY AND CONCLUSION

The pollution of ecosystems is one of the major environmental problems due to its impact on the biosphere and human health. The ecosystem is subject to numerous pollutions from direct discharges (urban and industrial effluents, etc.) and indirect discharges (fluvial and atmospheric inputs). The almost widespread pollution seriously threatens the existing natural resources in the environment. The xenobiotic substances likely to contaminate water are extremely numerous and diverse due to the multiple human activities that can be their source. Plastic wastes (microplastics) are particularly considered as priority contaminants of aqueous ecosystems due to their low solubilization in aqueous media and their adsorption to solid particles. In addition, microplastics present a significant toxicological risk even at low concentrations, in particular through their mutagenic and carcinogenic properties. In order to preserve and restore the environment and eliminate these pollutants, it is therefore necessary to develop reliable and effective pollution control methods. Physical and chemical methods are applied to immobilize the microplastics and destroy them to reduce their impact on health and the environment. Most of these techniques are expensive and environmentally unfriendly. However, a biological approach by bioremediation can be considered. It uses the purifying power of microorganisms in the environment, and has the advantage of being minimally invasive and less expensive. Current research is oriented towards the elimination of microplastics by bioprocesses, the interest of these new techniques of which lies above all in the non-polluting and economically profitable aspect. In this chapter we will discuss the different biological processes currently followed to rid micro plastics of wastewater and highlight the economic interest as we can benefit from the case studies.

Microplastics are small plastic particles with a diameter ranging from 0.1 μm to 5 mm. They can be of primary or secondary origin. They are ubiquitous and therefore affect both wildlife and humans. We are even exposed to microplastics through the air we inhale. Due to their high resistance to degradation, they persist for years in the environment. No solution to extract them from the environment has yet been found to date and the global overconsumption of plastic is aggravating the accumulation of these particles in these natural environments. The chemical substances entering into their composition as well as the pollutants adsorbed then released by the microplastics generate harmful eco-toxicological effects for the health of animals and humans. However, there is a lack of experimental data to assess toxicity in humans and no threshold has yet been established to ensure that there is no risk. Before being able to certify that microplastics pose a danger to biota and humans, it is necessary to improve the quality and internationally standardize the methods used to assess their exposure, risks, and effects. In the meantime, it is strongly advised to reduce the accumulation of these microplastics in the environment. For this, solutions to reduce the accumulation of macroplastics in the oceans are on the way, laws are put in place to reduce plastic consumption and research on alternatives to plastic is launched.

Changing the current situation of plastics pollution requires, first of all, a consideration of the basic principles in relation to the subject of waste management:

- The principle of prevention: since the cost of this operation is lower than the waste disposal costs.
- The polluter-pays principle: since polluting activities and the costs resulting from measures to prevent, reduce and combat pollution are borne by the polluter.

- Precaution: principle according to which the absence of scientific certainty should not lead a decision-maker to postpone the adoption of preventive measures against the potential health or environmental risk.
- The ecosystem approach: ensures that the collective pressures of human activities are taken into account.
- The principle of public participation as well as the principle of integration and environment sustainability considerations must be included in economic development.
- Legislation and the law prohibiting the use of single-use plastic bags.
- Education and awareness: develop awareness programs for citizens and especially students to change consumption patterns and choose products made from biodegradable or recycled materials instead of single-use disposable products;
- Replace plastic bags, containers, and film with glass, an inert material which, unlike plastic, does not release any contaminants;
- Be a responsible citizen by disposing of all waste (cigarette butts, wrappers, plastic toys, etc.) appropriately to avoid pollution of beaches and the environment.
- Scientific research must play a role in this process
- The organizational, financial and technical framework for waste collection by the municipality must be optimized.

REFERENCES

APEL (2019). Le lac Saint-Charles. [Online] www.apel-maraisdunord.org/apel/territoire/lac-saint-charles/

Baztan, J., Carrasco, A., Chouinard, O., Cleaud, M., Gabaldon, J.E., Huck, T., Jaffes, L., Jorgensen, B., Miguelez, A., Paillard, C., Vanderlinden, J.-P. (2014). Protected areas in the Atlantic facing the hazards of micro-plastic pollution: first diagnosis of three islands in the Canary Current. *Mar. Pollut. Bull.* 80, 302e311. [Online] http://dx.doi.org/10.1016/j.marpolbul.2013.12.052.

Bhuju, R. (2020). REMOVAL OF MICROPLASTICS FROM FRESH WATER USING PLANT DERIVED POLYSACCHARIDES,Tarleton State University ProQuest Dissertations Publishing, (2020). 27838126. Published by ProQuest LLC (ProQuest).

Cole, M, Lindeque, P, Halsband, C, Galloway, & Tamara, S. (2011), Microplastics as contaminants in the marine environment: a review. In: *Marine pollution bulletin*, vol. 62, n°12, p. 2588–2597. DOI: 10.1016/j.marpolbul.2011.09.025

Collard, F., B. Gilbert, G. Eppe, E. Parmentier, and K. Das, (2015). Detection of Anthropogenic Particles in Fish Stomachs: An Isolation Method Adapted to Identification by Raman Spectroscopy. *Archives of Environmental Contamination and Toxicology*, 69, 331–339, 74. DOI:10.1007/s00244-015-0221-0.

Da Costa, J.P., Santos, P.S.M., Duarte, A.C., & Rocha-Santos, T. (2016). (Nano)plastics in the environment – Sources, fates and effects. *Science of The Total Environment*, 566–567, 15–26. DOI: 10.1016/j.scitotenv.2016.05.041

DALBERG ADVISORS (2019). A report for WWF Pollution plastique: à qui la faute? Identification des défaillances systémiques et présentation du scénariozéroplastique dans la nature en 2030. www.wwf.fr/sites/default/files/doc-2019-03/20190305_Rapport_Pollutionplastique_a_qui_la_faute_WWF.pd

Dekiff, J.H., Remy, D., Klasmeier, J., & Fries, E. (2014).Occurrence and spatial distribution of microplastics in sediments from Norderney. *Environmental Pollution*, 186, 248–256

Delacuvellerie, A., Valentine, C., Gobert, S., Benali, S., & Wattiez, R., (2019). The plastisphere in marine ecosystem hosts potential specific microbial degraders including Alcanivoraxborkumensis as a key player for the low-density polyethylene degradation. *Journal of Hazardous Materials*. Volume 380, 120899. https://doi.org/10.1016/j.jhazmat.2019.120899

Dris, R., Gasperi, J., Rocher, V., Saad, M., Renault, N., & Tassin, B., (2015). Microplastic contamination in an urban area: a case study in Greater Paris. *Environ. Chem.* 12, 592–599

EFSA CONTAM Panel (European Food Safety Authority Panel on Contaminants in the Food Chain), (2016). Statement on the presence of microplastics and nanoplastics in food, with particular focus on seafood. *EFSA Journal* 2016; 14(6):4501, 30 pp.

European Commission. (2017). Sources Explanationlist. Accessed on (2019), on European Commission: https ec.europa.euinfositesinfofilessources_explanation_list.pdf

(IMO/FAO/UNESCOIOC/UNIDO/WMO/IAEA/UN/UNEP/UNDP Joint Group of Experts on the Scientific Aspects of Marine Environmental Protection).Rep.Stud. GESAMP No. 93, 220 p.

Welle, F., Franz, R., (2018). Microplastic in bottled natural mineral water – literature review and considerations on exposure and risk assessment, Food Additives & Contaminants: Part A, 35:12, 2482–2492.

Guasch, H., Bernal, S., Bruno, D., Almroth, B.C., Cochero, J., Corcoll, N., ...Martí, E. (2022). Interactions entre microplastiques et biofilms benthiques dans les écosystèmesfluviaux: lacunesdans les connaissances et tendances futures. *Freshwater Science*, 41(3), 442–458. doi.org/10.1086/721472,Dépôtinstitutionnel.

Haq, I., & Kalamdhad, A.S. (2021). Phytotoxicity and cyto-genotoxicity evaluation of organic and inorganic pollutants containing petroleum refinery wastewater using plant bioassay. *Environ. Technol. Innov* 23:101651.

Haq, I., & Kalamdhad, A.S. (2022). Advanced and Ecofriendly Technologies for the Treatment of Industrial Wastewater to Constrain Environmental Pollution. In: Arora, S., Kumar, A., Ogita, S., & Yau, Y.Y. (eds) *Innovations in Environmental Biotechnology*. Springer, Singapore. https://doi.org/10.1007/978-981-16-4445-0_22.

Haq, I., & Kalamdhad, A.S. (2023). Enhanced biodegradation of toxic pollutants from paper industry wastewater using Pseudomonas sp. immobilized in composite biocarriers and its toxicity evaluation. *Bioresour. Technol. Repor*, 24:101674.

Haq, I., Kalamdhad, A.S., & Malik, A. (2024) Bioremediation of Petroleum Refinery Wastewater Using Bacillus subtilis IH-1 and Assessment of Its Toxicity. *Arch. Microbiol*, 206, 296.

Hidalgo-Ruz Valeria, Lars Gutow, Richard C. Thompson, & Martin Thiel. (2012). Microplastics in the Marine Environment: A Review of the Methods Used for Identification and Quantification. *Environmental Science & Technology,* 46 (6): 3060–75. [Online] https://doi.org/10.1021/es2031505

Lambert, S., Sinclair, C., & Boxall, A. (2013). Occurrence, Degradation, and Effect of Polymer-Based Materials in the Environment. Reviews of Environmental Contamination and Toxicology, Volume 227, 1–53. DOI: 10.1007/978-3-319-01327-5_1

Lambert, S., Sinclair, C., & Boxall, A., (2014). Occurrence, degradation, and effect of polymerbased materials in the environment. *Reviews of Environmental Contamination and Toxicology*, 227, 1–53. DOI: 10.1007/978-3-319-01327-5_1

Mintenig, S.M., Kooi, M., Erich, M.W., Primpke, S., Redondo- Hasselerharm, P.E., Dekker, S.C., Koelmans, A.A., & van Wezel, A.P. (2020). A systems approach to understand microplastic occurrence and variability in Dutch riverine surface waters. *Water Research* 176, 115723.https://doi.org/10.1016/j.watres.2020.115723

PlasticsEurope. (2018). An analysis of European plastics production, demand and waste data. The Association of Plastics Manufacturers in Europe and EPRO.[Online] www.plasticseurope.org/en/resources/market-data.

Rajhi, H. (2012) The Bio Production of Hydrogen by Acid Fermentation". Department of Molecular Biology, Center of Molecular Biology (CBM), Autónoma University of Madrid (UAM), Madrid -Spain., under Reference (URI: http://hdl.handle.net/10486/12889).

Rajhi, H., & Bardi, A. (2023). Chapter: Phytoremediation of endocrine disrupting pollutants in industrial wastewater", published in the book (Pandey-CDBB-Endocrine Disrupting Pollutants) Book Series Elsievre 2023 – Bioressource Technology. Phytoremediation of endocrine disrupting pollutants in industrial wastewater.10.1016/B978-0-323-91902-9.00002-X

Rodríguez-Seijo, Andrés, & Ruth Pereira. (2017). Morphological and Physical Characterization of Microplastics. In *Comprehensive Analytical Chemistry*, 75:49–66. Elsevier. [Online] https://doi.org/10.1016/bs.coac.2016.10.007

Toussaint, B., Raffael, B., Angers-Loustau, A., Gilliland, D., Kestens, V., Petrillo, M., Rio Echevarria I., & Van den Eede, G. (2019). Review of micro- and nanoplastic contamination in the food chain, *Food Additives & Contaminants: Part A*, 36(5):639–673

UNEP and GRID-Arendal, (2016). Marine Litter Vital Graphics. United Nations Environment Programme and GRID-Arendal. Nairobi and Arendal. www.unep.org, www.grida.no, Consulté le 12 février 2019.

Vermaire Jesse C., Carrington Pomeroy, Sofia M. Herczegh, Owen Haggart, & Meaghan Murphy. (2017). Microplastic Abundance and Distribution in the Open Water and Sediment of the Ottawa River, Canada, and Its Tributaries. Édité par Daniel E. Schindler. FACETS 2 (1): 301–14. [Online] https://doi.org/10.1139/facets-2016-0070.

Vianello, A., Boldrin, A., Guerriero, P., Moschino, V., Rella, R., Sturaro, A., & Da Ros, L. (2013). Microplastic particles in sediments of Lagoon of Venice, Italy: first observations on occurrence, spatial patterns and identification. *Estuar.Coast.Shelf Sci.* 130, 54e61. [Online] http://dx.doi.org/10.1016/j.ecss.2013.03.022

Wright, S.L., & Kelly, F.J. (2017). Plastic and human health: a micro issue? *Environ. Sci. Technol.*, 51, 6634–6647.

Zuccarello, P., Ferrante, M., Cristaldi, A., Copat, C., Grasso, A., Sangregorio, D., Fiore, M., & Oliveri Conti, G. (2019). Exposure to Microplastics associated to plastic bottles mineral water consumption: the first quantitative study. Water Research. *Environmental and Food Hygiene Laboratories*, Department "G.F. Ingrassia", University of Catania.

7 Recent Development in Usage of Black Soldier Fly Larvae in Composting Process and Its Efficiency as a Potential Degrader

Sujata Phurailatpam and Meena Khwairakpam

7.1 INTRODUCTION

The worldwide waste generation is expected to reach 3.4 billion tones by 2050, with biodegradable materials accounting for up to 44% of total accumulated waste. However, with a higher proportion in low- to middle-income countries; the majority of this accumulated waste is currently disposed of in landfills (37%) and open dumps (33%) (Lopes et al., 2022). Management of organic waste can be done by various methods. Some of the methods which are mainly adopted for organic waste are open dumping, composting, and anaerobic digestion. However, there are a number of disadvantages related to the dumping of waste in landfills, including taking upwastes, the proliferation of hazardous organisms, and the wasting of precious space, the generation of foul smell, and lead to the emissions of greenhouse gas. Nevertheless, there are certain disadvantages associated with dumping of waste in landfills, including the use of space, the proliferation of harmful organisms, the formation of unpleasant odors, and the emission of greenhouse gases.
(FAO, 2013). Moreover, it takes a long time for the wastes to totally breakdown using these common waste management techniques). To effectively address the issue of waste buildup, more environmentally friendly techniques are needed. One of the emerging organic waste management is using the Black Soldier Fly larvae (BSFL) (Kinasih et al., 2020; C. Lalander et al., 2019; Shumo, Khamis, et al., 2019). *Hermetiaillucens,* or Black Soldier Fly (BSF) is an insect belonging to the family Stratiomyidae and order Diptera. The BSF resembles a wasp and has distinctive white and black color patterns all over its body (Tomberlin et al., 2002). The BSF has historically been found all across the Nearctic area of the globe. However, due to globalization and technological improvements, this species is currently widespread in most tropics and temperate areas (Oliveira et al., 2015). BSF larvae may break down organic waste, and their entire life cycle functions as a decomposer. It is widely

 DOI: 10.1201/9781003379386-7

recognized that treating organic waste with insect larvae is more efficient than traditional composting since it shortens the processing time and the residual substrates have better fertilizing qualities (Awasthi et al., 2019).

Converting organic waste using BSF larvae has the potential to aid both countries aiming for more sustainable resource exploitation and developing countries where garbage is disposed of in archaic and unsustainable ways, such as in open landfills. Furthermore, in order to increase costs for natural resources and livestock feed, the BSFL may capitalize on the nutritional content of organic waste (Awasthi et al., 2020). It attracts many researchers all over the world to study these. There are several benefits to using BSFL to turn manure and other organic wastes into valuable products, including the possibility of turning manure into safe compost, producing biofuels from larval fat and leftovers from rearing, and using larval biomass as a rich source of protein for cattle (Chia et al., 2021). Female BSFs can lay hundreds of eggs. BSFL development and reproduction are similar to other insects in that they are dependent on a complex interplay of abiotic and biotic parameters, such as ambient temperature, humidity, light, nutritional value (caloric content) of the substrate feed, and microbial colonization, among others (Sideris & Tsagkarakis, 2017; Yuvaraj et al., 2021).

Furthermore, organic waste may not always contain nutrients in a constant amount. It is worth noting that the environmental advantages of insect frass generation are directly tied to the type of substrate used to feed the larvae, with lesser impacts recorded when non-utilized waste streams are employed instead of typical items such as soybean meal (Smetana et al., 2019). Insects have a significant levels of crude protein ranging from 34 to 74% dry matter (DM) (Gasco et al., 2020). On a dry matter basis, most whole-insects contain 42 to 63.3% crude protein when insect meal is defatted, with up to 74% observed (Alfiko et al., 2022). Insects are an important alternative food source being researched (Siva Raman et al., 2022). In this chapter we will discuss the perspective of using BSF as a potential degrader and useful products generated from this fly. The chapter focuses mainly on the application of the *Hermetiaillucens* and its use in the circular economy of the developing country.

7.2 ASPECTS OF INVERTEBRATES IN COMPOSTING

In backyard compost heaps and other small-scale outdoor composting system, soil invertebrates are anticipated to help the breakdown of organic matter. Along with bacteria, fungi, and other microbes, invertebrates also form a complex food web that is difficult to understand. The pyramid foundation's foundation, or energy source, is composed of organic materials that include plant and animal waste. Several types of invertebrates, such as millipedes, sow bugs, snails, and slugs, feed plant matter like foliage or other plant residues. These invertebrates break down parts of the plant, increasing the surface area that fungus, bacteria, and actinomycetes may act on. These organisms are subsequently consumed by invertebrates like mites and springtails.

Tunnelling and feeding activities of the earthworm, potworm, nematods etc., help to aerate the compost and increase the surface area of organic materials for bacteria to operate on. More food is supplied to the web for additional decomposers as each decomposer dies or excretes.

TABLE 7.1
Invertebrates helpful in composting

Name of the microbes and invertebrates	Feeding substrate of the microbes and invertebrates during the composting
1. Bacteria	Sugar and starch
2. Fungi	Cellulose
3. Actinomycetes	Cellulose, lignin, chitin, protein
4. Potworm	Bacteria and fungi
5. Protozoa	Mycelia, the thread like strand produced by fungi
6. Earthworm	Undecomposed plant material
7. Mites	Found in all levels of compost food web i) primarily on leaves and rotten wood ii) secondarily on bacteria and fungi
8. Pseudoscorpion	Nematods, mites, springtails, small larvae and worm
9. Spider	Insect and other small invertebrates in compost pile
10. Springtails	Fungi, nematods and detritus
11. Housefly	Fresh and rotting food
12. Ants	Fungi, food scraps, other insects and seeds
13. Beetle	Insect, Snail, slugs and other small animals
14. Eagwig	Act as predator and eat detritus
15. Millipede	Decaying vegetation

(https://compost.css.cornell.edu/invertebrates.html)

Some of the invertebrates that are helpful in composting are listed in Table 7.1.

7.3 IMPROVEMENT IN WASTE MANAGEMENT THROUGH BLACK SOLDIER FLY

The main target of waste management is to lessen the quantity of waste produced, which will cut the costs associated with disposal and the effects on environmental and health issues (Periathamby et al., 2009). Poor waste management systems result in air and water pollution from careless open dumping and unrestricted burning, which are problems that are common in low- and middle-income countries (LMIC). When it comes to solid waste, between 50 and 70 percent of it is made up of organic materials, while the rest is made up of nonorganic items like glass, paper, metals, and polymers. Large volumes of non-organic waste are collected by waste pickers for recycling, whereas a sizable fraction of organic waste is left at the dumpsites to degrade (Komakech et al., 2015). Methane and carbon dioxide are produced as byproducts of the breakdown of organic waste, which results in the build-up of greenhouse gases that cause climate change (Isibika et al., 2019). Additionally, the poor sanitation at landfills promote the growth of rat, fly, and vermin vectors that disseminate illnesses like cholera and malaria throughout the surrounding area (Chowdhury et al., 2017). One way to lessen the quantity of organic waste dumped in landfills is by composting.

A tried-and-true way for dealing with organic matter is composting that may drastically cut down on waste generation. The efficacy of composting can be enhanced by the conversion of organic waste using saprophage (CORS) systems, which feed organisms (saprophages) with decomposing organic waste. Vermicomposting, in which worms and other microbes transform organic waste into nutrient-rich humus, is the most well-known use of CORS (Amrul et al., 2022). As converters of organic waste, *Hermetiaillucens* Linnaeus, Black Soldier Fly (BSF), has been proposed. Researchers have concentrated on a recently developed BSF-based approach for treating organic waste (De Smet et al., 2018; Zurbrügg et al., 2018). Food wastes (Nguyen et al., 2015), agro-industrial byproducts (Amrul et al., 2022) (cross), and dairy manure (Rehman et al., 2017) are only a few examples of the organic-rich waste that BSF larvae (BSFL) ravenously eat. The kind of organic material, the amount of nutrients present, and the ratio of waste to biomass are only a few of the variables that affect how economically viable waste treatment is (C. Lalander et al., 2019). According to research by Ibadurrohman et al. (2020), BSFL has an amazing capacity (75%) to recycle biological wastes, producing 4 kg of wastes yielded 800 g of larval biomass. Because conversion of organic waste using BSF has less negative environmental effects, like emissions of greenhouse gas and lower ecological footprint when manufacturing protein feed and other nutritional supplements, it is a great recycling technology and comprehensive solution. This method's application is possible for low- and middle-income nations because of its minimal capital investment (Van Huis A., Van Itterbeeck J., & Klunder H., 2013). Composting of BSF larvae has been the subject of extensive investigation.

7.3.1 Food Waste Management through Black Soldier Fly

Numerous studies have noted the important function that BSFL plays when dealing with biological waste (Purkayastha & Sarkar, 2022) and the valorization of diverse biodegradable wastes (Adebayo et al., 2021; Kawasaki et al., 2020). It is discovered that the larval biomass created by the biotransformation of organic wastes is a rising source of fat or oil (C. H. Kim et al., 2021) and proteins (Kinasih et al., 2020; Shumo, Osuga, et al., 2019). It is being used more frequently in a variety of sectors for animal feed, production of biodiesel, biopolymers (chitin), and frass produce as soil conditioner (Purkayastha & Sarkar, 2022).

Notably, whole or processed BSF larvae or pupae can be added to the diets of fish, poultry, pets, and pigs, offering as potential substitutes for typical feed elements like soybean- and fish-based meal (Surendra et al., 2020). As a result, the traditional feed components that BSF produce may substitute can be saved for various purposes, such as people utilization, so promoting food security. Additionally, the functional substances found in BSFL, like antimicrobial peptides, may significantly improve the nutritional value of animal diets (Surendra et al., 2020). However, a study by Wynants et al. (2019) highlights the need for extra care to be taken while growing BSFL since harmful organisms including *Bacillus cereus* and Salmonella sp. can be found in samples of residues and/or larvae, which can then spread to the animals to whom they are being fed. The researchers suggest using the appropriate decontamination technology in this situation (Wynants et al., 2019).

7.3.2 Biowaste Conversion using Black Soldier Fly

The management of organic waste is inadequate that results in contamination of the environment and produces no useful byproducts. As a result, managing organic waste using insects has acquired attention and considerable promise in recent years (C. H. Lalander et al., 2015). Some of the insects being examined most for the conversion of organic waste are mealworms, common houseflies, and BSF, which might be used as substitute sources of protein for feeding in animal (van Huis et al., 2020). These insects, especially BSFL larvae, are rapacious eaters of decomposing organic matter, so they have been used to successfully reduce waste streams such as animal manure (such as swine, cow, chicken, etc.), human waste, waste from farm slaughterhouses, vegetable waste, fruit, and waste from the palm oil industry (Beskin et al., 2018; Gold et al., 2020; C. Lalander et al., 2019; Raksasat et al., 2021). The possibilities has been given further thought of BSFL to decompose organic waste than to any other fly species. BSF provides a solution for the challenges of inadequate global waste management, urban unemployment, and rising need for sustainable animal feed in contemporary society (Zurbrügg et al., 2018; Dortmans et al., 2017). The benefit of black soldier fly in conversion of organic waste as a waste converter (environmental and functional health benefits) are capable of waste reduction in a less amount of time, carbon footprint will be low, more feed to body mass conversion, does not caused any disease like housefly, as well as decreasing pathogen and other pest populations. Moreover, from a commercial standpoint, it generates valuable resources at a low cost while using less water and land than other alternative protein sources. Furthermore, Beskin et al., 2018 showed that BSFL can lower the pungent components in dairy, swine, and poultry manures by up to 87% or more. Thus, utilizing BSFL as an ecologically beneficial approach of handling animal manure as opposed to existing traditional methods, proves an added advantage. BSF is a desirable remediation method to turn biowaste into high-quality goods because of all of the aforementioned benefits. In conclusion, Bortolini et al. (2020) reported that the advantage of BSF system in degradation of organic waste into larval biomass and leaving no remnant at the finished of the process, because the remnant (frass) from the process can be used as soil conditioner and it will help in the circular economy of the country.

7.3.3 Reduction Rate by Black Soldier Fly

Waste reduction and bioconversion rate are two indicators that are frequently used to evaluate the effectiveness of the type of waste that BSFL handles (Gold et al., 2020; C. Lalander et al., 2019). Whereas the biomass conversion rate is normally measured by dividing the total feed delivered by the larvae weight growth (the end larval weight less the beginning larval weight), waste reduction is a figure on a dry mass basis that is determined by dividing the amount of feed used in the experiment by the amount of residue (Gold et al., 2020).

Numerous variables, including feed composition (amounts of nutrients that may be digested, pH, and moisture levels), feeding rates, and others, affect the biomass conversion rate (Banks, 2014). The process of converting organic waste is, however, greatly influenced by the Larval population size and feeding schedule (Banks, 2014;

Dortmans et al., 2017). The bioconversion rate is also dependent on the water level of the supporting material. The ideal moisture content for the optimum bioconversion ranges from 70–80%. High protein content and the availability of digestible carbon both significantly increase the value of biomass production (C. Lalander et al., 2019). By combining different waste streams to overcome the drawbacks of isolated streams, the effectiveness of organic waste management using BSFL may be boosted (Ermolaev et al., 2019).

7.3.4 BSF is Not a Disease Vector

The flies that are drawn to food can often spread pathogenic microorganisms. Common house flies are an annoyance to homes and businesses close to stored and raw biodegradable waste. These nuisance flies are the main focus for management of fly and control initiatives (Siddiqui et al., 2022). The rearing of BSF near our habitat area does not cause any disease to human (Dortmans et al., 2017). The glucose that was stored in the larval body's fat is what the BSF adults rely on to survive because they avoid human homes and meals (Sheppard et al., 2002; Van Huis, A., Van Itterbeeck, J., Klunder, H., 2013; van Huis et al., 2020). The larvae's large fat reserves diminish or completely eliminate the need for adults to eat and spread any dangerous diseases (Van Huis, A., Van Itterbeeck, J., Klunder, H., 2013; van Huis et al., 2020). House flies appear to be more prone to spread illness than BSF. Therefore, the law could favor BSF over the house fly, which might be a major problem if manures and excrement are used for feeding substrates for Black Soldier Fly (van Huis et al., 2020).

7.3.5 Organic Waste Treatment Including Food Waste using BSFL

BSFL has been used to treat a variety of substrates. If livestock excreta is not handled, it does not degrade rapidly and efficiently, polluting the environment. Previously, high temperature composting was the primary method for treating animal manure; moreover, people start paying attention to manure composting utilizing BSFL in the early 1970s.

There is a difference both in the rate of substrate intake across equally nutrient-rich substrates and the production of biomass from larvae. BSFL are capable of consuming a significant amount of organic waste; in general, 17 g of five-day-old larvae may eat up to 0.7 kg of biowaste a day (Pintowantoro et al., 2021). The microorganism population of the substrate was boosted by the *Hermetiaillucens* biotransformation of chicken manure, raising the fraction of dominating Firmicutes bacteria in poultry manure after 15 days, from 56% prior to pre-treatment to 98% post-treatment (Liu et al., 2021).

BSFL may be impacted not only by the microbial composition of the substrate, but also by the physical and chemical properties of the larvae. The quantity and make-up of microbial communities might differ under the control of where they are grown and the qualities of the biodegradable waste. This suggests that microbial communities may impact not only the efficacy of degradation processes, but also the possibility of discovering novel metabolic pathways important for understanding the treatment of biowaste as well as livestock manure involves biochemical kinetics. Appreciable organic waste treatment efficacy may be achieved by using fecal sludge as a feeding substrate for BSFL enriched with 30% biowaste at a feeding rate of 200 mg per larva per day.

7.4 BLACK SOLDIER FLY LIFE CYCLE

BSF, also known as *Hermetiaillucens*, is a member of the phylum Arthropoda, class Insecta, order Diptera, and family Stratiomyidae. It is widely distributed in tropical and warmer climates between the latitudes of 45° N and 40° S. (R & K, 2013; Doelle et al. 2016; Purkayastha et al., 2017). Egg, larva, prepupae, pupae, and fly are the major five stages of black soldier fly life as shown in Figure 7.1. The average lifespan of an adult BSF ranges from 8–20 days (Tomberlin et al., 2009; Fly & Hermetia, n.d.). Females flies choose dry areas over damp area for egg laying, and it takes them around 20–30 minutes to lays eggs. Each egg weighs approximately 0.028 plus or minus 0.001 mg, and each egg cluster typically contains 324 to 998 eggs (Chia et al., 2018). After four days of oviposition, the fertilized eggs hatch (Zhang et al., 2010). After hatching from the eggs, the BSFL feeds rapaciously on different types of biowaste and accumulates a substantial amount of fat and protein in its body for later use. The BSFL typically feeds based on the surroundings, for 2–4 weeks (relative humidity and temperature), food sources, and other factors. They subsequently mature into prepupae, which ceased feeding on the substrate, their guts are emptied, and move away from the food in search of a dry area (Diener et al., 2009)

. The Black Soldier Fly molts prior to the pupal stage, and there are six instars during its life (W. Kim et al., 2010). Prepupae develop into adults in approximately

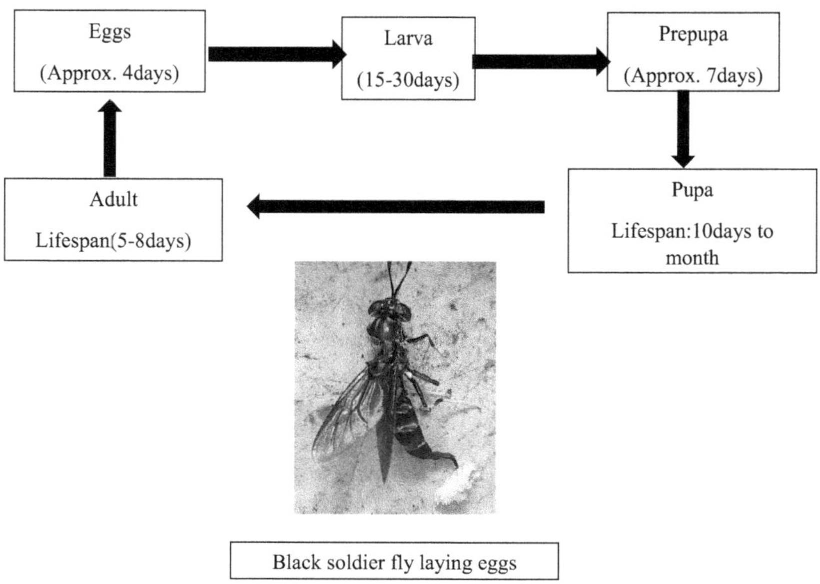

FIGURE 7.1 The life cycle of black soldier fly.

two weeks, however under environmental stress, it may take up to five months (Furman et al., 1959).

7.4.1 Factors Influencing the Black Soldier Fly Larvae's Growth and Development

Temperature, humidity, sunshine, moisture content, pH, and other environmental factors affect the development and growth of *Hermetiaillucens*.

7.4.1.1 Humidity and Temperature on the Black Soldier Fly

The eurythermic Black Soldier Fly species can endure temperatures between 15 and 47°C, but it is also extremely sensitive to changes in temperature (Park, 2016). Temperature changes have an impact on the different phases of BSFL. *Hermetiaillucens* larvae exposed at high thermal develop eggs faster than those grown at lower temperature. The larvae's developmental time varied significantly between temperature ranges and raising surfaces. The effect of temperature and the brood substrate on the larvae's development period interact significantly. The mature BSF survival of both males and females is significantly influenced by temperature. In comparison to greater severe temperatures, adult flies survive longer in intermediate temperatures (Chia et al., 2018; de Sousa Antunes et al., 2021). Temperature and relative humidity have a big impact on how a species develops, mates, and lays its eggs (Tomberlin et al., 2009; Barry, 2017).

According to research by several authors, the majority of oviposition (about up to 99.6%) takes place between 27.5 and 37.5°C and at a 60% of relative humidity (Bradley et al., 1984; Sheppard et al., 2002). Other research by several authors has demonstrated that keeping the site's relative humidity at roughly 60% and its temperature at 27°C are ideal for mating and egg-laying (Sheppard et al., 2002).

7.4.1.2 Sunlight

In the natural world, exposure to direct sunshine is crucial for growth, development, reproduction, and survival. For tests conducted indoors, artificial lighting sources are necessary. Nearly 85% of mating events require sunlight with an intensity of 110 μ mol m^{-2} s^{-1}(Park, 2016). Artificial light sources were found to have an influence on species' reproduction event in one research. When rearing animals outside of their natural habitat, where sunlight is the key factor, this precise procedure is advantageous and successful. The species' mating and oviposition are favorably associated with quartz iodide lamps (135 μ mol m^{-2} s^{-1}) as artificial light sources (Zhang et al., 2010). Beyond 700 nm, insects cannot see the light. Consequently, adult breeding events are most suited to the wavelength range between 450 and 700 nm(Briscoe and Chittka, 2001; Zhang et al., 2010).

7.4.1.3 Influence of Moisture Content on Black Soldier Fly

The amount of water content in the waste has a big impact on BSFL's survival and development (Cheng et al., 2017). A high moisture content slows down degradation and causes agglomeration. Thick materials may also accompany the resulting residue, complicating further processing (Diener et al., 2011). According to Ferronato &

Torretta (2019), when the larva are fed in chicken dung having the moisture content ranges between 40% and 60% reached the highest survival rate. Similar to this, Banks (2014) found in his research that pupae develop faster when 85% of the fecal sludge's water content is present.

7.4.1.4 pH Influence on the Black Soldier Fly

The pH level has a significant impact on the lifespan of the black soldier fly. The developmental process of the larvae of BSF require a pH higher than 6, as demonstrated by several investigations on these insects (Meneguz et al., 2018). According to Green & Popa (2012) BSF larvae may raise the pH of liquid organic leachate to 9.M. Alattar (2012) also found that the density of the larvae is the only factor that controls how well the organic medium can regulate the pH. Ma et al. (2018) researched the influence of various levels of pH on the development of larva in more detail and demonstrated that the performance for the growth of larva is highest at pH level above 6 to 10. when the pH is 4 or 2 the weight of the larva is raised. Additionally, it has been demonstrated that larvae on alkaline substrates have the ability to raise the pH from 8 to 8.5 but not in severely acidic substrates.

7.5 ADVANTAGES OF BLACK SOLDIER FLY TECHNOLOGY

Many countries of the world are seeing an increase in the usage of BSFL for the rapid and low-cost stabilization of organic waste. While the same aim may be achieved in a variety of other methods or with different insects, the attraction of utilizing the BSF prepupae may have economic value derived from it as a whole or by manufacturing a number of valuable goods from it through the agency of *Hermetiaillucens* larva, as shown in Figure 7.2. Even the BSFL's exterior shell is formed of chitin, which

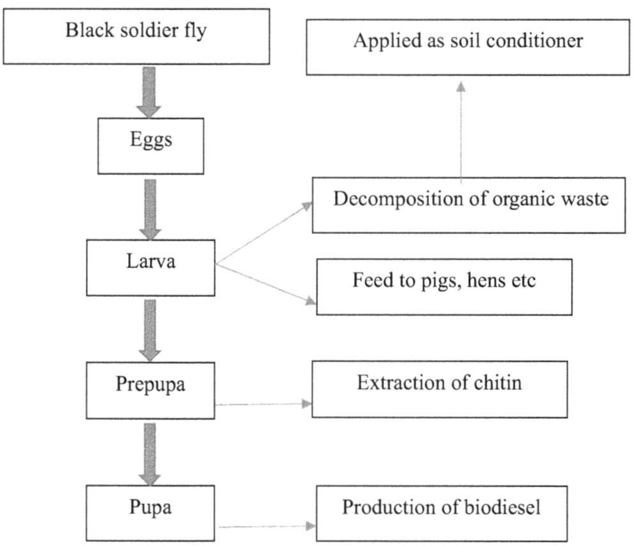

FIGURE 7.2 The advantages of black soldier fly product.

may be removed and used as a biodegradable polymer. Some of the uses of BSF are discussed below.

7.5.1 BSF Larva used as Animal Feed

Typically, as they are high in protein, soybean and fish meal are used often as feed materials in livestock farming units. The protein content of BSF prepupae can be used as an alternative for animal feed (Lee et al., 2018; Su et al., 2019). Farming/rearing of BSF prepupae uses less space as compare to other farming, and it is possible to stack them vertically to increase manufacture on the equal amount of ground, if required (Zurbrügg et al., 2018). They are significantly less expensive than soybean and fish meal since they are produced on wastes. After defatting, BSF prepupae have the same amino acid composition as soybean meal (Surendra et al., 2016; Caligiani et al., 2018). However, the food source used to produce BSFL depends on composition and quantity (Caligiani et al., 2018; Marco et al., 2015; Onsongo et al., 2018). Since it increases the protein content of BSF feed, the defatted BSF prepupae has also been considered to be a workable alternative, providing additional opportunity for animal feed supplementation (Cullere et al., 2018; Nery et al., 2018; Schiavone et al., 2017). Compared to conventional animal feed, the introduction of BSF prepupae increased the economic benefit by a considerable amount. Onsongo et al. (2018) found an economical ratio of 16.0% and a return on investment of 25.0% for replacing fish meal made of soybean in the basal and finisher diets containing 42 and 55.5% BSF prepupae in broiler chicken, respectively. The characteristics of the eggs produced by Lohmann brown classic laying hens given BSFL food were similar to those of eggs produced using soybean as a diet (Secci et al., 2018); however, contrasting findings have also been observed by Ruhnke et al. (2018). The productivity of BSF prepupae meal given to European seabass (45% replacement of fish meal) was the same as that of fish fed 100% fish meal. Similar development patterns were seen in rainbow trout fed either 100% fish food or BSF prepupae in place of 25% of their normal fish food (Sciences et al., 2007).

Regardless of the feeding substrate, the crude protein level drops as the BSFL emerges, particularly in the sixth larval stage, just before the transition to prepupae (Purkayastha & Sarkar, 2022). The enzymatic catalysis of protein during the development of the chitin layer of the exoskeleton causes such a change (Wong et al., 2019), a process called sclerotization. Moreover, it is well known that BSF with a greater chitin level makes it more difficult for animals like poultry, fish, and swine to digest its lipids (Kroeckel et al., 2012; Schiavone, De Marco, et al., 2017). Soit's important to gather BSF prepupae in a timely manner. BSF prepupae is a preferable option for animal feed, however in order to produce a sustainable feed source, intensive animal studies are required. The stringent regulations that are in force in some countries regarding using insects produced on waste material as animal feed are another significant, if not limiting, holdback.

7.5.2 Use of Fat as Biodiesel

Many researchers are looking into the idea of using other sources of energy in place of oil and its derivatives as a result of the increased awareness of energy difficulties and environmental challenges related to burning fossil fuels. Biodiesel is appealing due to its low emissions of carbon monoxide and particulate matter, renewable nature, high degree of degradability, lack of toxicity, and lack of hydrocarbon release. According to ASTM International (International Association for Testing and Materials), long-chain fatty acid mono-alkyl esters derived from renewable vegetable oils or animal fats are used to create the fuel known as "biodiesel" that satisfies ASTM D6751's specifications (Mishra & Goswami, 2018).

Biodiesel stands out among them as being particularly intriguing because of several factors, with its excellent degradability and low toxicity, its ability to replace diesel fuel in many uses, such as boilers and internal combustion engines, with only minor performance changes necessary, its nearly release of zero sulfur dioxide, aromatic compounds, and other environmentally harmful chemical substances, and its relatively negligible net carbon dioxide contribution (Shahid & Jamal, 2011). The significance of energy is rapidly increasing daily as a result of global modernization and industrialization (Mishra & Goswami, 2018). Because of industrialization and globalization, the value of energy is steadily increasing. Most importantly, the production of renewable energy is rapidly rising as a result of fossil fuels depletion, the growth in the price of petroleum, and major other effects. The primary sources of fossil fuels are acknowledged to be petroleum, coal, and natural gas, which have been used to supply our energy needs (Atabani et al., 2012). One of the most discussed and studied alternatives to minimize emissions of greenhouse gas, alternative fuels are being sought for is biodiesel. Even while study on and usage of plant-based feedstock, particularly edible vegetable oil, for the generation of biodiesel has been substantial, there are various questions and justifications for its use regarding their economic stability and food security (Adewale et al., 2015; Rezania et al., 2019).

The transesterification reaction is a chemical process used to create biodiesel. Triacylglycerols react with short-chain monohydric alcohols in the presence of an appropriate catalyst—usually an alkali, an acid, or an enzyme—at high temperatures to produce fatty acid alkyl esters (FAAE) and glycerol. The production and purification of biodiesel from the *Hermetiaillucens* is shown in Figure 7.3. In order to provide economically and environmentally sustainable feedstock, the manufacturing of biodiesel from insects has become more popular (Hasnol et al., 2020; Närvänen et al., 2019). Creating biodiesel from insects follows a comparable procedure to generating it from other biological sources of fat (Teresa L. Alleman, 2016). The amount of fat obtained from insects differ according to the species, the kind of food waste used, and the stage of development; the larval stage has the largest amount of fat (Manzano-Agugliaro et al., 2012). *Hermetiaillucens*, sometimes known as the "black soldier fly," is a common fly of the Stratiomyidae family that inhabits tropical and temperate climates. The development period of *Hermetiaillucens* is 21 days, that is significantly longer than the development period of house flies (5 days). In contrast to house flies, it indicates that a single black soldier fly larva may devour more food scraps and produce bigger pupae (Feed & Food, 2017; Newton et al., 2015). The average lipid content of BSFL and their pupae is 29%, with ranges between 11 and 42% (dry basis) (Feed & Food, 2017). Since BSFL naturally leave the substrate to go to a high, clean area before entering the pupa stage, there is no need to separate

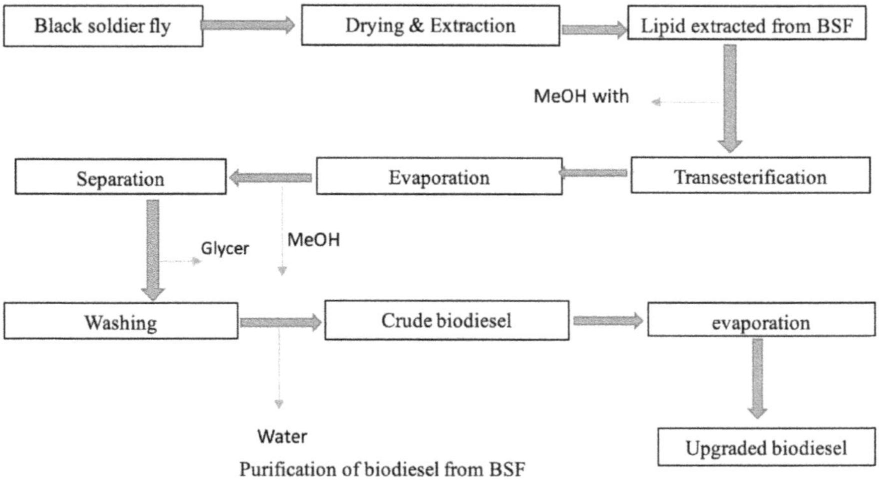

FIGURE 7.3 Flowchart showing extraction and purification of biodiesel from *Hermetiaillucens.*

food waste from BSFL further. Such innate tendency, also known as self-harvesting, can assist in removing a time-consuming gathering process. Some (Zheng et al., 2012) examples of insects that are utilized to make biodiesel include agricultural waste from fast food are converted by black soldier fly larvae with extra microbes (Rid-X), yielding 43.8 g of biodiesel from 1 kg of waste. Additionally, due to the high quantities of lauric acid (>60%) in their fatty acid profiles, the oil obtained from BSFL was proven to be less suitable than oils of locust for usage in skincare products. Odor, color, phospholipid, and free fatty acid removal were shown to be essential to make all of the studied insect oils (Siddiqui et al., 2022). In conclusion, there are several processes involved for the bioconversion process of food waste based on insect that may be utilized to extract sources of fuel, offering an alternative to our limited supply of fossil fuels.

7.5.3 Use of Black Soldier Fly Frass as Fertilizer

Reusing nitrogen (N) and phosphorus (P) sources rather than using external sources in the food chain aids in mitigating the effects of excess N and P on the environment. Agriculture's primary components are nutrients, water, and light. However, as the population of the globe grows and more people choose diets high in nutrients, the quality of the air, land, and water is being put under tremendous strain. Growing crops and raising feed from animals, an expanding global population has grown dependent on the availability of N and P fertilizer, which has significant social and economical advantages (Eickhout et al., 2006; G. P. Robertson & Vitousek, 2009). However, very little of the N and P used in agronomy makes it into people's mouths (Schmitt & de Vries, 2020). Most of these nutrients are either wasted in the food system or lost into the environment (Jenkins et al., 2016). Finding sources of N and P that are generated endogenously from the food system and returned to the food system rather than sourced exogenously is crucial for resolving the excess N and P impacts and keeping within our planetary limitations.

A growing market for insects as food and feed is now emerging, with one of the main products being a compost-like substance known as frass, which is formed by providing insects thriving in exceptionally high densities with bacteria with reduced organic food.

Byproducts from our food system are fed to insects, which results in a frass that contains N and P that is naturally occurring instead of being externally contributed to the agriculture system. *Hermetiaillucens* and *Tenebrio molitor* are a couple of the species for which study on frass has been conducted (Osimani et al., 2018; Xiao et al., 2018). The procedure for creating and using insect frasses has three benefits. First, this method typically has a higher economic value than traditional composting because additionally, first before providing these to crops, manufacturers immediately upgrade a sizeable percentage of the nutrients, particularly N and P, in the waste organics into insect components, such lipids and proteins that may be utilized in the food system (Chia et al., 2019). Second, because they contain significant amounts of insect by products, like chitin via shed skins (i.e. exuviae).Contrary to conventional composts, insect frasses are shown to change the soil microbiology and plant activity. Plants growing in contact with them may experience major advantages as a result of these impacts (Poveda et al., 2019). Last but not least, frass manufacturing is as efficient as other methods in terms of a variety of environmental impact indicators as standard biogas production and composting and produces a lot lesser greenhouse gas than chemical N fertilizer (Smetana et al., 2019). Microorganisms play a crucial part in the generation of BSF. Wherein microbes and larvae degrade down meal resources symbiotically, which is less obvious but perhaps more fascinating. The characteristics of the frass are affected by this.

The health of the soil can be improved by an abundance of helpful bacteria, which can be found in large numbers in frass (Xiao et al., 2018). Economically, as compared to production of biogas and compost from the organic waste, production of the frass and other valuable products can be more profitable from the farming or rearing of BSF. Insect frass used as fertilizer has suitable chemical and physical characteristics with other industrial goods(Salomone et al., 2017). For instance, in one research, cabbages cultivated with BSF frass had the same rate of growth and chemical composition as cabbages treated with commercial fertilizers (Choi et al., 2009). Similar to that, both insect frass and compost additives had the same effect on onion production (Zahn, 2017). This could be as a result of the additional ammonia (NH_4^+) from nitrogen found in insect frass, which has been shown to rise five times more than in plants that weren't fertilized (Green & Popa, 2012). Composting insect frass has other advantages, such as reducing insecticides and pathogenic microbes(C. Lalander et al., 2016). In terms of nitrogen, phosphorus, potassium (NPK), and germination index values, BSFL-aided composting produces organic fertilizer of higher quality than conventional composting, however there is variation in the quality of fras depending on the parameters of the waste stream (Liu et al., 2019). The BSFL improves the quality of the compost and decreases the amount of heavy metals in the waste stream (Attiogbe et al., 2019). However, there are issues that some insects' frass may contain heavy metals (Diener et al., 2015). So, it is really important to consider what kind of substrate is feeding to the BSFL.

7.5.4 Use of Chitin

Chitin, the primary structural element of the exoskeleton of arthropods and the cell wall of fungus, is the next most prevalent polysaccharide after cellulose on Earth (J. Robertson, 2004). It is a polymer with expanding marketable possibilities. It is used in agriculture to combat biologic infections (Manjula et al., 2001), in tissue regeneration) for wound dressing, in the removal use as a hydrating element in pharmaceuticals, the removal of colors and contaminants from contaminated water, and the production of paper (Felse and Panda, 1999) Chitin creates a network of structured fibers and has a crystalline structure. Its chemical structure is remarkably similar to that of mucopolysaccharides (heparin and hyaluronic acid). Those organisms that have it gain stiffness and resilience. Additionally, chitin may be transformed into chitosan, a polymer with greater sensitivity and solubility due to free amino groups, which has a larger variety of uses (Yeul & Rayalu, 2013).

Black soldier fly or *Hermetiaillucens* L. is one of the species that is most often produced for bio-conversion of insect-based. This fly's larval stages may feed on a various range of organic materials and transform them into useful products including proteins and fat for animal feed, lipids for the creation of biodiesel and cosmetics, and frass for improving soil (Jucker et al., n.d.). Several exoskeletons from the larvae are shed during the transition to the pupae stage. Pupal exuviae are produced as pupae develop into adult flies (Tomberlin et al., 2002). The exoskeleton of BSFLs, in particular, is mostly made of chitin. Throughout each stage of development the BSF's chitin composition varies (Hahn et al., 2022). The chitin in the frass functions as a biostimulant and supports the growth of beneficial bacteria in the soil (Schmitt & de Vries, 2020). Exuviae of the pupa stage are easiest to get from Black Soldier Fly farms because they self-collect in the most easily available and necessary containers for the development of pupae into adults. As a result, it is necessary to design a method for removing chitin from pupal exuviae. The majority of research to date has been on improving the extraction of chitin from fishing waste, for which numerous techniques are documented in the literature. Chitin from insects is still mostly purified in laboratories using techniques described for crustaceans. These techniques include consecutive processes to remove minerals, proteins, pigments, and catecholamine chemicals, which are identifiable by the black color of the insect cuticle. According to (Hahn et al., 2020), it represents an extensive overview of techniques for insect chitin and chitosan purification.

Demineralization, deproteinization, and bleaching phases of the purification process were carried out under various circumstances to find the ideal ones as shown in Figure 7.4. The effectiveness of demineralization was examined in correlation to the kind of acid employed, its fatality, and the quantity of the final washing. Temperature, molarity, and incubation time were taken into consideration for optimizing the alkali-based deproteinization process. Regarding the chemical employed, time, and temperature of the final bleaching procedure, an evaluation was made.

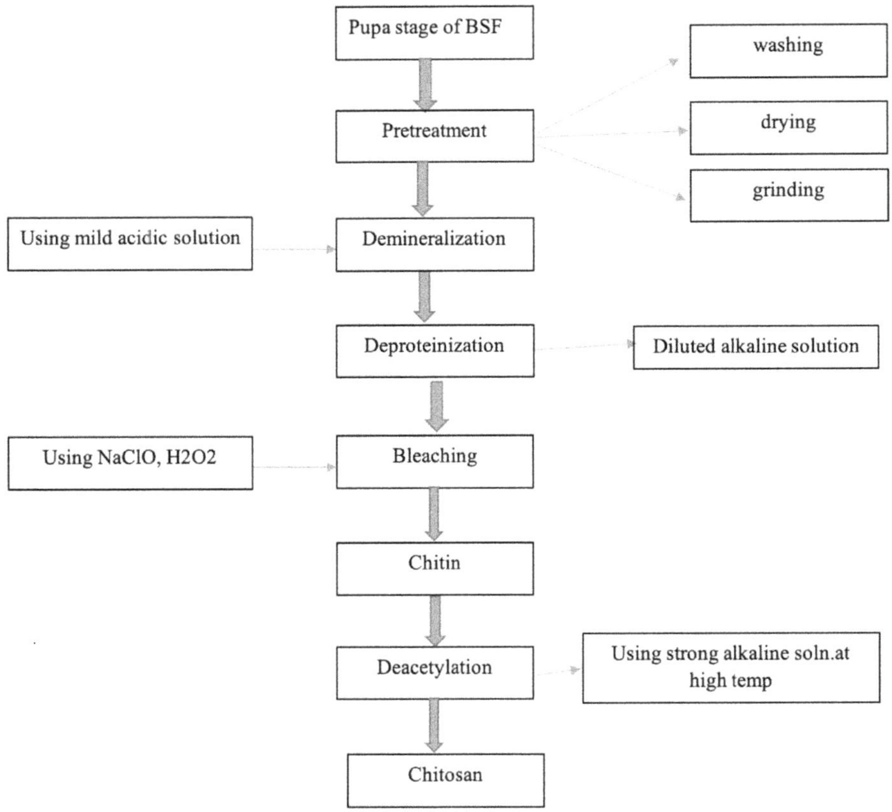

FIGURE 7.4 The extraction and purification of chitin from black soldier fly.

7.6 FUTURE PERSPECTIVE

To concentrate on a more narrow area, the BSF waste treatment system might be run on a household scale. According to the Khazanah Research Institute (KRI), home waste accounts for 44.5% (6.1 million tons) of all solid waste produced annually and is the main source of municipal solid waste (MSW). Between 0.5 and 0.8 kg of leftover food is typically thrown out by Malaysian homes each day, and with population expansion, this quantity is anticipated to rise (M. A. Alattar et al., 2016). Food waste is rising, which might make landfill problems worse, such bad aromas, formation of harmful leachate, emissions of greenhouse gas, and vermin infestations (Song et al., 2021). Addressing the underlying sources of the issues, including putting up a black soldier fly at homebased system to recycle organic waste into usable items like organic fertilizers for household gardening or for revenue, is one option to solve these difficulties. The home BSF system employs unconsumed food used as a feeding surface for larvae, which eliminates the need for transportation expenses. However, there are a number of difficulties using the BSF system at home, including odor and space restrictions.

To encourage anybody who is eager to take part in the effort to decrease food waste, it is necessary to create a household BSF system that uses little space, less time, and minimal technical knowledge. In order to design an efficient system for waste treatment that takes into account the crucial aspects of the surroundings in residences, it is necessary to do more study to identify the ideal environmental conditions for BSF larvae growth. It should also be mentioned that the BSF bin is convenient and handy, and that the larva are self-pickable and adaptive to the area environment condition

7.7 SUMMARY AND CONCLUSION

Recent composting experiments using Black Soldier Fly (BSF) larvae have shown their degrading capabilities. BSF larvae, also known as *Hermetiaillucens*, transform organic waste into useful resources, making them popular in waste management. The BSF larvae are introduced into organic waste, such as food waste or agricultural residues, during decomposition. These voracious eaters rapidly consume the refuse and decompose it into simpler compounds. Their highly efficient digestive system can metabolize various organic matter, including proteins, carbohydrates, and lipids. The ability of BSF larvae to convert refuse into nutrient-rich frass is one of the larvae's primary benefits in decomposition. Frass is the byproduct of the larvae's digestion of organic material. High levels of nitrogen, phosphorus, and potassium are present in this potent fertilizer. This nutrient-rich manure can be used directly as a soil amendment, reducing the need for synthetic fertilizers. Further, it has been determined that the BSF larvae decomposition process is environmentally friendly. It substantially reduces emissions of greenhouse gases, including methane, a potent contributor to climate change. Additionally, the process minimizes the odor problems associated with conventional decomposition techniques, making it more socially acceptable in urban areas.

A new waste treatment technology using BSFL, which is used to process waste, may create products with value added that are beneficial for the market and help develop circular economy models of waste management that are both economically and environmentally sustainable. The innovation for treating garbage with BSFL remains in its beginning stage, while being seen to offer significant promise for turning waste into commercial developments. Bioconversion insect-based businesses and the governing bodies responsible for safeguarding need to communicate more effectively, which calls for research. More research on the security of items made from insects will probably result in regulation that is biologically sound. Bioconversion of insect-based food waste has the capability to be an effective instrument for reducing food waste, generating employment, and providing an environmentally benign source of protein to help feed the world's expanding population if the necessary safeguards are in place.

REFERENCES

Adebayo, H. A., Kemabonta, K. A., Ogbogu, S. S., Elechi, M. C., & Obe, M. T. (2021). Comparative assessment of developmental parameters, proximate analysis and mineral compositions of black soldier fly (Hermetiaillucens) prepupae reared on organic waste substrates. *International Journal of Tropical Insect Science*, *41*(3), 1953–1959.

Adewale, P., Dumont, M. J., & Ngadi, M. (2015). Recent trends of biodiesel production from animal fat wastes and associated production techniques. *Renewable and Sustainable Energy Reviews*, *45*, 574–588. https://doi.org/10.1016/j.rser.2015.02.039

Alattar, M. (2012). Biological Treatment of Leachates of Microaerobic Fermentation. *Dissertations and Theses*, 78. http://archives.pdx.edu/ds/psu/7959

Alattar, M. A., Alattar, F. N., & Popa, R. (2016). *Plant Science Today fly larvae food scrap processing residues on the growth. 3*, 57–62.

Alfiko, Y., Xie, D., Astuti, R. T., Wong, J., & Wang, L. (2022). Insects as a feed ingredient for fish culture: Status and trends. *Aquaculture and Fisheries*, *7*(2), 166–178. https://doi.org/10.1016/j.aaf.2021.10.004

Amrul, N. F., Ahmad, I. K., Basri, N. E. A., Suja, F., Jalil, N. A. A., & Azman, N. A. (2022). A Review of Organic Waste Treatment Using Black Soldier Fly (Hermetia illucens). *Sustainability (Switzerland)*, *14*(8), 1–15. https://doi.org/10.3390/su14084565

Atabani, A. E., Silitonga, A. S., Badruddin, I. A., Mahlia, T. M. I., Masjuki, H. H., & Mekhilef, S. (2012). A comprehensive review on biodiesel as an alternative energy resource and its characteristics. *Renewable and Sustainable Energy Reviews*, *16*(4), 2070–2093. https://doi.org/10.1016/j.rser.2012.01.003

Attiogbe, F. K., Ayim, N. Y. K., & Martey, J. (2019). Effectiveness of black soldier fly larvae in composting mercury contaminated organic waste. *Scientific African*, *6*. https://doi.org/10.1016/j.sciaf.2019.e00205

Awasthi, M. K., Chen, H., Awasthi, S. K., Duan, Y., Liu, T., Pandey, A., Varjani, S., & Zhang, Z. (2019). Application of metagenomic analysis for detection of the reduction in the antibiotic resistance genes (ARGs) by the addition of clay during poultry manure composting. *Chemosphere*, *220*, 137–145. https://doi.org/10.1016/j.chemosphere.2018.12.103

Awasthi, M. K., Duan, Y., Liu, T., Awasthi, S. K., & Zhang, Z. (2020). Relevance of biochar to influence the bacterial succession during pig manure composting. *Bioresource Technology*, *304* (December 2019), 122962. https://doi.org/10.1016/j.biortech.2020.122962

Banks, I. J. (2014). *To Assess the Impact of Black Soldier Fly (Hermetiaillucens) Larvae on Faecal Reduction in Pit Latrines. Doctoral thesis, London School of Hygiene Tropical Medicine.* 1–231. http://researchonline.lshtm.ac.uk/1917781/

Beskin, K. V., Holcomb, C. D., Cammack, J. A., Crippen, T. L., Knap, A. H., Sweet, S. T., & Tomberlin, J. K. (2018). Larval digestion of different manure types by the black soldier fly (Diptera: Stratiomyidae) impacts associated volatile emissions. *Waste Management*, *74*, 213–220. https://doi.org/10.1016/j.wasman.2018.01.019

Bortolini, S., Macavei, L. I., Hadj Saadoun, J., Foca, G., Ulrici, A., Bernini, F., Malferrari, D., Setti, L., Ronga, D., & Maistrello, L. (2020). Hermetia illucens (L.) larvae as chicken manure management tool for circular economy. *Journal of Cleaner Production*, *262*, 121289. https://doi.org/10.1016/j.jclepro.2020.121289

Bradley, S. W., Booth, D. C., & Sheppard, D. C. (1984). Parasitism of the Black Soldier Fly by Trichopria sp. (Hymenoptera: Diapriidae) in Poultry Houses. *Environmental Entomology*, 13, 451–454.https://doi.org/10.1093/ee/13.2.451

Barry, T. (2017). Evaluation of the economic, social, and biological feasibility of bioconverting food wastes with the black soldier fly (Hermetia illucens) [electronic resource].

Briscoe, A.D., & Chittka, L. (2001). The evolution of color vision in insects. *Annu. Rev. Entomol.* 46, 471–510.

Caligiani, A., Marseglia, A., Leni, G., Baldassarre, S., Maistrello, L., Dossena, A., & Sforza, S. (2018). Composition of black soldier fly prepupae and systematic approaches for extraction and fractionation of proteins, lipids and chitin. *Food Research International*, *105*, 812–820. https://doi.org/10.1016/j.foodres.2017.12.012

Cheng, J. Y. K., Chiu, S. L. H., & Lo, I. M. C. (2017). Effects of moisture content of food waste on residue separation, larval growth and larval survival in black soldier fly bioconversion. *Waste Management*, *67*, 315–323. https://doi.org/10.1016/j.wasman.2017.05.046

Chia, S. Y., Mbi, C., Id, T., Khamis, F. M., Mohamed, A., Salifu, D., Id, S. S., Fiaboe, K. K. M., Niassy, S., Loon, J. J. A. Van, Id, M. D., & Ekesi, S. (2018). *Threshold temperatures and thermal requirements of black soldier fly Hermetia illucens: Implications for mass production*. 1–26.

Chia, S. Y., Tanga, C. M., Loon, J. J. A. Van, & Dicke, M. (2019). ScienceDirect Insects for sustainable animal feed: inclusive business models involving smallholder farmers. *Current Opinion in Environmental Sustainability*, *41*, 23–30. https://doi.org/10.1016/j.cosust.2019.09.003

Chia, S. Y., Tanga, C. M., Osuga, I. M., Alaru, A. O., Mwangi, D. M., Githinji, M., Dubois, T., Ekesi, S., van Loon, J. J. A., & Dicke, M. (2021). Black Soldier Fly Larval Meal in Feed Enhances Growth Performance, Carcass Yield and Meat Quality of Finishing Pigs. *Journal of Insects as Food and Feed*, *7*(4), 433–447. https://doi.org/10.3920/JIFF2020.0072

Choi, Y., Choi, J., Kim, J., Kim, M., Kim, W., Park, K., & Bae, S. (2009). *Potential Usage of Food Waste as a Natural Fertilizer after Digestion by Hermetia illucens (Diptera: Stratiomyidae)*. *19*(1), 171–174.

Chowdhury, F. R., Nur, Z., Hassan, N., Seidlein, L., & Dunachie, S. (2017). Pandemics, pathogenicity and changing molecular epidemiology of cholera in the era of global warming. *Annals of Clinical Microbiology and Antimicrobials*, *16*(1), 1–6. https://doi.org/10.1186/s12941-017-0185-1

Cullere, M., Tasoniero, G., Giaccone, V., Acuti, G., Marangon, A., & Dalle Zotte, A. (2018). Black soldier fly as dietary protein source for broiler quails: Meat proximate composition, fatty acid and amino acid profile, oxidative status and sensory traits. *Animal*, *12*(3), 640–647. https://doi.org/10.1017/S1751731117001860

De Smet, J., Wynants, E., Cos, P., & Van Campenhout, L. (2018). Microbial Community Dynamics during Rearing of Black. *Applied and Environmental Microbiology*, *84*(9), 1–17.

de Sousa Antunes, L. F., de Souza, R. G., de Sousa Vaz, A. F., dos Santos Ferreira, T., & Correia, M. E. F. (2021). Evaluation of millicomposts from different vegetable residues and production systems in the lettuce seedling development. *Organic Agriculture*, *11*(3), 367–378. https://doi.org/10.1007/s13165-020-00342-y

Diener, S., Zurbrügg, C., & Tockner, K. (2009). Conversion of organic material by black soldier fly larvae: Establishing optimal feeding rates. *Waste Manag. Res.* 27, 603–610. https://doi.org/10.1177/0734242X09103838

Diener, S., Zurbrügg, C., Gutiérrez, F. R., Nguyen, D. H., Morel, A., Koottatep, T., & Tockner, K. (2011). Proceedings of the executive summary WasteSafe 2011: 2nd International Conference on Solid Waste Management in Developing Countries, Khulna, Bangladesh February 13-15, 2011. *Proceedings of the WasteSafe 2011 – 2nd International Conference on Solid Waste Management in the Developing Countries*, *52*(February), 275.

Diener, S., Zurbrügg, C., & Tockner, K. (2015). Bioaccumulation of heavy metals in the black soldier fly, Hermetia illucens and effects on its life cycle. *Journal of Insects as Food and Feed*, *1*(4), 261–270. https://doi.org/10.3920/JIFF2015.0030

Doelle, K., Oliveira, F. R., & Smith, R. P. (2016). Gold nanoparticle uptake and distribution in the digestive tract of Hermetia illucens stratiomyidae: Diptera (L.1758) based on transmission electron microscopy. *Annual Research and Review in Biology*, *9*(3), 1–8. https://doi.org/10.9734/ARRB/2016/23907

Dortmans, B., Diener, S., Verstappen, B., & Zurbrügg, C. (2017). Black Soldier Fly Biowaste Processing. In *Black soldier fly biowaste processing. A step-by step guide. Swiss Agency for Development and Cooperation (SDC) and the Swiss State Secretariat for Economic Affairs (SECO).*

Eickhout, B., Bouwman, A. F., & van Zeijts, H. (2006). The role of nitrogen in world food production and environmental sustainability. *Agriculture, Ecosystems and Environment*, *116*(1–2), 4–14. https://doi.org/10.1016/j.agee.2006.03.009

Ermolaev, E., Lalander, C., & Vinnerås, B. (2019). Greenhouse gas emissions from small-scale fly larvae composting with Hermetia illucens. *Waste Management*, *96*, 65–74. https://doi.org/10.1016/j.wasman.2019.07.011

Feed, A., & Food, H. (2017). *Review of Black Soldier Fly (Hermetia illucens) as Animal Feed and Human Food*. https://doi.org/10.3390/foods6100091

Felse, P. A., & Panda, T. (1999). Regulation and cloning of microbial chitinase genes. *Appl. Microbiol. Biotechnol. 51*, 141–151. https://doi.org/10.1007/s002530051374

Ferronato, N., & Torretta, V. (2019). Waste mismanagement in developing countries: A review of global issues. *International Journal of Environmental Research and Public Health*, *16*(6). https://doi.org/10.3390/ijerph16061060

Fly, S., & Hermetia, L. (n.d.). *crossm Microbial Community Dynamics during Rearing of Black. ii.*

FAO. (2013). Food wastage footprint. In *Fao*. www.fao.org/nr/sustainability/food-loss-and-waste/en/

Furman, D. P., Young, R. D. and Catts, E. P. (1959). Hermetiaillucens (Linnaeus) as a Factor in the Natural Control of Musca domestica Linnaeus. *Journal of Economic Entomology*, *52*, 917–921. https://doi.org/10.1093/jee/52.5.917

Gasco, L., Acuti, G., Bani, P., Dalle Zotte, A., Danieli, P. P., De Angelis, A., Fortina, R., Marino, R., Parisi, G., Piccolo, G., Pinotti, L., Prandini, A., Schiavone, A., Terova, G., Tulli, F., & Roncarati, A. (2020). Insect and fish by-products as sustainable alternatives to conventional animal proteins in animal nutrition. *Italian Journal of Animal Science*, *19*(1), 360–372. https://doi.org/10.1080/1828051X.2020.1743209

Gold, M., Cassar, C. M., Zurbrügg, C., Kreuzer, M., Boulos, S., Diener, S., & Mathys, A. (2020). Biowaste treatment with black soldier fly larvae: Increasing performance through the formulation of biowastes based on protein and carbohydrates. *Waste Management*, *102*, 319–329. https://doi.org/10.1016/j.wasman.2019.10.036

Green, T. R., & Popa, R. (2012). Enhanced ammonia content in compost leachate processed by black soldier fly larvae. *Applied Biochemistry and Biotechnology*, *166*(6), 1381–1387. https://doi.org/10.1007/s12010-011-9530-6

Hahn, T., Tafi, E., Paul, A., Salvia, R., Falabella, P., & Zibek, S. (2020). Current state of chitin purification and chitosan production from insects. *J. Chem. Technol. Biotechnol. 95*, 2775–2795. https://doi.org/10.1002/jctb.6533

Hahn, T., Tafi, E., von Seggern, N., Falabella, P., Salvia, R., Thomä, J., Febel, E., Fijalkowska, M., Schmitt, E., Stegbauer, L., & Zibek, S. (2022). Purification of Chitin from Pupal Exuviae of the Black Soldier Fly. *Waste and Biomass Valorization*, *13*(4), 1993–2008. https://doi.org/10.1007/s12649-021-01645-1

Hasnol, S., Kiatkittipong, K., Kiatkittipong, W., Wong, C. Y., Khe, C. S., Lam, M. K., Show, P. L., Oh, W. Da, Chew, T. L., & Lim, J. W. (2020). A review on insights for green production of unconventional protein and energy sources derived from the larval biomass of black soldier fly. *Processes*, *8*(5). https://doi.org/10.3390/PR8050523

Ibadurrohman, K., Gusniani, I., Hartono, D. M., & Suwartha, N. (2020). The potential analysis of food waste management using bioconversion of the organic waste by the black soldier fly (Hermetia illucens) larvae in the cafeteria of the faculty of engineering, universitas Indonesia. *Evergreen*, *7*(1), 61–66. https://doi.org/10.5109/2740946

Isibika, A., Vinnerås, B., Kibazohi, O., Zurbrügg, C., & Lalander, C. (2019). Pre-treatment of banana peel to improve composting by black soldier fly (Hermetia illucens (L.), Diptera: Stratiomyidae) larvae. *Waste Management*, *100*, 151–160. https://doi.org/10.1016/j.wasman.2019.09.017

Jenkins, W., Tucker, M. E., & Grim, J. (2016). Routledge handbook of religion and ecology. In *Routledge Handbook of Religion and Ecology*. https://doi.org/10.4324/9781315764788

Jucker, C., Lupi, D., Moore, C. D., Leonardi, M. G., & Savoldelli, S. (n.d.). *Nutrient Recapture from Insect Farm Waste: Bioconversion with Hermetia illucens (L .) (Diptera: Stratiomyidae). 2006.*

Kawasaki, K., Kawasaki, T., Hirayasu, H., Matsumoto, Y., & Fujitani, Y. (2020). Evaluation of fertilizer value of residues obtained after processing household organic waste with black soldier fly larvae (Hermetia illucens). *Sustainability (Switzerland)*, *12*(12). https://doi.org/10.3390/SU12124920

Kim, C. H., Ryu, J., Lee, J., Ko, K., Lee, J. Y., Park, K. Y., & Chung, H. (2021). Use of black soldier fly larvae for food waste treatment and energy production in asian countries: A review. In *Processes* (Vol. 9, Issue 1, pp. 1–17). MDPI AG. https://doi.org/10.3390/pr9010161

Kim, W., Bae, S., Park, H., Park, K., Lee, S., Choi, Y., Han, S., & Koh, Y. (2010). *[Note] The Larval Age and Mouth Morphology of the Black Soldier Fly, Hermetia illucens (Diptera: Stratiomyidae)*. *21*(2), 185–187.

Kinasih, I., Suryani, Y., Paujiah, E., Ulfa, R. A., Afiyati, S., Adawiyah, Y. R., &Putra, R. E. (2020). Performance of Black Soldier Fly, Hermetia illucens, Larvae during valorization of organic wastes with changing quality. *IOP Conference Series: Earth and Environmental Science*, *593*(1). https://doi.org/10.1088/1755-1315/593/1/012040

Komakech, A. J., Sundberg, C., Jönsson, H., & Vinnerås, B. (2015). Life cycle assessment of biodegradable waste treatment systems for sub-Saharan African cities. *Resources, Conservation and Recycling*, *99*, 100–110. https://doi.org/10.1016/j.resconrec.2015.03.006

Kroeckel, S., Harjes, A. G. E., Roth, I., Katz, H., Wuertz, S., Susenbeth, A., & Schulz, C. (2012). When a turbot catches a fly: Evaluation of a pre-pupae meal of the Black Soldier Fly (Hermetia illucens) as fish meal substitute – Growth performance and chitin degradation in juvenile turbot (Psetta maxima). Aquaculture 364–365, 345–352. https://doi.org/10.1016/j.aquaculture.2012.08.041

Lalander, C., Diener, S., Zurbrügg, C., & Vinnerås, B. (2019). Effects of feedstock on larval development and process efficiency in waste treatment with black soldier fly (Hermetia illucens). *Journal of Cleaner Production*, *208*, 211–219. https://doi.org/10.1016/j.jclepro.2018.10.017

Lalander, C. H., Fidjeland, J., Diener, S., Eriksson, S., &Vinnerås, B. (2015). High waste-to-biomass conversion and efficient Salmonella spp. reduction using black soldier fly for waste recycling. *Agronomy for Sustainable Development*, *35*(1), 261–271. https://doi.org/10.1007/s13593-014-0235-4

Lalander, C., Senecal, J., Gros Calvo, M., Ahrens, L., Josefsson, S., Wiberg, K., & Vinnerås, B. (2016). Fate of pharmaceuticals and pesticides in fly larvae composting. *Science of the Total Environment*, *565*, 279–286. https://doi.org/10.1016/j.scitotenv.2016.04.147

Lee, J., Kim, Y., Park, Y., Yang, Y., Jung, B., & Lee, B. (2018). *Black soldier fly (Hermetia illucens) larvae enhances immune activities and increases survivability of broiler chicks against experimental infection of Salmonella Gallinarum*. 736–740. https://doi.org/10.1292/jvms.17-0236

Liu, T., Awasthi, S. K., Qin, S., Liu, H., Awasthi, M. K., Zhou, Y., Jiao, M., Pandey, A., Varjani, S., & Zhang, Z. (2021). Conversion food waste and sawdust into compost employing black soldier fly larvae (diptera: Stratiomyidae) under the optimized condition. *Chemosphere*, *272*, 129931.https://doi.org/10.1016/j.chemosphere.2021.129931

Liu, T., Kumar, A. M., Chen, H., Duan, Y., & Awasthi, S. K. (2019). *Performance of black soldier fl y larvae (Diptera: Stratiomyidae) for manure composting and production of cleaner compost. 251* (September). https://doi.org/10.1016/j.jenvman.2019.109593

Lopes, I. G., Yong, J. W., & Lalander, C. (2022). Frass derived from black soldier fly larvae treatment of biodegradable wastes. A critical review and future perspectives. In *Waste Management* (Vol. 142, pp. 65–76). Elsevier Ltd. https://doi.org/10.1016/j.wasman.2022.02.007

Ma, J., Lei, Y., Rehman, K. U., Yu, Z., Zhang, J., Li, W., Li, Q., Tomberlin, J. K., & Zheng, L. (2018). Dynamic Effects of Initial pH of Substrate on Biological Growth and Metamorphosis of Black Soldier Fly (Diptera: Stratiomyidae). *Environmental Entomology*, *47*(1), 159–165. https://doi.org/10.1093/ee/nvx186

Manjula, K. & Podile, A. R. (2001). Chitin-supplemented formulations improve biocontrol and plant growth promoting efficiency of *Bacillus subtilis* AF 1. *Canadian Journal of Microbiology*. **47**(7): 618–625. https://doi.org/10.1139/w01-057

Manzano-Agugliaro, F., Sanchez-Muros, M. J., Barroso, F. G., Martínez-Sánchez, A., Rojo, S., & Pérez-Bañón, C. (2012). Insects for biodiesel production. *Renewable and Sustainable Energy Reviews*, *16*(6), 3744–3753. https://doi.org/10.1016/j.rser.2012.03.017

Marco, M. De, Martínez, S., Hernandez, F., Madrid, J., Gai, F., Rotolo, L., Belforti, M., Bergero, D., Katz, H., Dabbou, S., Kovitvadhi, A., & Zoccarato, I. (2015). Nutritional value of two insect larval meals (Tenebrio molitor and Hermetia illucens) for broiler chickens: Apparent nutrient digestibility, apparent ileal amino acid digestibility and apparent metabolizable energy. *Animal Feed Science and Technology*, *209*, 211–218. https://doi.org/10.1016/j.anifeedsci.2015.08.006

Meneguz, M., Gasco, L., & Tomberlin, J. K. (2018). Impact of pH and feeding system on BSF larval development. *PLoS ONE*, *13*(8), 1–15.

Mishra, V. K., & Goswami, R. (2018). A review of production, properties and advantages of biodiesel. *Biofuels*, *9*(2), 273–289. https://doi.org/10.1080/17597269.2017.1336350

Närvänen, E., Mesiranta, N., Mattila, M., & Heikkinen, A. (2019). Food waste management: Solving the wicked problem. In *Food Waste Management: Solving the Wicked Problem*. Springer International Publishing. https://doi.org/10.1007/978-3-030-20561-4

Nery, J., Gasco, L., Dabbou, S., & Schiavone, A. (2018). *Protein composition and digestibility of black soldier fly larvae in broiler chickens revisited according to the recent nitrogen-protein conversion ratio*. *4*(3), 171–177. https://doi.org/10.3920/JIFF2018.0006

Newton, G. L., Lacy, R. C., Kozánek, M., & Helena, C. (2015). *The use of fly larvae for organic waste treatment ˇ ic. 35*, 68–80. https://doi.org/10.1016/j.wasman.2014.09.026

Nguyen, T. T. X., Tomberlin, J. K., & Vanlaerhoven, S. (2015). Ability of Black Soldier Fly (Diptera: Stratiomyidae) Larvae to Recycle Food Waste. *Environmental Entomology*, *44*(2), 406–410. https://doi.org/10.1093/ee/nvv002

Oliveira, F., Doelle, K., List, R., & Reilly, J. R. O. (2015). *Assessment of Diptera: Stratiomyidae, genus Hermetia illucens (L ., 1758) using electron microscopy*. *3*(5), 147–152.

Onsongo, V. O., Osuga, I. M., Gachuiri, C. K., Wachira, A. M., Miano, D. M., Tanga, C. M., Ekesi, S., Nakimbugwe, D., & Fiaboe, K. K. M. (2018). *Insects for Income Generation Through Animal Feed: Effect of Dietary Replacement of Soybean and Fish Meal With Black Soldier Fly Meal on Broiler Growth and Economic Performance*. *111*(May), 1966–1973. https://doi.org/10.1093/jee/toy118

Osimani, A., Milanović, V., Cardinali, F., Garofalo, C., Clementi, F., Pasquini, M., Riolo, P., Ruschioni, S., Isidoro, N., Loreto, N., Franciosi, E., Tuohy, K., Petruzzelli, A., Foglini, M., Gabucci, C., Tonucci, F., & Aquilanti, L. (2018). The bacterial biota of laboratory-reared edible mealworms (Tenebrio molitor L.): From feed to frass. *International Journal of Food Microbiology*, *272*(March), 49–60. https://doi.org/10.1016/j.ijfoodmicro.2018.03.001

Park, H. H. (2016). Black Soldier Fly Larvae Manual. *Student Showcase*, *14*, 1–13. http://scholarworks.umass.edu/sustainableumass_studentshowcase/14

Periathamby, A., Hamid, F. S., & Khidzir, K. (2009). Evolution of solid waste management in Malaysia: Impacts and implications of the solid waste bill, 2007. *Journal of Material Cycles and Waste Management*, *11*(2), 96–103. https://doi.org/10.1007/s10163-008-0231-3

Pintowantoro, S., Setiyorini, Y., Aljauhari, A. M., Abdul, F., & Nurdiansah, H. (2021). Black soldier fly biowaste treatment and its recycle waste to produce chitosan. *IOP Conference Series: Earth and Environmental Science*, *649*(1). https://doi.org/10.1088/1755-1315/649/1/012004

Poveda, J., Jiménez-gómez, A., Saati-santamaría, Z., Usategui-martín, R., & Rivas, R. (2019). Mealworm frass as a potential biofertilizer and abiotic stress tolerance- inductor in plants. *Applied Soil Ecology*, *142*(March), 110–122. https://doi.org/10.1016/j.apsoil.2019.04.016

Purkayastha, D., & Sarkar, S. (2022). Sustainable waste management using black soldier fly larva: a review. *International Journal of Environmental Science and Technology*, *19*(12), 12701–12726. https://doi.org/10.1007/s13762-021-03524-7

Purkayastha, D., Sarkar, S., Roy, P., & Kazmi, A. A. (2017). Isolation and morphological study of ecologically-important insect "Hermetia illucens" collected from Roorkee compost plant. *Pollution*, *3*(3), 453–459. https://doi.org/10.7508/pj.2017.03

Gujarathi, G. R., & Pejaver, M. K. (2013). Occurrence of Black Soldier Fly Hermetia illucens (Diptera: Stratiomyidae) in Biocompost. *Research Journal of Recent Sciences*, *2*(4), 65–66.

Raksasat, R., Kiatkittipong, K., Kiatkittipong, W., Wong, C. Y., Lam, M. K., Ho, Y. C., Oh, W. Da, Suryawan, I. W. K., & Lim, J. W. (2021). Blended sewage sludge–palm kernel expeller to enhance the palatability of black soldier fly larvae for biodiesel production. *Processes*, *9*(2), 1–13. https://doi.org/10.3390/pr9020297

Rehman, K. ur, Rehman, A., Cai, M., Zheng, L., Xiao, X., Somroo, A. A., Wang, H., Li, W., Yu, Z., & Zhang, J. (2017). Conversion of mixtures of dairy manure and soybean curd residue by black soldier fly larvae (Hermetia illucens L.). *Journal of Cleaner Production*, *154*, 366–373. https://doi.org/10.1016/j.jclepro.2017.04.019

Rezania, S., Oryani, B., Park, J., Hashemi, B., Yadav, K. K., Kwon, E. E., Hur, J., & Cho, J. (2019). Review on transesterification of non-edible sources for biodiesel production with a focus on economic aspects, fuel properties and by-product applications. *Energy Conversion and Management*, *201*(July), 112155. https://doi.org/10.1016/j.enconman.2019.112155

Robertson, G. P., & Vitousek, P. M. (2009). Nitrogen in agriculture: Balancing the cost of an essential resource. *Annual Review of Environment and Resources*, *34*, 97–125. https://doi.org/10.1146/annurev.environ.032108.105046

Robertson, J. (2004). 11119. *The American Mathematical Monthly*, *111*(10), 915. https://doi.org/10.2307/4145104

Ruhnke, I., Normant, C., Campbell, D. L. M., Iqbal, Z., Lee, C., Hinch, G. N., & Roberts, J. (2018). Impact of on-range choice feeding with black soldier fl y larvae (Hermetia illucens) on fl ock performance, egg quality, and range use of free-range laying hens. *Animal Nutrition*, *4*(4), 452–460. https://doi.org/10.1016/j.aninu.2018.03.005

Salomone, R., Saija, G., Mondello, G., Giannetto, A., Fasulo, S., & Savastano, D. (2017). Environmental impact of food waste bioconversion by insects: Application of Life Cycle Assessment to process using Hermetia illucens. *Journal of Cleaner Production, 140*, 890–905. https://doi.org/10.1016/j.jclepro.2016.06.154

Schiavone, A., De Marco, M., Martínez, S., Dabbou, S., Renna, M., Madrid, J., Hernandez, F., Rotolo, L., Costa, P., Gai, F., & Gasco, L. (2017). Nutritional value of a partially defatted and a highly defatted black soldier fly larvae (Hermetia illucens L.) meal for broiler chickens: Apparent nutrient digestibility, apparent metabolizable energy and apparent ileal amino acid digestibility. *Journal of Animal Science and Biotechnology*, *8*(1), 1–9. https://doi.org/10.1186/s40104-017-0181-5

Schmitt, E., & de Vries, W. (2020). Potential benefits of using Hermetia illucens frass as a soil amendment on food production and for environmental impact reduction. *Current Opinion in Green and Sustainable Chemistry*, *25*, 100335. https://doi.org/10.1016/j.cogsc.2020.03.005

Sciences, B., Extension, T. C., Dt, D., Uk, U. B., Science, D., Science, V., Fish, H., & Experimental, C. (2007). *Fly Prepupae as a Feedstuff for Rainbow Trout, Oncorhynchus mykiss*. *38*(1), 59–67.

Secci, G., Bovera, F., Nizza, S., Baronti, N., Gasco, L., Conte, G., Serra, A., Bonelli, A., & Parisi, G. (2018). Quality of eggs from Lohmann Brown Classic laying hens fed black soldier fl y meal as substitute for soya bean. *Animal, The International Journal of Animal Biosciences*, *12*(10), 2191–2197. https://doi.org/10.1017/S1751731117003603

Shahid, E. M., & Jamal, Y. (2011). Production of biodiesel: A technical review. *Renewable and Sustainable Energy Reviews*, *15*(9), 4732–4745. https://doi.org/10.1016/j.rser.2011.07.079

Sheppard, D. C., Tomberlin, J. K., Joyce, J. A., Kiser, B. C., & Sumner, S. M. (2002). Rearing methods for the black soldier fly (diptera: Stratiomyidae). *Journal of Medical Entomology*, *39*(4), 695–698. https://doi.org/10.1603/0022-2585-39.4.695

Shumo, M., Khamis, F. M., Tanga, C. M., Fiaboe, K. K. M., Subramanian, S., Ekesi, S., Huis, A. Van, & Borgemeister, C. (2019). Influence of temperature on selected life-history traits of black soldier fly (Hermetia illucens) reared on two common urban organic waste streams in Kenya. *Animals*, *9*(3). https://doi.org/10.3390/ani9030079

Shumo, M., Osuga, I. M., Khamis, F. M., Tanga, C. M., Fiaboe, K. K. M., Subramanian, S., Ekesi, S., van Huis, A., & Borgemeister, C. (2019). The nutritive value of black soldier fly larvae reared on common organic waste streams in Kenya. *Scientific Reports*, *9*(1), 1–13. https://doi.org/10.1038/s41598-019-46603-z

Siddiqui, S. A., Ristow, B., Rahayu, T., Putra, N. S., Widya Yuwono, N., Nisa', K., Mategeko, B., Smetana, S., Saki, M., Nawaz, A., & Nagdalian, A. (2022). Black soldier fly larvae (BSFL) and their affinity for organic waste processing. *Waste Management*, *140* (December 2021), 1–13. https://doi.org/10.1016/j.wasman.2021.12.044

Sideris, V. F., & Tsagkarakis, A. E. (2017). Immature Development Time of <i>Hermetia illucens</i> L. in Different Varieties of Feed. *Advances in Entomology*, *05*(03), 109–114. https://doi.org/10.4236/ae.2017.53011

Siva Raman, S., Stringer, L. C., Bruce, N. C., & Chong, C. S. (2022). Opportunities, challenges and solutions for black soldier fly larvae-based animal feed production. *Journal of Cleaner Production*, *373*(August), 133802. https://doi.org/10.1016/j.jclepro.2022.133802

Smetana, S., Schmitt, E., &Mathys, A. (2019). Resources, Conservation & Recycling Sustainable use of Hermetia illucens insect biomass for feed and food: Attributional and consequential life cycle assessment. *Resources, Conservation & Recycling*, *144*(February), 285–296. https://doi.org/10.1016/j.resconrec.2019.01.042

Song, S., Wei, A., Ee, L., Koon, J., Tan, N., Chin, J., Chiam, Z., Arora, S., Ngai, W., Tiang, H., & Tan, W. (2021). Upcycling food waste using black soldier fl y larvae: Effects of further composting on frass quality, fertilising effect and its global warming potential. *Journal of Cleaner Production*, *288*, 125664.https://doi.org/10.1016/j.jclepro.2020.125664

Su, C., Nguyen, H. C., Bui, T. L., & Huang, D. (2019). Enzyme-assisted extraction of insect fat for biodiesel production. *Journal of Cleaner Production*, *223*, 436–444. https://doi.org/10.1016/j.jclepro.2019.03.150

Surendra, K. C., Olivier, R., Tomberlin, J. K., Jha, R., & Kumar, S. (2016). Bioconversion of organic wastes into biodiesel and animal feed via insect farming. *Renewable Energy*, *98*, 197–202. https://doi.org/10.1016/j.renene.2016.03.022

Surendra, K. C., Tomberlin, J. K., van Huis, A., Cammack, J. A., Heckmann, L. H. L., & Khanal, S. K. (2020). Rethinking organic wastes bioconversion: Evaluating the potential of the black soldier fly (Hermetia illucens (L.))(Diptera: Stratiomyidae) (BSF). *Waste Management*, *117*, 58–80. https://doi.org/10.1016/j.wasman.2020.07.050

Teresa L. Alleman, R. L. M. (2016). Biodiesel handling and use guide. In *Biodiesel Handling and Use Guide (Fifth Edition)* (Issue November).

Tomberlin, J. K., Adler, P. H., & Myers, H. M. (2009). *Development of the Black Soldier Fly (Diptera: Stratiomyidae) in. Hogsette 1992*, 930–934.

Tomberlin, J. K., Sheppard, D. C., & Joyce, J. A. (2002). Selected life-history traits of black soldier flies (Diptera: Stratiomyidae) reared on three artificial diets. *Annals of the Entomological Society of America*, *95*(3), 379–386. https://doi.org/10.1603/0013-8746(2002)095[0379:SLHTOB]2.0.CO;2

Van Huis, A., Van Itterbeeck, J., Klunder, H., et al. (2013). Edible Insects: Future Prospects for Food and Feed Security. In *Food and Agriculture Organization of the United Nations* (Vol. 97, Issue 18). https://library.wur.nl/WebQuery/wurpubs/fulltext/258042

van Huis, A., Oonincx, D. G. A. B., Rojo, S., & Tomberlin, J. K. (2020). Insects as feed: house fly or black soldier fly? *Journal of Insects as Food and Feed*, *6*(3), 221–229. https://doi.org/10.3920/JIFF2020.X003

Wong, C. Y., Rosli, S. S., Uemura, Y., Ho, Y. C., Leejeerajumnean, A., Kiatkittipong, W., Cheng, C. K., Lam, M. K., & Lim, J. W. (2019). Potential protein and biodiesel sources from black soldier fly larvae: Insights of larval harvesting instar and fermented feeding medium. *Energies*, *12*(8). https://doi.org/10.3390/en12081570

Wynants, E., Frooninckx, L., Crauwels, S., Verreth, C., De Smet, J., Sandrock, C., Wohlfahrt, J., Van Schelt, J., Depraetere, S., Lievens, B., Van Miert, S., Claes, J., & Van Campenhout, L. (2019). Assessing the Microbiota of Black Soldier Fly Larvae (Hermetia illucens) Reared on Organic Waste Streams on Four Different Locations at Laboratory and Large Scale. *Microbial Ecology*, *77*(4), 913–930. https://doi.org/10.1007/s00248-018-1286-x

Zahn, N. H. (2017). The effects of insect frass created by Hermetia illucens on spring onion growth and soil fertility Author: Nils Holger Z AHN Supervisor: Dr. Richard Q UILLIAM The effects of insect frass created by Hermetia illucens on spring onion growth and soil fert.

Xiao, X., Mazza, L., Yu, Y., Cai, M., Zheng, L., Tomberlin, J. K., Yu, J., van Huis, A., Yu, Z., Fasulo, S., & Zhang, J. (2018). Efficient co-conversion process of chicken manure into protein feed and organic fertilizer by Hermetia illucens L. (Diptera: Stratiomyidae) larvae and functional bacteria. *Journal of Environmental Management*, *217*, 668–676. https://doi.org/10.1016/j.jenvman.2018.03.122

Yeul, V. S., & Rayalu, S. S. (2013). *Unprecedented Chitin and Chitosan: A Chemical Overview*. 606–614. https://doi.org/10.1007/s10924-012-0458-x

Yuvaraj, A., Thangaraj, R., Karmegam, N., Ravindran, B., Chang, S. W., Awasthi, M. K., & Kannan, S. (2021). Activation of biochar through exoenzymes prompted by earthworms for vermibiochar production: A viable resource recovery option for heavy metal contaminated soils and water. *Chemosphere*, *278*, 130458. https://doi.org/10.1016/j.chemosphere.2021.130458

Zhang, J., Huang, L., He, J., Tomberlin, J. K., Li, J., Lei, C., Sun, M., Liu, Z., & Yu, Z. (2010). An artificial light source influences mating and oviposition of black soldier flies, Hermetia illucens. *Journal of Insect Science*, *10*(202), 1–7. https://doi.org/10.1673/031.010.20201

Zheng, L., Li, Q., Zhang, J., & Yu, Z. (2012). Double the biodiesel yield: Rearing black soldier fly larvae, Hermetia illucens, on solid residual fraction of restaurant waste after grease extraction for biodiesel production. *Renewable Energy*, *41*, 75–79. https://doi.org/10.1016/j.renene.2011.10.004

Zurbrügg, C., Dortmans, B., Fadhila, A., Vertsappen, B., & Diener, S. (2018). *FROM PILOT TO FULL SCALE OPERATION OF A WASTE-TO-PROTEIN TREATMENT FACILITY*. *01*, 18–22.

8 Scientific Insights of Biosorption

A Sustainable Technique for Removal of Heavy Metals

Bhaswati Devi, Netra Prova Baruah, and Arundhuti Devi

8.1 INTRODUCTION

Heavy metals have been identified to be present in the category of the most critical contaminants in aqueous resources because of their non-biodegradable, carcinogenic, interminable nature and also cause adverse effects on health and economy (Ahamed et al., 2021; Aragay et al., 2011). They are also classified as toxic radionuclides or radioactive isotopes which not only contaminate surface water (infecting lakes, pools and seas) but also pollute groundwater, eventually advancing towards being a serious threat for all life forms on earth. According to biochemistry, heavy metals have been defined as metal elements depending on their behavior of accepting electron pairs, i.e., Lewis acids. Their specific gravity is more than 5.00 and atomic weights range from 62.50–200.60. They are non-degradable in nature and effortlessly store up in plants, aquatic life forms and infiltrate the human body by means of drinking water, through food chains or by skin contact. A few heavy metals act as micronutrients in the human body and when present in excess, they reconstruct the functions of essential enzymes thereby generating a multiplicity of illnesses and even fatality. For example, Cd entering into the bodies of human beings cause hypertension, lung cancer, and renal dysfunction. Similarly, excess of Zn intake leads to anemia, appetite loss and diarrhea whereas Hg gives rise to Minamata disease (Panqing et al., 2023). Thus, it is indispensable to effectively treat and process the effluents present in wastewater comprising heavy metals prior to their release into the environment (Haq et al., 2017; Haq et al., 2022). The heavy metal ions particularly Cd (II), Cr (III), Cu (II), Hg (II), Pb (II), Zn (II), etc. emanating from industries involved in battery manufacturing, coal, fertilizers, metal plating, metal refining, mining operations, pesticides, steel tempering, tanneries, textiles, etc. are the chief pollutants present in their wastewater effluents and their percentage is increasing with rapid industrialization (Thakur et al.,2022;Singh et al., 2020; Haq et al., 2016a b; 2017, 2018, 2020, 2022a b c; Haq and Kalamdhad, 2021; Haq and Kalamdhad, 2022, 2023; Haq et al., 2024). Their presence poses considerable risks to human health such as brain, kidney, liver, and

DOI: 10.1201/9781003379386-8

lung damages when aggregated in internal organs exceeding a standard confined limit. Consequently, there is an urgent necessity for decimating these harmful and environmentally hazardous metal ions before they are slipped into the aqueous bodies. Some standard techniques or approaches used for the minimization of heavy metal ions and ensure restoring the environment are adsorption, chemical coagulation, chemical precipitation, electrodeposition, flocculation, ion exchange, nanofiltration, reverse osmosis, solvent extraction, and ultrafiltration (Thakur et al., 2022; Bolisetty et al., 2019). The remediation of these wastewaters originating from anthropogenic sources has turned out to be a matter of solicitude to the health of citizens as billions of people consume these contaminated waters either directly or indirectly for drinking purpose. To achieve the objective of making clean water available to the public for drinking, innovative bio-based technologies have been developed by making use of authentic natural products which are highly recommended to be embraced in applicable cases. Adsorption has been ascertained to be a brilliant approach for the treatment of industrial effluents, extending significant assets like availability, efficiency, feasibility, reasonably priced, and profitability. Biosorption of the toxic heavy metals has proven to be very promising in the effluent treatment and contaminant removal process from aqueous media (Demirbas, 2008). A few common mechanisms responsible for the process of biosorption to take place effectively are adsorption-complexation on pores and surface, chemisorption, complexation, condensation onto the biological surface, ion- exchange, microprecipitation, and surface adsorption (Inamuddin et al., 2021). The process of biosorption rests on particular parameters like structure of cell wall embodied by the biomass, concentration of the metal ions, contact time of pollutants with biomass, pH of aqueous medium, and the types of metal ions. This chapter emphasizes the discussion of biosorption, the types of novel biosorbents developed, advantages and disadvantages of this approach, its mechanism, effect of pH and temperature parameters, the different types of adsorption isotherms and its applications in practical field. Sorption is a method which is conveniently implemented due to its technical feasibility over an extensive range of pH and temperature, consumption of reagent is much less, generation of less amount of sludge and simplicity in fabrication as well as usage (Singh et al., 2020). It is essential to extend scientific exploration on systems involving real wastewater treatment to substantiate the possibility of utilizing the class of most promising and widely accepted materials to ensure environmental sustainability as studies conducted on existing systems of pollutants are very limited.

8.2 HEAVY METALS

One of the most unwanted yet prevalent environmental pollutants is the class of heavy metals which are leading to serious public health complications nowadays. The heavy metals released into the environment find their ways from the discharge of wastewater effluents, sewage sludge, industrial and agricultural wastes, mining, water pipe corrosion, dumping, electrolysis, electroplating, electro-osmosis, fuel and energy producing, metal surface treating, atomic energy installations, etc. (Sayago & Ballesteros, 2023). All the naturally occurring elements possessing atomic number between 63.5 and 200.6 and elemental density greater than 5 g cm^3 are classified as "heavy metals." Some of the top priority heavy metals with high degree of toxicity includes *arsenic, cadmium, chromium, lead and mercury* (Hegazy, 2023). Bioaccumulation of heavy metals in the internal organs may pose severe health hazards, notably brain, liver, kidney, and lung damage. In an attempt to modulate a harmony between the human welfare and sustainable biospheres, some standards and regulations are being framed across the world to set a permissible range for the safe discharge of heavy metals into the ecosystem. Due to the non-biodegradable and persistent nature, heavy metals are considered to be the most robust and enduring environmental pollutants. Hence, getting rid of heavy metals from the outflows prior to the emanation into the ecosystem has been a great challenge. These heavy metals enter the bodies of humans and other living beings through exposure routes like oral ingestion and skin contact, and accordingly pose a critical menace to human health. Heavy metals such as copper, iron, manganese, and zinc are nutritionally beneficial for a healthy life but they have a tendency to assemble and constitute complexes, particularly with ligands having biological origin consisting of nitrogen, oxygen, and sulfur. This causes alteration in the protein's molecular structure, hydrogen bond breaking, and even occurrence of enzyme inhibition. They remain in ecological systems and expose predators of top-level to extreme levels of heavy metal contamination. This is the purpose why safe limits and permissible maximum levels of contaminant have been formulated for drinking water by many organizations worldwide. Table 8.1 outlines the benchmark and guidelines for the presence of heavy metals in drinking water suggested by the Environmental Protection Agency (EPA) and the World Health Organization

TABLE 8.1
The standard guidelines related to the presence of Heavy Metals suggested by the EPA and WHO on the basis of scientific studies and toxicity data

Heavy Metal	EPA limit (mg/L)	WHO limit (mg/L)	Toxic effect
As	0.010	0.010	Depression of bone marrow, bronchitis, hepatomegaly, dermatitis, hemolysis.
Cd	0.005	0.003	Bronchitis, bone marrow disorder, cancer, gastrointestinal disorder, renal dysfunction, hypertension, weight loss, Itai-itai, lung insufficiency.
Cu	1.3	2	Diarrhoea, dizziness, neurotoxicity.
Hg	0.002	0.001	Corrosive to eyes, skin and muscles; dermatitis, protoplasm poisoning, kidney damage, nervous system damage.
Ni	0.04	0.07	Conjunctivitis, eczema, the mucous membranes of the upper airways suffers inflammation, contact allergy dermatoses.
Pb	0.015	0.010	Anaemia, anorexia, brain damage, gastrointestinal damage, kidney damage, liver, loss of appetite, mental retardation in children.
Zn	5	3	Metal fume fever (short term), gastrointestinal distress, appetite loss, anaemia.

(WHO) (Zareh et al., 2022; Al-Badaani, 2023; Aragay et al., 2011; Hołyńska-Iwan et al., 2023).

However, a plethora of technologies such as electroplating, ion exchange, precipitation, membrane filtration, etc. have been developed so far for the decontamination of heavy metals. Although these techniques are quite efficient for heavy metal decontamination, they too have some sort of limitations such as need for high energy and chemicals, hazardous sludge formation, high costs when used on a wide scale, use of sophisticated instruments, low efficiency for metal concentration below 100 mgL^{-1}, generation of toxic by-products, need for trained manpower, not accessible for common people, etc. (Haq et al., 2022). As the current wastewater treatment situation in developing nations is increasingly dire, it requires further attention by promoting and inventing affordable as well as environmentally friendly technology so that heavy metal release into the ecosystem can be adequately controlled. In this context, biosorption has evolved into a promising technique due to its varied advantages such as low costs, high metal removal efficiency even at low concentrations, no chemical requirement, ease of operation, easy recovery, etc. Over the past few decades, the adsorption technology has been reckoned to be the most efficient, competent and eco-friendly approach for heavy metal removal (Haq et al., 2020; Haq et al., 2022; Fadl, 2023).

8.3 BIOSORPTION

The term "sorption" is defined by both physisorption and chemisorption which are often used alternatively. "Absorption" is the uniform assimilation or diffusion of one state into those of another state in order to merge with them to form a solution (e.g., absorption of liquids by gases or solids). On the other hand, the term "adsorption" is frequently used to describe the deposition of ions and molecules at the surface or interface of any two phases, such as the interface between liquid and liquid, gas and liquid, gas and solid, or liquid and solid. In this instance, the surface where the substances get adsorbed is called "adsorbent" and the substance that has accumulated at the surface is known as "adsorbate." The term "biosorption" can be defined as a kind of adsorption where a biological framework is used as an adsorbent. The process involves the prompt and persistent adherence of ions and molecules present in aqueous solutions to functional groups that exist on the surface of lignocellulosic materials. It is a relatively novel technique that has shown to be particularly effective at removing pollutants from wastewater. It is the ability of specific biomolecules (or types of biomasses) to bind and concentrate particular ions or other molecules from aqueous solutions. The potential of biological materials for rapid passive metal sequestration by non-growing biomass/adsorbents from sewage through physiologically regulated or physico-chemical processes of uptake is known as biosorption (Pokethitiyook and Poolpak, 2016). Because of the fact that lignocellulosic biosorbents bind metal ions through chemical functional groups like carboxyl, amino, or phenolics, raw lignocellulosic biosorbents have generally undergone numerous modifications to boost their sorption capabilities (Singh et al., 2020). In recent years, significant effort has been made to create new adsorbents and enhance current adsorbents. Many researchers have looked into the viability of utilizing inexpensive agricultural wastes

(Charazińska et al., 2022). Microbial biomass has become a viable option in recent years for the development of cost-effective and environmentally friendly wastewater treatment methods. Since biotechnology has the potential to be applied to the control and removal of metal pollution, it has received considerable interest and is slowly becoming an increasingly prominent subject in the field. Biosorption is an economically and technically viable alternative for the removal of heavy metals from even the most diluted water solutions. It has been demonstrated that algae, bacteria, fungi, and yeasts are potential metal biosorbents. The biosorption technology's cost advantage would ensure a significant invasion of the massive market of heavy metal contaminated industries. The removal of heavy metal ions effectively from industrial contaminated water has been made possible using modern-day approaches provided by contemporary research (Charazińska et al., 2022). Some of these methods include advanced oxidation process, adsorption employing natural and man-made adsorbents, electrocoagulation, implementation of magnetic field, membrane processes, and many other noteworthy techniques. Biosorption is the capacity of biological substances to collect heavy metals from contaminated water or wastewater by means of metabolically mediated or through physico-chemical trajectory of uptake. It can be defined as the characteristic feature of specific biomolecules or biomass to attach and condense selected or preferred ions or molecules from liquid solutions. Earlier studies have shown that algae, bacterial consortia, fungi, and yeast work as potential and ideal metal biosorbents. It is a prompt phenomenon of phlegmatic metal sequestration brought into action by the non-expanding biomass or bio-adsorbents. Its cost compatible nature, high efficiency, depreciation in biological or chemical sludge, no necessity of supplementary nutrient, biosorbent regeneration, and plausibility of metal recovery are a few salutary features in contrast with traditional techniques (Hegazy, 2023). It is comprised of a solid phase (biological material) called the biosorbent, adsorbent, or simply sorbent, and a liquid/aqueous phase called solvent and is usually water, which contains the adsorbate (dissolved metal) and needs to be adsorbed. This adsorbent has a high affinity for the adsorbate species which causes the former to attract the latter species and is bound by different mechanisms. The phenomenon of adsorption persists till an equilibrium is confirmed between the bound adsorbate and its remaining fraction in the solution (Charazińska et al., 2022; Zareh et al., 2022). The concept of using biomaterials for the efficient uptake of heavy metals, taking place through physico–chemical interactions between cellular constituents of biological species and the metal ion counterparts, has been comprehensively investigated for the last few decades. The idea of using microbes as biosorbents has emerged as an economical and propitious technology in the recent few decades because of their ability to remove and recover heavy metals from aquatic systems. This entire process of metal biosorption follows a complicated mechanistic approach and is influenced by category of biomaterials used, significance of the biomass—whether living or non-living, chemical features of metal solution and ambient conditions like pH and temperature (Das et al., 2008). The approach of metal biosorption by living cells is a two-stage process: at first, the interactions between metals and functional groups appearing on the cell surface lead to the process of adsorption. The metal ion counterparts move across the cell wall of the biomaterial before finally acquiring entry to the cell cytoplasm as well as the cell membrane. The active sites for binding the metal ions are provided by the assortment of proteins and polysaccharides contained in the cell wall. The different types of microbes possess differences in the composition of their cell walls thereby resulting in significant variation in the amount and type of metal ions they bind (Abbas et al., 2014). Some examples of significant natural adsorbents established on agricultural wastes are black gram, bark of neem tree, rice husk, tea wastes, Turkish coffee, walnut shell, etc. (Inamuddin et al., 2021; Jahanban-Esfahlan et al., 2020).

8.4 MECHANISM OF BIOSORPTION

The various mechanisms involved in the process of biosorption have been demonstrated in Figure 8.1 (Inamuddin et al., 2021; Atkins et al., 2014).

1. **Physical adsorption:** In the process of physical adsorption (or contracted to physisorption), the adsorbate and the substrate maintains a van der Waals interaction. These van der Waals interactions, even though being of long range, are weak and the order of magnitude of the energy liberated on physisorption

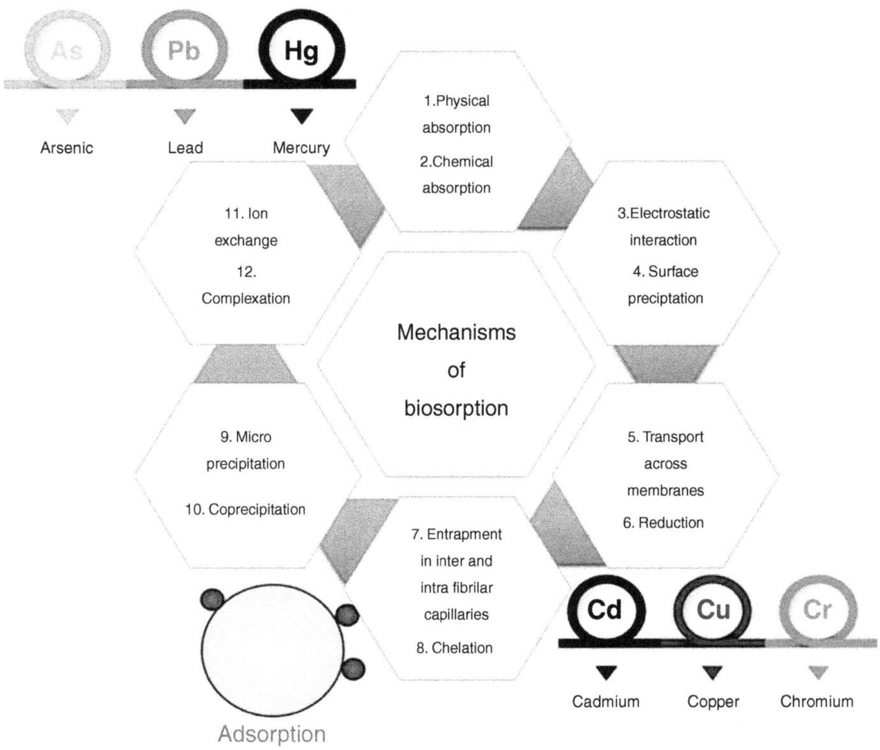

FIGURE 8.1 Pictorial illustration of the different mechanisms involved in the phenomenon of adsorption.

of a particle is synonymous with that of condensation enthalpy. In this case, a molecule undergoing physisorption preserves its identity even though there is a possibility of its distortion in the proximity of the surface and this takes place because of the insufficiency of the slight enthalpy change to direct to bond breaking. The typical values of enthalpy of physisorption, as found in literature, lie in the region of -20 $kJmol^{-1}$ (Atkins et al., 2014).

2. **Chemical adsorption:** It is also termed as chemisorption where the atoms or molecules constitute a chemical (generally covalent) bond resulting in adherence to the surface, and is also liable to identify sites which augment their coordination number with the substrate. The typical values of enthalpy of chemisorption lie in the region of -200 $kJmol^{-1}$ and is much greater than that for physisorption and also the distance between the closest adsorbate atom and the surface is shorter in the former case (chemisorption) than in the latter (physisorption) (Atkins et al., 2014).

8.5 TYPES OF BIOSORBENTS

8.5.1 Agricultural Wastes

The main components of agricultural by-products are typically cellulose and lignin, although they may also contain additional polar functional groups of lignin, such as aldehydes, ketones, alcohols, phenolic, carboxylic, and ether groups. Heavy metals have the tendency to bind to these groups leading to complex formation (Demirbas, 2008).

Lignin has an apparent indefinite molecular weight and a three-dimensional aromatic polymer structure. The secondary cell wall is where the cellulose is primarily found. In hardwoods, lignin is covalently bonded to xylans, while in softwoods, it is coupled to galactoglucomannans. Lignin is not soluble in water, while being structurally cleavable to a relatively small molecular weight. Lignin has a high surface area (180 m^2g^{-1}) and an excellent resistance to chemical reactions. The polymeric lignin's molecular weight ranges from 2000 to 15,000 $gmol^{-1}$. These characteristics of lignin indicate that it may be employed as an adsorption medium to filter out heavy metals from waste fluids (Haq et al., 2016).

Cellulose is a magnificently pure organic polymer that is only made up of anhydroglucose units. These anhydroglucose units are connected by -(1,4)-glycosidic connections. This linkage establishes cellobiose as the repeating unit for cellulose chains. Before being fermented into ethanol, cellulose needs to be hydrolyzed to glucose (Al-Badaani, 2023).

Hemicelluloses, unlike cellulose, are made up of several monosaccharide units. Hemicellulose polymer chains are also amorphous and feature short branches. Hemicelluloses have an amorphous shape, which makes them only partially soluble or expandable in water. Hemicelluloses serve as the cement in binding the cellulose micells and fiber together and are mostly generated from the chains of pentose sugars.

The removal of heavy metal ions using inexpensive, widely accessible adsorbents: agricultural wastes like hazelnut shells, tea waste and coffee, red fir, peanut hull, olive cake, sawdusts, and maple, different bark samples, charcoal pine needles, pinus bark, coconut husk, modified cellulosic materials, palm kernel husk and peanut skins, chemically modified cotton, corncobs and its modified form, rice hulls, apple wastes, coffee grounds, tea leaves, bark, modified bark, wool fibers, wool, cactus leaves, almond shells, modification of sugar beet pulp, banana and orange peels, modified lignin, palm fruit bunch, modified sunflower stalk, maize leaf, and different agricultural by-products were employed and examined (Haq et al., 2016; Demirbas, 2008; Nguyen et al., 2013).

8.5.2 Algal Biosorbents

Algal biodiversity is characterized by phylogenetic relationships, life stages, morphological types, environments used by various species, or chemical variety. Their habitats span from wide oceans, where planktonic microalgal species live, to rocky coasts, where marine macroalgae, seaweeds, and benthic microalgae may be present. The group of noticeable filamentous algae inhabits freshwater ecosystems such rivers, ditches, lakes, and ponds. The benthic microalgae that live on bedrock in different moist and transiently moist marine, freshwater, and terrestrial settings are less visible. The diversity of algae taxonomy with diverse habitats suggested that distinct species had evolved analogous metabolic pathways to carry out fundamental tasks including protection from biotic and abiotic threats. Microalgae can be colonial or unicellular algae and can take the appearance of filaments. The majority of them include chlorophyll, however other types of common algae appear brown or red instead of green because they also contain other pigments like carotenoids. One or more chloroplasts, membrane structures that contains the pigments necessary for photosynthetic activity, are found in algae cells. The chemical and structure of algae's cell walls exhibit wide variation. In many instances, the cell wall is made up of a web of cellulose fibrils that are typically changed by the addition of additional polysaccharides such as pectin, alginic acids, or fucinic acid. Some algae have walls made of calcium carbonate deposition, with chitin present on occasion (Pokethitiyook and Poolpak, 2016; González et al., 2011).

Different types of algae have been utilized and studied in biosorption as biosorbents for metal removal. The biomass cell's constituents, particularly the cell surface and the spatial organization of the cell wall, play a key role in metal biosorption. Teichoic acids, lipoteichoic acids, and peptidoglycan are all crucial chemical elements of bacterial surface structures. It has been demonstrated that a number of polysaccharides, such as cellulose, chitin, alginate, glycan, etc., found in the cell walls of fungi or algae, are crucial for some types of biomass' ability to bind metals. It has been discovered that some functional groups, particularly carboxyl groups, can bind metal ions. According to research, functional groups including O-, N-, S-, or P- directly contribute to the binding of certain metals (Wang and Chen, 2009). Algal cell wall polysaccharides contain carboxyl, hydroxyl, sulphate, and amino groups that serve as binding sites for metals (Lesmana et al., 2009; Inamuddin et al., 2021). Additionally, algal acid treatment can partially dissolve polysaccharide molecules in the cell wall's outer layer, producing extra binding sites (often amino groups) (Pokethitiyook and Poolpak, 2016); González et al., 2011).

8.5.3 Bacterial Biosorbents

Prokaryotes, which include bacteria as their core group of unicellular creatures, are widely distributed throughout the environment, particularly in soil and water. They have a straightforward morphology and can be found in the spherical or coccus, rod-shaped or bacillus-shaped, and spiral-shaped or spirillum forms. Both in size and structure, bacteria are diverse. Bacteria cells typically measure 2.0–6.0 m long by 1.1–1.5 m broad.

Gram staining was used to classify bacteria. Gram-positive bacteria stained purple, while Gram-negative bacteria stained pink or red. The cell walls of Gram-positive bacteria are stronger due to the thicker peptidoglycan layer, despite the fact that the surface of Gram-negative cells is significantly more complex chemically and physically than that of the other. The peptidoglycan layer in the cell wall of gram-negative bacteria is 2–7 nm thick, while the outer membrane is 7–8 nm thick. The complex macromolecular lipopolysaccharides that Gram negative bacteria produce also come in a variety of forms. Each LPS is maintained in the outer membrane by ionic and hydrophobic contacts as well as relatively weak cohesive forces, and it is possible to use surface-active chemicals to separate it from the cell surface. The negative surface charge of Gram-negative bacteria is what causes the net negative charge of LPSs. It has been determined that the phosphate groups in phospholipids and LPSs are the main locations for metal ions to bind.

The small size, ubiquity, and capacity for regulated growth of bacteria make them intriguing candidates for use as biosorbents. They can also survive in a variety of different environmental conditions. In accordance with the species, bacteria may retain the ability to biosorb a variety of elements or, in certain situations, may be element-specific (Pokethitiyook and Poolpak, 2016; Dawwam et al., 2023).

8.5.4 Fungal Biosorbents

Fungal biomass has drawn a lot of attention as a biosorbent due to the large proportion of cell-wall material, which enhances the variety of functional groups associated with metal binding and hence increases the potential of fungi to sequester metal. The fungal kingdom is extremely diverse, with species developing as unicellular yeasts and/or branching hyphae that produce a surprising variety of spores and other reproductive structures. Each time, the mechanical strength of the cell wall—which plays a variety of vital roles in how the fungus interacts with its surroundings—determines the shape and integrity of the fungus (Dhankhar and Hooda, 2011).

Approximately 80% of the dry weight of fungi's cell walls are made up of polysaccharides, which are then joined in diverse ways by a large variety of distinct proteins that are frequently extensively glycosylated. Due to the wide range of functional groups present and the large proportion of cell-wall material, fungi exhibit excellent metal-binding capacity. Fungus has a short multiplication cycle compared to various biosorbents like plant products or algal biomass; it is simple to cultivate at a big scale. Furthermore, it is simple to grow utilizing low-cost growth media and simple fermentation processes. In addition to this, biomass yield is also fairly high. Thirdly, fungal biomass is easily accessible as industrial waste products, such as Saccharomyces cerevisiae (trash from the brewing sector) and Aspergillus niger (waste from the production of citric acid) (Zareh et al., 2022). Numerous industrial fermentations, as well as the food, brewing, and distilling industries, produce different forms of fungal biomass, which are also the subject of ongoing research. As opposed to other forms of microbial biomass, this gives the fungal biosorbents an economic advantage. Fourth, a large fraction of fungal biosorbents is not pathogenic, making them typically considered harmless and well-liked by the general populace when used practically. In addition, S. cerevisiae, a fungus, is a perfect model organism for identifying the mechanism of biosorption in metal ion removal, particularly the molecular interaction between metal with microbe (Dhankhar and Hooda, 2011).

Researchers have looked into the biosorption capacity of fungi such *Rhizopus, Aspergillus, Streptoverticillum,* and *Saccharomyces.* A stoichiometric interaction between the reactive chemical groups in the cell wall and the metal appears to be the first stage in the process of metal binding, which is followed by an inorganic deposition of higher metal concentrations. Metal translocation results in intracellular buildup, which is reliant on cellular metabolism. This indicates that only living cells can support this type of biosorption. It is frequently linked to fungi's active defense mechanism, which responds when hazardous metals are present (Dhankhar and Hooda, 2011; Pokethitiyook and Poolpak, 2016).

8.5.5 Industrial Waste Biosorbents

High economic and/or ecological interest can be found in recycling landfill garbage to encourage reuse or recycling possibilities and the manufacture of marketable products. There are now just a few novel ideas for the eco-friendly, risk-free, and beneficial handling of hazardous industrial waste sludge documented in research journals. Industrial sludge and/or sludge ash can be transformed into lightweight artificial aggregates, bricks, tiles, concrete and cement-like materials, and other usable materials to solve disposal issues and lower the solid waste content.

Tannery sludge was employed as a biomaterial due to its accessibility and low cost for the elimination of heavy metal ions. It is made up of both live and dead microbial components as a natural complex biosorbent with several chemical functional groups at the cell walls of biomasses that could demonstrate a greater and faster rate of metal ion sequestration.

8.5.6 Natural Resources

In the category of the various biomaterials obtainable, the polysaccharides most frequently used for coating Magnetic nanoparticles (MNPs) subsume agarose, alginate, carrageenan, dextran, gum Arabic, heparin, pullulan and starch. Among the aforementioned materials, cellulose (CL) and chitosan (CS) are prominent owing to their characteristics like biocompatibility, biodegradability, chemical, and thermal stability. CL is the most rampant natural polymer which is renewable; CL based beads, films and resins have been extensively used in heavy metal adsorption. CS is proclaimed to maintain the leading sorption potential for Cu^{2+} ions, assisted by abundant NH_2

and OH functional groups in CS functioning as the sites for chelation. Its outstanding adsorption attributes can be ascribed to:

(i) Superior hydrophilic behavior because of ample number of OH groups.
(ii) A massive number of functional groups like acetamido, NH_2, OH possessing powerful chemical reactivity are present.
(iii) The bendable structure of the chain of polymer.

But certain disadvantageous features of CS like high crystallinity, poor mechanical strength and weaker chemical resistance confine its usage as a functional adsorbent. However, designing MNPs by applying coating methods using lesser chemicals is favored to modify the overall preparation procedure into a green process thereby ensuring less toxicity and opening more environmentally compatible applications (Gautam & Chattopadhyaya et al., 2016). Also, there are several other naturally occurring adsorbents like clay (Kaolinite e.g. pyrophyllite, serpentine, sepiolite, vermiculite; micas- e.g. illitite; smectites- e.g. montmorillonite and saponite), siliceous materials (alunite, dolomite, perlite, glasses and silica beads) and zeolites (clinoptilolite, zeolite with manganese oxide- coating, natural Chinese zeolite) (Singh et al., 2020).

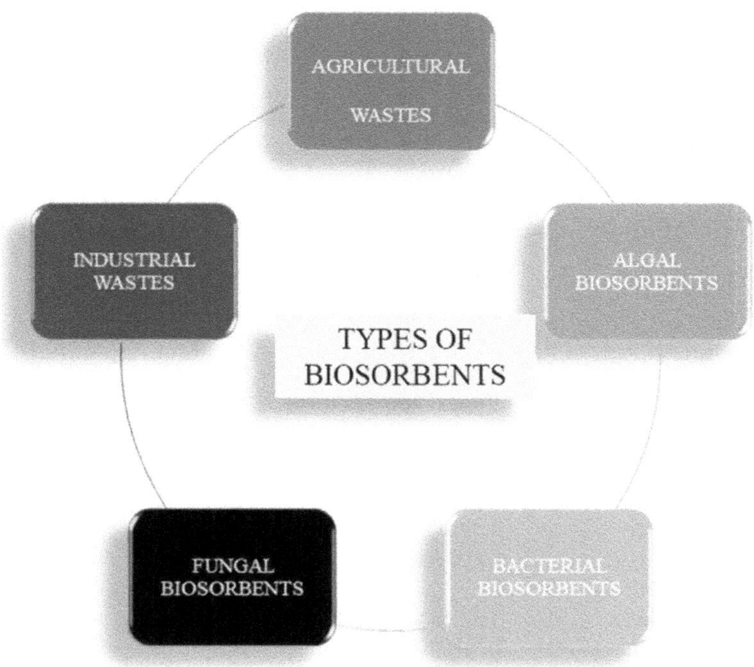

FIGURE 8.2 Diagrammatic representation of the different types of biosorbents developed.

8.5.7 Others

A few examples of adsorbents having high adsorption capacity for the effective removal of heavy metals prepared using natural polymers are cellulose-chitosan (CL-CS) nanoporous composite microspheres (NMCMs), β-cyclodextrin (CD)/ metal or polymer- grafted MNPs, carbon-disulfide modified chitosan (CMCS)-Fe_3O_4, Cu- chitosan/nano-Al_2O_3 for removal of As (III), chitosan/Fe (III)/ PVA novel biopolymeric nanofibers to eliminate bacterial growth as well as heavy metal contamination and provide safe drinking water, chitosan-poly(vinyl alcohol)/bentonite (CTS-PVA/BT) nanocomposite with high selectivity for Hg (II) and many more (Gautam & Chattopadhyaya et al., 2016; Al-Badaani, 2023) (see Figure 8.2).

8.6 FACTORS AFFECTING BIOSORPTION

8.6.1 Effect of pH

The solubility of the metal ions, the concentration of the counter ions on the functional groups of the adsorption/biosorption (i.e., the surface charge of the adsorption/ biosorption), and the degree of ionization of the adsorbate during reaction are all influenced by the acidity (pH), which is a crucial parameter for adsorption/biosorption of metal ions from aqueous solution. Since it affects not only the adsorbate's solution chemistry but also the performance of the functional groups in the adsorbent and biosorbent, as well as the rivalry between adsorbate and sorbate ions.

Since the hydrogen ion itself is a difficult competing adsorbate, changes in pH have an impact on efficacy. Table 8.2 lists the functional groups in biosorbents that are present and are in charge of sorbent binding, along with their acidity constants.

TABLE 8.2
Representation of essential functional groups associated in metal ion biosorption

Name of binding group	Structural formula	Ligand atom	pKa
Amide	-C=O-NH2	N	-
Amine	-NH2	N	8–11
Carbonyl (ketone)	C=O<	O	-
Carboxyl	-C=O-OH	O	1.7–4.7
Hydroxyl	-OH	O	9.5–13
Imidazole	-C-N-H>CH H-C-N	N	6.0
Imine	>NH	N	11.6–12.6
Phosphodiester	>P=O-OH	O	1.5
Phosphonate	OH-P=O-OH	O	0.9–2.1
		O	6.1–6.8
Secondary amine	>NH	N	13
Sulfhydryl (thiol)	-SH	S	8.3–10.8
Sulfonate	O-S=O	O	1.3
Thioether	S<	S	-

Over the pH range of 3.0–10.0, the surface charge is mainly negative. The overall surface charge on the cells of microorganisms turns positive below a pH of 3.0 (Zareh et al., 2022).

The solubility of metals sufficiently declines at higher solution pH levels to permit precipitation, which may impede the sorption process. By adjusting the pH level, the activity of binding sites can also be modified. For instance, pH values between 3 and 6 have been reported to be favorable for biosorption when metal ions are absorbed by bacterial biomass. During the biosorption of metals, a protonated bacterial biomass releases H+ ions, which in turn lowers the pH of the solution. Since the majority of the reaction usually takes place in the initial stage, these pH shifts happen quickly in the beginning and take a while to reach equilibrium. Until equilibrium is attained, the pH value should be maintained for the entire contact period.

In batch tests, cadmium removal by dead mycelial biomass at various pH ranges from 3 to 8 was investigated. The pH value affects the electrostatic binding of metal ions to corresponding functional groups, which affects the adsorption of metal ions. The outcome showed that pH values of 4 are ideal for the greatest adsorption of several metal species. At high pH values, the amount of metal ions adsorbed is considerably reduced by the production of metal hydroxide and other metal complexes.

The pH level during the absorption of cations on the surface of biomass had an impact on the biosorption of heavy metals onto Pseudomonas aeruginosa. Additionally, studies have demonstrated that the industrial biomass Pseudomonas aeruginosa is effective in removing heavy metals at pH levels between 3 and 5 (Pokethitiyook and Poolpak, 2016) (see Table 8.2).

8.6.2 Effect of Temperature

The impact of temperature on metal uptake has been extensively studied by researchers. The solubility of metal ions is also impacted by the change in solution temperature, in addition to the diffusion rate of metal ions (Park et al., 2010). The adsorption capacity of a certain biosorbent is influenced by temperature in accordance with the surface functional groups of that biosorbent. However, a common finding of much research is that temperature has a limited impact and only does so over a narrow temperature range (Sahmoune et al., 2011). Depending on whether the biosorption process is exothermic or endothermic, temperature may have distinct effects on the biosorption process. Adsorption and ion exchange are exothermic processes by nature, thus as the temperature rises, so does the rate of their reactions.

The temperature dependence of metabolizing cells, in contrast to temperature independent non-metabolizing cells, has a significant impact on the biosorption process. On the other hand, cell wall damage at very high temperatures may result in a decrease in biosorption. Most metal uptake rose when temperature climbed between 4 and 13°C, but only a small amount of metal uptake reduced as temperature increased between 20 and 40°C.

Adsorption may indicate chemisorption if it increases with temperature, but a reduction in adsorption with rising temperature may indicate physical adsorption. The uranium adsorption by prawn shell flakes showed that with increase in temperature, the efficiency of uranium adsorption decreases and that at room temperature, the highest adsorption efficiency occurs before the adsorption efficiency minimizes to reach the least values at 70 °C, so at room temperature, the evaluation of biosorption is still suitable because it is fit for replication (Pokethitiyook and Poolpak, 2016).

8.6.3 Effect of Initial Concentration

A high metal uptake is caused by initial higher metal ion concentration that have certain impacts on biosorption. The explanation is that at decreasing metal concentrations, the ion to accessible surface area ratio drops, and hence the fractional sorption becomes unrelated to metal concentration at first. In contrast, at high metal concentrations, the availability of sorption sites declines in comparison to the ion numbers, and the subsequent metal removal is heavily dependent on the original metal concentration (Pokethitiyook and Poolpak, 2016).

8.6.4 Effect of Biosorbent Loading

The amount of biosorbent has a significant influence on the amount of metal biosorption. With the increase in biomass surface area there is an increase in the number of binding sites which causes metal biosorption to rise as biomass increases. In contrast, due to complicated circumstances, increasing the concentration of biosorbent reduces the amount of metal sorption per unit of biomass weight. Low metal uptake is caused by a lack of accessible metal to completely cover the accessible exchangeable sites, which has a substantial impact on biosorption at high concentrations of biosorbent.

8.6.5 Effect of Contact Time

An involuntary physical adsorption of metal occurs very quickly at the cell surface after metal ions make contact with the biosorbent. This behavior suggests that the external surface of the biosorbent experiences rapid initial adsorption, subsequently undergoing a slower interior diffusion process that may be the rate-limiting step. This is significant because one of the criteria for a cost-effective wastewater treatment plant application is equilibrium time (Pokethitiyook and Poolpak, 2016).

8.7 ADVANTAGES AND DISADVANTAGES OF BIOSORPTION

The advantages and disadvantages of the biosorption (nanotechnology) approach have been delineated below in Table 8.3 (Gautam &Chattopadhyaya et al., 2016; Inamuddin et al., 2021).

8.8 APPLICATIONS OF BIOSORPTION

A few examples of some of the most popular biosorbents are: BIO-FIX, a biosorbent fabricated using biomass from a miscellany of sources (algae, cyanobacterium-*Spirulina*, yeast and plants- *Lemma* sp. And *Sphagnum* sp. blended with guar gum and xantham) and is a creation of the US Bureau of Mines, AMT-BIOCLAIM™ which is a biosorbent of algal origin made by Advanced Mineral Technologies and

TABLE 8.3
Representation of the advantages and disadvantages of the biosorption technology in tabular form

ADVANTAGES	DISADVANTAGES
Biosorbent can be regenerated	Applicability is constraint to laboratory
Discharge of hazardous chemical sludge is restricted	Estrangement of tiny particles of spent biomass from the effluent after treatment (biomass granulation or immobilization can help overcome this problem)
Ease of maintenance	Loss of mass post regeneration, especially porous adsorbents take longer regeneration duration
Low-priced biosorbents and hence does not need much investment of capital	Low-grade mechanical characteristics
Microsize polymer adsorbents have better regeneration capacity and surface-modified adsorbents have greater affinity to certain metals	Presence of particulate content creates problems in functioning
Separation of beneficial constituents of valued metals is possible	The type of biosorbent used regulates the performance
Superior removal efficiency	Porous adsorbents have their adsorption sites positioned inside the pores and due to the defiance of pore diffusion, porous adsorbents take longer time to attain adsorption equilibrium

AlgaSORB™ which is also a product of algal origin. The granules of BIO-FIX are designed to resolve the problem of heavy metal removal from the wastewater discharged as a result of mining and other industrial activities. It composition chiefly comprises of algae, Lemna sp., Sphagnum sp., Spirulina sp. and yeasts Spirulina sp. disseminated over porous granules. The outstanding features of this biosorbent besides its efficiency in treatment of effluents containing heavy metals are: its stability and regeneration capacity even after 200 cycles of usage. Another biosorbent BV-SORBEX$_{TM}$ is a product of the B.V.SORBEX$_{TM}$ meant for heavy metal removal is an amalgamation of several algae namely, *Ascophyllum nodosum, Chondrus crispus, Palmyra pamata, Halimeda opuntia* and *Sargassum natans*. The company is known for manufacturing reliable products effective in an ample range of pH and also in solutions containing calcium and magnesium, with the benefit of ease of regeneration. AlgaSORB™, a noteworthy creation of Bio-Recovery System Inc. (USA) developed by incorporation of Chlorella vulgaris-freshwater algae, efficiently expels metal ions from solutions with concentration 1–100 mg/L and has the advantage of withstanding around 100 or more cycles of regeneration.

A recent study was conducted by Sasan Narkesabad et al. (2023) on the synthesis of a novel bioadsorbent and exploring its effectiveness in heavy metals adsorption from aqueous solutions. The aim of the study was the synthesis and characterization of amidoximated biocomposite *Luffa cylindrica* (AO-LC) followed by evaluation of its heavy metal adsorption performance from the aquatic medium (Sasan Narkesabad et al., 2023). It is known from earlier scientific studies that amidoxime group enhances the adsorbing capability of adsorbents due to its elevated efficiency in the approach of adsorption of a heterogeneous class of contaminants, primarily focusing on heavy metals. Owing to the permeable lattice and convoluted hierarchy of *Luffa cylindrica*, its fibers can be utilized for performing decontamination of Cd, dyes, Ni, oils, surfactants and other contaminants from aqueous bodies (Haq et al., 2018). After successfully grafting 3-(trimethoxysilyl)propyl methacrylate (MPS) and Polyacrylonitrile (PAN) onto the surface of LC, AO-LC was able to adsorb heavy metals, Pb^{2+}> Ag^{+}> Cu^{2+}> Cd^{2+}> Co^{2+}> Ni^{2+}, in an efficacious manner. The adsorption potential observed was 18.88 mg/L and percentage of metal ion removal was found to be 99.07% with adsorption equilibrium attained within initial 20 minutes of treatment (Sasan Narkesabad et al., 2023).

Another example of an effectual attempt to use biomass for the construction of a biosorbent by Sayago et al. (2023), helped elucidate the heavy metal decontamination capacity of the chosen biomaterial along with its plausible mechanism of function (Sayago & Ballesteros Ballesteros, 2023). The development of a treatment mechanism for an aqueous system contaminated with Cr(VI) was made possible by using cellulose xanthogenate from *Eichhornia crassipes* (Inamuddin et al., 2021; Das et al., 2008).

8.9 BIOSORPTION ISOTHERMS

There exists a dynamic equilibrium between the adsorbed gas and the free gas in the phenomenon of adsorption. The surface fractional coverage (θ) is dependent on the pressure of the overlying or excessive gas. Thus, adsorption isotherm is the deviation of θ with pressure at a selected temperature (Atkins et al., 2014; Singh et al., 2020; Fadl, 2023; Ayawei et al., 2017) (see Figure 8.3).

8.9.1 Single Isotherm Models

a. **Freundlich model**: The empirical formula put forward by Freundlich and Kuster (1907)to express the adsorption process from solution is given as: $q_e = KC_e^{1/n}$ where, the Freundlich constants are K (mg/g) $(1/mg)^{1/n}$ and nrelated to the capacity and intensity of adsorption respectively (Al-Badaani et al., 2023; Ayawei et al., 2017).
b. **Langmuir model**: A new isotherm model for adsorption of liquid or gas on solid was published by Irving Langmuir in the year 1918, based on the supposition that presence of a saturated monolayer of solute molecules on the surface of adsorbent results in maximum adsorption with constant adsorption energy and restriction of adsorbate transmigration on the adsorbent's surface (Al-Badaani et al., 2023). The Langmuir isotherm model is given as:

$$Q_e = \frac{q_m b C_e}{1 + bC_e}$$

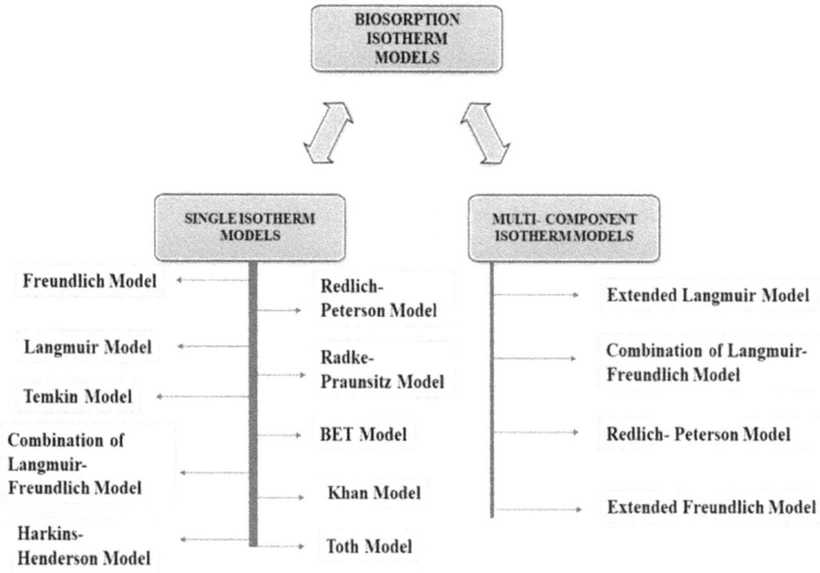

FIGURE 8.3 Different types of biosorption isotherm models.

There are four main assumptions of Langmuir adsorption isotherm model:

i. The adsorbent possesses a uniform surface.
ii. There is absence of interaction among adsorbed molecules on adjacent sites.
iii. The mechanism for every adsorption is consistent.
iv. The adsorption of molecules occurs at distinct sites on the surface of the adsorbent.

The dimensionless equilibrium parameter or separation constant/factor "R_s" is given by:

$$R_s = \frac{1}{1+bC_o}$$

Where, if $R_s > 1$ then the type of isotherm is unfavorable, $R_s = 0$ yields a linear isotherm, $0 < R_s < 1$results in favorable type of isotherm and finally if, $R_s < 0$, an irreversible type of adsorption isotherm is obtained.

c. **Temkin model**: The adsorption isotherm expression is given as:

$$q_e = \frac{RT}{b}\ln\left(K_T C_e\right)$$

where, K_T (1/mg) = equilibrium binding constant congruous with the maximum bending energy, andB_1 (KJ/mol) is associated with the heat of adsorption (Al-Badaani et al., 2023; Ayawei et al., 2017). The equation outlined below gives the linear representation of the form above:

$$q_e = B_1 lnK_T + B_1 lnC_e$$

where, B_1 = RT/b
R = 8.314 KJ/mol.K is the universal constant, and
T = absolute temperature in K

d. BET model

Stephan Brunauer, Paul Emmett, and Edward Teller derived a method to assess isotherms which describe a multilayer adsorption which is one of the most widely used and accepted method called the Braunauer-Emmett-Teller (BET) model (Yurdakal et al., 2019). It is given as:

$$\frac{V}{V_{mon}} = \frac{c(p/p^0)}{1-(p/p^0)\left[1-(1-c)\frac{p^0}{p}\right]}$$

To obtain the volume of a monolayer (V_{mon}) and the parameter of the model *c*, there is a linearized equation: $\frac{p}{V(p^0-p)} = \frac{1}{V_{mon}c} + \frac{c-1}{V_{mon}c}\frac{p}{p^0}$

where,

p^0 = vapor pressure on top of a layer of adsorbate whose thickness is more than one molecule and is similar to a liquid which is bulk and pure,
V_{mon} = adsorbate's monolayer coverage volume,
c = constant whose value rests on the enthalpy of desorption ($\Delta_{des}H°$).

If the value of $\Delta_{des}H°$ is high compared to that of enthalpy of vaporization of the liquid adsorbate ($\Delta_{vap}H°$), then the value of c is also large. The BET isotherm fits experimental results or observations rationally/moderately well over regions of pressure which are restricted, but on the other hand, at low pressures, the extent of adsorption is typically underestimated and in cases of high pressures, the same is overestimated (Pokethitiyook and Poolpak, 2016).

e. Harkins-Henderson model

Multilayer adsorption process as well as the adsorbent's heterogeneous distribution of pores is explained by this model. The equation expressing the Harkins-Henderson model is given as$q_e = \frac{K_h^{1/n_h}}{C_e^{1/n_h}}$

where,

K_h (mg/g) (mg/L) $1/n_h$ and n_h are the model parameters associated with capacity of adsorption and intensity of adsorption respectively.

f. **Redlich-Peterson model**

g. The Redlich-Peterson adsorption isotherm infers that formation of monolayer takes place in the process of adsorption along with simultaneous multi-site intermolecular interactions (Ayawei et al., 2017); Vilela et al., 2019). The representation of the nonlinear form of Redlich-Peterson isotherm model is given by the following equation: $q_e = \frac{K_R C_e}{1 + a_R C_e^{\beta}}$

where, a_R(mg^{-1}) and K_R (Lg^{-1}) = Redlich-Peterson constants, and
β = an exponent having value between 0 and 1.

h. **Toth model**: This isotherm model is expressed as:

$$q = \frac{q_m p}{\left(\left(\frac{1}{K_t}\right) + p^t\right)^{1/t}}$$

where, K_t and t = Toth isotherm constants.
K_t is associated with the binding energy and q_m= capacity of monolayer adsorption.

The Toth isotherm is one of the most extensively used isotherm models to express data of both gas and liquid phase equilibrium. Mathematically, it enables prediction of the monolayer adsorption ability of a surface which is homogeneous in nature and will coincide with the prediction of surface area given by Langmuir isotherm (Ayawei et al., 2017; Kumar et al., 2021). An equation involving three isotherm constants was also proposed by Toth:

$$q = \frac{q_m \chi^{1/t} p}{\left(\frac{1}{K_t} + p^t\right)^{1/t}} = \frac{q_m \chi^{1/t} p}{\left(1 + p^t\right)^{1/t}}$$

where, χ is a constant given by: $\chi = \frac{1}{K_t} + 1$

i. Radke-Prausnitz model

This adsorption isotherm model reduces to a linear form of isotherm at low concentration of adsorbate; it takes the form of Freundlich adsorption isotherm at elevated temperature and when MRP is equal to zero, it takes the form of Langmuir adsorption isotherm (Ayawei et al., 2017). The fit over a wide range of adsorbate concentrate provided by Radke-Prausnitz model is quite good, which is a noteworthy characteristic of this model thereby making it highly preferrable in most systems of adsorption at a low concentration of adsorbate (Ayawei et al., 2017; Saravanan et al., 2021).

$$q_e = \frac{q_{MRP} K_{RP} C_e}{(1 + K_{RP} C_e)^{MRP}}$$

where, q_{MRP} = capacity of the adsorbent (mg/g),
K_{RP} = equilibrium constant, and
MRP = model exponent.

j. Combination of Langmuir-Freundlich model

The assumption of combined Langmuir-Freundlich model of expressing adsorption isotherm, also commonly called the Sips model, is that the process of adsorption appears only in systems which are heterogeneous in nature (Ayawei et al., 2017). It is extensively used for adsorption involving a single component. The non-linear equation for Sips isotherm model is given as:

$$q_e = \frac{q_{ms} K_s C_e^{\beta_S}}{1 + K_S C_e^{\beta_S}}$$

where, q_{ms} (mg g^{-1}) = maximum Sips adsorption capacity,
K_S (L mg^{-1}) = Sips constant,
β_S = Sips exponent which expresses either the homogeneous or heterogeneous nature of the process of adsorption.

k. **Khan model**: This is a comprehensive model for biadsorbate adsorption from pure, uncontaminated dilute solutions. To acquire the isotherm parameters, many researchers have implemented non-linear methodologies (Abbas et al.,2014). The equation is given as:

$$Q_e = \frac{Q_{max} b_k C_e}{(1 + b_k C_e) a_k}$$

Where, a_k = Khan isotherm,
b_k = Khan isotherm model constant, and
Q_{max} = Khan isotherm maximum adsorption capacity expressed in mg g^{-1}

8.9.2 Multi-component Isotherm Models

A huge quantity of variables is involved in case of multi-component biosorption and this makes its experimental evaluation time-consuming. Hence, a lot of attention has been generated by the complication of predicting multi-component adsorption isotherm from the data of single component adsorption. For the delineation of competitive adsorption, a number of isotherms has been propounded. On the basis of parameters of single component isotherm as well as the elements of correction abstracted from the experimental competitive information or data, many of these isotherms have been established and a few of such models have been listed below (Abbas et al., 2014):

a. **Extended Langmuir model**
b. **Combination of Langmuir-Freundlich model**
c. **Redlich-Peterson model**
d. **Extended Freundlich model.**

8.10 SUMMARY AND CONCLUSION

Unsystematic industrialization, mining activities, rapid urbanization, and many other related factors steer to the discharge and dissemination of a wide variety of pollutants in groundwater as well as surface water which lead to the generation of severe health issues and thus are regarded as a matter of calamitous environmental consideration (Ye Panqing, 2023; Thakur et al., 2022). Due to such haphazard activities over the last few decades, majority of water bodies are transforming into contaminated sites with untreated effluents from industries which contain heavy metals and a wide range of toxic materials. These heavy metals are also raw materials and essential strategic resources in the domain of advanced and sophisticated functional materials. Some examples include the usage of copper compounds in fabrication of antifouling paints, electrical gadgets, fungicides, plating metals, producing wood pulp, and semiconductor devices. Nickel compounds find their usage in the field of manufacturing catalysts with high-performance (Hołyńska-Iwan et al., 2023). There also exists an important relationship of certain heavy metals essential for living organisms even though in trace amounts such as Co, Cu, Fe, Mn, Mo, and Zn, but their excessive unneeded levels can be harmful when they infiltrate to the food chain of humans (Thakur et al., 2022). The heavy metals emerging from industrial activities, mining, natural resources and so on are reckoned as inimical to public health when integrated into supplies of fresh water. Toxicological studies have observed that the degree of toxicity of heavy metals relies on its oxidation state. Even after the government has imposed strict laws, guidelines and regulations, many industries still permit unscientific and uncontrolled release of their effluents, which do not undergo adequate treatment into drinking water sources (lakes and rivers) and public sewers only to retain their operational costs.

There exist a range of techniques like adsorption, bio/electrokinetic remediation, chemical reduction precipitation, evaporation, floatation and oxidation are accessible in laboratory proportion for heavy metal removal, only a few technologies are efficiently initiated in the industrial scale. Among the above-mentioned methods, adsorption is a widely accepted, conventional yet effective technology to remove hazardous metal ions and also pathogenic bacteria from aqueous solutions. This chapter summarizes the mechanism, related parameters controlling the rate of adsorption and efficient use of green biosorbents procured from agricultural residues, a range of microbes and other natural resources (collectively called biomaterial/biomass) for the productive removal of multifarious metal ions from water sources (Gautam & Chattopadhyaya et al., 2016; Das et al. (2008).

REFERENCES

Abbas, S. H., Ismail, I. M., Mostafa, T. M., & Sulaymon, A. H. (2014). Biosorption of heavy metals: a review. *Journal of Chemical Science and Technology* 3(4): 74–102.

Ahamed, M. I., & Lichtfouse, E. (Eds.). (2021). *Water pollution and remediation: Heavy metals*. Springer.

Al-Badaani, A. A., Hifney, A. F., Adam, M. S., & Gomaa, M. (2023). Low-cost biosorption of Fe (II) and Fe (III) from single and binary solutions using Ulva lactuca-derived cellulose nanocrystals-graphene oxide composite film. *Scientific Reports* 13(1): 6422.

Aragay, G., Pons, J., & Merkoçi, A. (2011). Recent trends in macro-, micro-, and nanomaterial-based tools and strategies for heavy-metal detection. *Chemical Reviews* 111(5): 3433–3458.

Atkins, P., Atkins, P. W., & de Paula, J. (2014). *Atkins' Physical Chemistry*. Oxford university press.

Ayawei, N., Ebelegi, A., & Wankasi, D. (2017) Modelling and interpretation of adsorption isotherms. *Journal of Chemistry*, 2017, Article ID: 3039817. https://doi.org/10.1155/2017/3039817

Bolisetty, S., Peydayesh, M., & Mezzenga, R. (2019). Sustainable technologies for water purification from heavy metals: review and analysis. *Chemical Society Reviews* 48(2): 463–487.

Charazińska, S., Burszta-Adamiak, E., & Lochyński, P. (2022). The efficiency of removing heavy metal ions from industrial electropolishing wastewater using natural materials. *Scientific Reports* 12(1): 17766.

Das, N., Vimala, R., & Karthika, P. (2008). Biosorption of heavy metals–an overview. *Indian Journal of Biotechnology* 7(2): 159–169.

Dawwam, G. E., Abdelfattah, N. M., Abdel-Monem, M. O., Jahin, H. S., Omer, A. M., Abou-Taleb, K. A., & Mansor, E. S. (2023). An immobilized biosorbent from Paenibacillusdendritiformis dead cells and polyethersulfone for the sustainable bioremediation of lead from wastewater. *Scientific Reports* 13(1): 891.

Demirbas, A. (2008). Heavy metal adsorption onto agro-based waste materials: a review. *Journal of Hazardous Materials* 157(2-3): 220–229.

Dhankhar, R., & Hooda, A. (2011). Fungal biosorption – an alternative to meet the challenges of heavy metal pollution in aqueous solutions. *Environmental Technology* 32(5): 467–491.

Fadl, M. G. (2023). Prediction of heavy metal biosorption mechanism through studying isotherm kinetic equations. *Scientific Reports* 13(1): 1576.

Gautam, R. K., & Chattopadhyaya, M. C. (2016). *Advanced nanomaterials for wastewater remediation*. CRC press.

González, F., Romera, E., Ballester, A., Blázquez, M. L., Muñoz, J. Á., & García-Balboa, C. (2011). Algal biosorption and biosorbents. *Microbial Biosorption of Metals:* 159–178.

Haq, I., Kumar, S., Kumar, V., Singh, S.K., & Raj, A. (2016a). Evaluation of bioremediation potentiality of ligninolytic Serratia liquefaciens for detoxification of pulp and paper mill effluent. *Journal of Hazardous Materials* 305:190–199.

Haq, I., Kumari, V., Kumar, S., Raj, A., Lohani, M., & Bhargava, R.N. (2016b). Evaluation of the phytotoxic and genotoxic potential of pulp and paper mill effluent using Vigna radiata and Allium cepa. *Advanced Biology:* 8065736.

Haq, I, Kumar, S., Raj, A., Lohani, M., &Satyanarayana, G.N.V. (2017). Genotoxicity assessment of pulp and paper mill effluent before and after bacterial degradation using Allium cepa test. *Chemosphere* 169: 642–650.

Haq, I., Raj, A., & Markandeya (2018). Biodegradation of Azure-B dye by Serratia liquefaciens and its validation by phytotoxicity, genotoxicity and cytotoxicity studies. *Chemosphere* 196: 58–68.

Haq, I, Mazumder, P., & Kalamdhad, A.S. (2020). Recent advances in removal of lignin from paper industry wastewater and its industrial applications–A review. *Bioresource Technology* 312: 123636.

Haq, I., & Kalamdhad, A.S. (2021). Phytotoxicity and cyto-genotoxicity evaluation of organic and inorganic pollutants containing petroleum refinery wastewater using plant bioassay. *Environmental Technology and Innovations* 23: 101651.

Haq, I., Kalamdhad, A.S., & Pandey, A. (2022a). Genotoxicity evaluation of paper industry wastewater prior and post-treatment with laccase producing Pseudomonas putida MTCC 7525. *Journal of Cleaner Production* 342: 130981.

Haq, I., Singh, A., Kalamdhad, A.S. (2022b) Integrated technologies for the remediation of paper industry waste in a circular bioeconomy. In Editor(s): Sunita Varjani, Ashok Pandey, Mohammad J. Taherzadeh, Huu Hao Ngo, R.D. Tyagi, *Biomass, Biofuels, Biochemicals*, Elsevier, pp. 351–363, ISBN 9780323885119.

Haq, I., Roy, S., & Kalamdhad, A.S. (2022c). Characterization of Pulp and Paper Mill Wastewater and Its Toxicity Analysis Using Vigna Radiata. In *Environmental Degradation: Monitoring, Assessment and Treatment Technologies*; Springer International Publishing: Cham, Switzerland, pp. 261–269. ISBN 9783030941475.

Haq, I., & Kalamdhad, A.S. (2022) Advanced and Ecofriendly Technologies for the Treatment of Industrial Wastewater to Constrain Environmental Pollution. In: Arora, S., Kumar, A., Ogita, S., & Yau, Y.Y. (eds) *Innovations in Environmental Biotechnology*. Springer, Singapore. https://doi.org/10.1007/978-981-16-4445-0_22.

Haq, I., & Kalamdhad, A.S. (2023). Enhanced biodegradation of toxic pollutants from paper industry wastewater using Pseudomonas sp. immobilized in composite biocarriers and its toxicity evaluation. *Bioresource Technology Reports* 24:101674.

Haq, I., Kalamdhad, A.S., & Malik, A. (2024). Bioremediation of Petroleum Refinery Wastewater Using Bacillus subtilis IH-1 and Assessment of Its Toxicity. *Archives of Microbiology* 206:296.

Hegazy, G. E., Soliman, N. A., Ossman, M. E., Abdel-Fattah, Y. R., & Moawad, M. N. (2023). Isotherm and kinetic studies of cadmium biosorption and its adsorption behaviour in multi-metals solution using dead and immobilized archaeal cells. *Scientific Reports* 13(1): 2550.

Hołyńska-Iwan, I., Sobiesiak, M., Kowalczyk, W., Wróblewski, M., Cwynar, A., & Szewczyk-Golec, K. (2023). Nickel ions influence the transepithelial sodium transport in the trachea, intestine and skin. *Scientific Reports* 13(1): 6931.

Inamuddin. (2021). Toxic pollutants in the environment: Challenges in analytical chemistry-Volume III: Wastewater treatment using Biomass. *Current Analytical Chemistry* 17(7): 902–903. DOI: https://doi.org/10.2174/157341101707210615094649

Jahanban-Esfahlan, A., Jahanban-Esfahlan, R., Tabibiazar, M., Roufegarinejad, L., & Amarowicz, R. (2020). Recent advances in the use of walnut (Juglans regia L.) shell as a valuable plant-based bio-sorbent for the removal of hazardous materials. *RSC Advances* 10(12): 7026–7047.

Kumar, K. V., Gadipelli, S., Howard, C. A., Kwapinski, W., & Brett, D. J. (2021). Probing adsorbent heterogeneity using Toth isotherms. *Journal of Materials Chemistry A* 9(2): 944–962.

Lesmana, S. O., Febriana, N., Soetaredjo, F. E., Sunarso, J., & Ismadji, S. (2009). Studies on potential applications of biomass for the separation of heavy metals from water and wastewater. *Biochemical Engineering Journal 44*(1): 19–41. DOI: https://doi.org/10.1016/j.bej.2008.12.009

Nguyen, T. A. H., Ngo, H. H., Guo, W. S., Zhang, J., Liang, S., Yue, Q. Y., ...& Nguyen, T. V. (2013). Applicability of agricultural waste and by-products for adsorptive removal of heavy metals from wastewater. *Bioresource Technology* 148: 574–585.

Panqing, Y., Abliz, A., Xiaoli, S., & Aisaiduli, H. (2023). Human health-risk assessment of heavy metal–contaminated soil based on Monte Carlo simulation. *Scientific Reports* 13(1): 7033.

Park, D., Yun, Y. S., & Park, J. M. (2010). The past, present, and future trends of biosorption. *Biotechnology and Bioprocess Engineering*, *15*, 86–102. DOI: https://doi.org/10.1007/s12257-009-0199-4

Pokethitiyook, P., & Poolpak, T. (2016). Biosorption of heavy metal from aqueous solutions. *Phytoremediation: Management of Environmental Contaminants* 3: 113–141.

Sahmoune, M. N. (2019). Evaluation of thermodynamic parameters for adsorption of heavy metals by green adsorbents. *Environmental Chemistry Letters 17*(2): 697–704. DOI: https://doi.org/10.1007/s10311-018-00819-z

Saravanan, A., Karishma, S., Kumar, P. S., Varjani, S., Yaashikaa, P. R., Jeevanantham, S., ...& Reshma, B. (2021). Simultaneous removal of Cu (II) and reactive green 6 dye from wastewater using immobilized mixed fungal biomass and its recovery. *Chemosphere* 271: 129519.

Sasan Narkesabad, Z., Rafiee, R., & Jalilnejad, E. (2023). Experimental study on evaluation and optimization of heavy metals adsorption on a novel amidoximated silane functionalized Luffa cylindrica. *Scientific Reports* 13(1): 3670.

Sayago, U. F. C., & Ballesteros Ballesteros, V. (2023). Development of a treatment for water contaminated with Cr (VI) using cellulose xanthogenate from E. crassipes on a pilot scale. *Scientific Reports* 13(1): 1970.

Singh, S., Kumar, V., Datta, S., Dhanjal, D. S., Sharma, K., Samuel, J., & Singh, J. (2020). Current advancement and future prospect of biosorbents for bioremediation. *Science of the Total Environment* 709: 135895.

Thakur, A. K., Singh, R., Pullela, R. T., & Pundir, V. (2022). Green adsorbents for the removal of heavy metals from Wastewater: A review. *Materials Today: Proceedings* 57: 1468–1472.

Vilela, P. B., Matias, C. A., Dalalibera, A., Becegato, V. A., & Paulino, A. T. (2019). Polyacrylic acid-based and chitosan-based hydrogels for adsorption of cadmium: Equilibrium isotherm, kinetic and thermodynamic studies. *Journal of Environmental Chemical Engineering* 7(5): 103327.

Wang, J., & Chen, C. (2009). Biosorbents for heavy metals removal and their future. *Biotechnology Advances 27*(2): 195–226. DOI: https://doi.org/10.1016/j.biotechadv.2008.11.002

Yurdakal, S., Garlisi, C., Özcan, L., Bellardita, M., & Palmisano, G. (2019). (Photo) catalyst characterization techniques: adsorption isotherms and BET, SEM, FTIR, UV–Vis, photoluminescence, and electrochemical characterizations. In *Heterogeneous photocatalysis* (pp. 87–152). Elsevier.

Zareh, M. M., El-Sayed, A. S., & El-Hady, D. M. (2022). Biosorption removal of iron from water by Aspergillus niger. *npj Clean Water* 5(1): 58.

9 Phytoremediation of Microplastics from Industrial Wastewater

Mandeep Prakash Parhi, Varsha Gupta, A. A. Narmada Thathsarani Perera, Muskan Sharma, and Izharul Haq

9.1 INTRODUCTION

Plastic has become nearly ubiquitous in our day-to-day lives because of its strong and lightweight nature. Its corrosion-resistant, high thermal tolerance, and electrical insulation properties have multiple industrial, medical, and transportation applications (Bajt, 2021). As per Plastics Europe Market Research Group, the total plastic production has increased from two million metric tons in 1950 to 390.7 million metric tons in 2021, with most of the plastic produced in China alone (32%), followed by the rest of Asia (17%) and the world (51%). Most of the plastic production focuses on electronics, cosmetics, automotive, food/dairy, textiles, and clothing industries, with most widely used plastic types being polyethylene (PE), polystyrene (PS), polypropylene (PP), polyamide (PA), polyvinyl chloride (PVC) polyester (PEST), and polyethylene terephthalate (PET).

Most of these industries are set in urban areas, possibly near a large water source, and a large quantity of wastewater is generated from the industry-dense urban areas of these regions daily. This wastewater is known to contain a large number of microplastics (MPs; synthetic polymer less than 5 mm in size), which are then treated in a nearby wastewater treatment plant (WWTP) (Turan et al., 2021).

Industrial wastewater and WWTPs are known to be significant contributors to environmental MP pollution through effluent and sludge discharge (Ou et al., 2024). A typical WWTP receives diverse types of MPs which include microbeads, microfibers, pallets, and fragments which typically range from 1 to 1000 µm. These MPs mostly originate from cosmetics, nail polishes, face scrubbers, toothpaste, and medical equipment or formed due to the fragmentation of larger plastics such as plastic containers, panels, and synthetic fibers due to physical, chemical, or biological processes (Van et al., 2022). A conventional WWTP employs pretreatments, primary treatment, and secondary treatment. Various treatment processes, such as bar

DOI: 10.1201/9781003379386-9

screening, primary and secondary sedimentation, flocculation, and biofilm process/active sludge formation, are used. Additionally, tertiary treatments such as membrane filtration, sand filtration, and advanced oxidation, can be used to further improve effluent quality (Acarer, 2023). WWTPs can remove 25% to >99% of MPs depending upon the applied technology.

Literature review suggests that WWTPs in Vietnam are capable of removing 25.5% of MPs, whereas treatment plants in Wuhan, China can remove up to 65% of MPs from wastewater influent (Liu et al., 2019; Van et al., 2022). Additionally, some treatment plants present in Sydney, Australia, removed 66% of MPs and the ones present in the Charleston Harbor watershed, The United States are capable of removing 80 to 97.6%MPs from wastewater (Ziajahromi et al., 2017; Conley et al 2019). An estimated 78–98% of MPs are removed in the primary treatment alone and 7–20% in secondary treatment (Prata, 2018). Advanced wastewater treatment technologies such as membrane bioreactor can remove >99% of MPs, and sand filtration and dissolved air floatation can remove 97% and 95%, respectively (Prata, 2018). Despite the efficient removal of MPs by modern WWTPs, approximately 6.5×10^6 MPs are released into the environment every day after secondary treatment. This can be attributed to the large volume (10^8 L/day) wastewater containing small microbeads and fibers of size <0.5 mm is discharged into the environment (Murphy et al., 2016; Ou et al., 2024). Additionally, advanced treatment like coagulation requires the addition of metal ions to the water, which might have potential negative and toxic effects. Microbial degradation has low removal efficiency and even though membrane bioreactor has high removal efficiency, its performance drastically drops due to fouling of the surface and pores clogging up by plastics (Murphy et al., 2016). This necessitates in-situ bioremediation of the effluent containing MPs that has been discharged to the water system. Several field and laboratory studies have shown potential MPs adhesion to aquatic macrophytes which has opened the possibility of in-situ phytoremediation (Sharma, Gupta, and Gupta, 2024). Aquatic macrophytes have been extensively used for the past 30 years to remediate various organic pollutants, agrochemicals, heavy metals, pharmaceuticals and in recent years more and more emphasis is being given to MPs remediation (Rozman et al., 2022). This chapter focuses on phytoremediation of MPs using macrophytes in wastewater post-treatment, the mechanism used, and the cultivation strategies employed. This chapter will also examine interactions of various MPs with several species of macrophytes while discussing MPs long- and short-term impacts on the macrophytes. And lastly, we will discuss the limitation of phytoremediation.

9.2 CHARACTERIZATION OF MPS IN INDUSTRIAL WASTEWATER

The physio-chemical properties of MPs vary based on their origin. The wastewater originating from both domestic and industrial sources predominately carry polyamide (PA), polyethylene (PE), polyester (PEST), polycaprolactone (PCL), Ethylene acrylic acid (EAA), polypropylene (PP) and polyethylene terephthalate (PET), high density polyethylene (HDPE) (Acarer, 2023). Additionally, industrial wastewater is reported to carry plastic particles belonging to diallyl phthalate (DAP) and acrylonitrile styrene acrylate (ASA) (Franco et al., 2021). MPs in industrial wastewater are found to have diverse morphology such as spherical (granules, beads, pellets), lines (fibers, filaments), fragments, or foams with fiber being the most common shape. Franco et al. (2021) found that MPs found in the influent generally ranged from size 100 μm to 5000 μm with 53% being in the size range of 100–355 μm. In the effluent they found eight polymers, namely, HDPE, PVC, PE, EAA, ASA, DAP, PP, PCL, out of which HDPE, PVC, PE, and EAA, were more abundant (availability > 10%) (Sadia et al., 2022). Additionally, 71.8% of MPs were found to be in the size range of 100–355 μm in the effluent, indicating that this WWTP has less removal efficiency in case of small and lighter MPs. Another possible explanation can be fragmentation of larger MPs during treatment processes (Cabrera et al., 2023).

According to Tamburri et al. (2022), MPs can also be directly generated in the aquatic environment as a result of paint abrasion on ships and ropes or the breakdown of bigger plastic objects. This is on top of MPs found in wastewater. Such MPs contamination cannot be removed by physicochemical techniques. Therefore, biodegradation and in-situ phytoremediation are gaining so much attention nowadays. Bacteria are a popular candidate for MPs remediation and degradation along with fungi and algae. Phytoremediation on the other hand has been less popular when compared with bioremediation; however, it has started to gain great attention recently.

9.3 AQUATIC PHYTOREMEDIATION

According to certain studies, macrophytes may absorb very tiny MPs (Mateos-Cárdenas et al., 2021). However, it appears that adhesion, which is the process by which MPs adsorb on plant biomass, is the primary mode of interaction between MPs and macrophytes (Kalčíková et al., 2017). Aquatic phytoremediation by definition is bioremediation of pollutants by specifically using macrophytes. Microalgae and other microorganisms, also play crucial roles in aquatic phytoremediation (Fletcher et al., 2020). Macrophytes can be classified into three growth forms: floating, submerged, and emergent. Floating macrophytes float on the water surface with submerged roots which can be free floating or rooted to the soil. It includes the genera of *Lemna*(duckweeds) and *Nymphaea* (water lilies). Submerged macrophytes grow below the water surface, either attached to the substrate or unrooted. It includes the genus of *Ceratophyllum* (hornworts). And finally, macrophytes that are present on the edges of water bodies and anchored to a substrate. Major portions of their shoots grow above the water level and they include genus *Typha* (reedmace) and *Phragmites* (common reed) (Newete and Byrne, 2016). MPs are generally known to aggregate around the root of the plan and are then adsorbed (Figure 9.1). This is known as rhizostabilization. But as per a field study done by Rozman et al. (2022), weak adsorption was predominant so it was unclear if rhizostabilization was possible. The interaction of MPs with root is also size-dependent. Small MPs and nano plastics are easily taken up. Giorgetti et al. (2020) observed that nanoparticles aggregation was promoted by root exudate, which inhibited uptake of positively charged PS-NH2. This suggested that, in addition to size, surface charge also plays a role in uptake of plastic particles. After uptake, the MPs are translocated throughout the plant body (Figure 9.1). A study was done using fluorescent microbeads to investigate if and

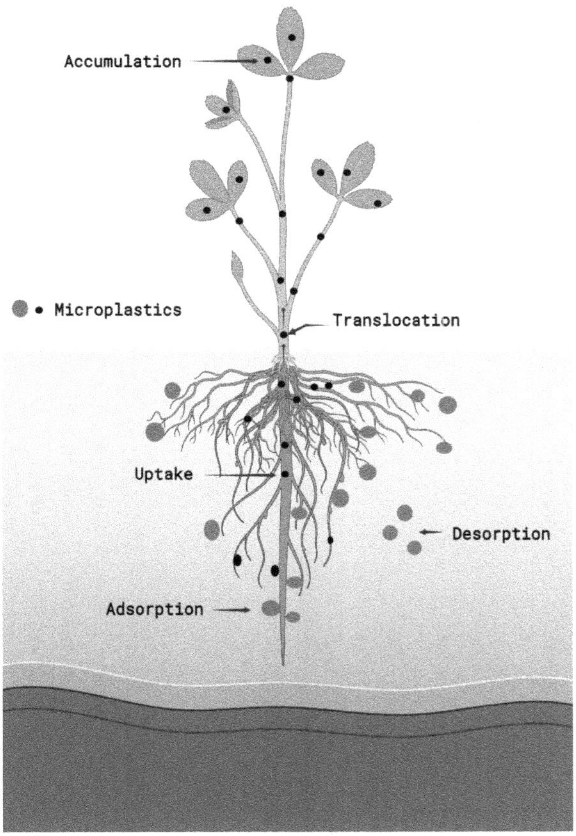

FIGURE 9.1 Fate of microplastics during phytoremediation.

how PS beads are translocated throughout *Triticum aestivum* (Li et al., 2020). They observed the bead movement was through apoplastic pathways from root to upper part of the plant. This confirmed that phytoextraction does take place in aquatic environments and is a vital mechanism of MPs phytoremediation.

9.3.1 Phytoremediation of Microplastics

According to recent research, both in the natural environment (Esiukova et al., 2020) and in laboratory settings (Kalčíková et al., 2017), MPs cling significantly to the biomass of aquatic macrophytes. By trapping and stabilizing MPs in water bodies, macrophytes can subsequently eliminate them through their biomass. There is also plenty of data to support the concept that different macrophytes are not considerably affected by ecologically relevant (and even greater) quantities of MPs (Capozzi et al., 2018). Some of the removing species and their ability to remove the microplastics by phytoremediation are shown in Table 9.1.

TABLE 9.1
Macrophytes used for MPs phytoremediation and its inference

Macrophytes	Conditions for phytoremediation of MPs	Observation of phytoremediation	Inference on phytoremediation	References
Urban tidal wetland	Analysis of MPs abundance in the tidal wetland and their distribution	It was observed that vegetation-free mudflats had abundance of microfibers, whereas vegetated wetlands had more deposition of microfragments. MPs accumulation was significant in the wetland.	Wetland serve as sink and source of MPs	Helcoski et al., (2020)
Lemna minor (L.)	Plants were exposed to PE MPs fragments of 150 μm at concentration of 9600 MP/L over 12 weeks	Growth, photosynthetic activity, and chlorophyll content remain unchanged. The roots became tolerant to MPs and although the adsorption was swift, desorption was higher.	Fast and weak adsorption was prevalent as the saturation was reached by day 7. MPs adhered were not degraded and environmental factors may alter adsorption and increase desorption	Rozman et al., (2022)
Lemna minor (L.)	This was a lab study in which experimental plant were grown on a medium containing 50,000 MPs per ml	PE MPs of six ranges 10–45 m were strongly adsorbed on the plant's surface. Up to 7 PE MPs were absorbed throughout the surface of the plan over 7 days. No impact on photosynthesis and growth.	The adsorption of MPs resulted in uptake of MPs by plant and consequently along the food chain.	Mateos-Cárdenas et al., (2019)
Phragmites australis	The PE MPs were co-exposed with Cu or Cd	Increased uptake of Cu and Cd. Presence of MPs did not impede phytoremediation of the heavy metal.	MPs Co-exposure had no effect on Cu and Cd phytoremediation.	Manjate et al., (2020)

9.4 MACROPHYTES ESTABLISHMENT SYSTEMS

Macrophyte establishment systems are essentially approaches to plant macrophytes in order to maximize phytoremediation of pollutants, in this case phytoremediation of MPs. We are going to discuss three planting systems namely, Constructed wetlands system (CWS), Floating treatment wetlands (FTW), and microalgae-based bioremediation (MBR).

9.4.1 Constructed Wetlands

Recent studies have focused on the potential of treatment wetlands in MPs removal from the released effluent (Cabrera et al., 2023). Treatment wetlands can either be a natural wetland system or constructed wetland systems (CWS). Natural wetlands are known to have macrophytes with high MP interception. MPs can be found in leaves, roots, or adhering to the surface of plants as biofilms (Cabrera et al., 2023; Duan et al., 2021). Additionally, local macroinvertebrate (mainly earthworm) populations found in the CWS and natural wetlands also play a crucial role in MPs distribution and break down. The gut microflora of these earthworms is also found to degrade MPs polymers into simpler and basic monomers (Lwanga et al., 2018). Field study done by Wang et al. (2020) showed that a CWS in Belgium can potentially remove 88% of MPs ranging up to 5600 μm in size. Similarly, In China Zhou et al. (2022) showed >89% of MPs ranging from 20–5000μm were removed by the CWS.

Constructed wetland systems (CWS) have been used for treatment of wastewater, wastewater polishing, sewage, and stormwater for a long time (Fletcher et al., 2020). A CWS is defined as an artificial natural wetland system by the US Environmental Protection Agency. It typically consists of several irregularly shaped or rectangular basins or cells coupled in a sequence and encircled by vegetation, rocks, or other materials (USEPA, 2000). A CWS with its low operation and maintenance requirements, serves as a nature-based waste water treatment solution for small communities or as tertiary treatment in bigger WWTPS (Cabrera et al., 2023). CWS, based on flow of water, can be classified as Free Water Surface flow wetlands (FWS), subsurface flow wetlands (SSF), and hybrid basins that incorporate surface and subsurface flows (Fletcher et al., 2020).

FWS wetlands are defined as wetland systems where the water surface is exposed to the atmosphere. A typical FWS CWS is generally a low-tech gravity-fed system which contains emergent and floating macrophytes, growing from a shallow basin with 20–30 cm root soil with 20–40 cm water depth, allowing the influent to flow through slowly. This will enable the capture uptake, and accumulations of MPs in the body of floating macrophytes (Fletcher et al., 2020).

SSF wetlands are more complex CWS in design, in which a channel or bed is designed sub-surface with substrate of stone, earth or gravel matrix, through which the influent water is passed. Emergent macrophytes grow over the substrate matrix, above the surface. This results in water flowing through the matrix and coming into direct contact with plant roots (Fletcher et al., 2020; Wang et al., 2020). SSF CWS are further divided into horizontal and vertical SSF CWS (HSSF CWS and VSSF CWS) based on the direction of flow through the substrate matrix (Perdana et al., 2018).

SSF CWS have several advantages over FWS CWS. In this method, the water surface is kept below the media surface, which minimizes the risk of odor3 along with other human, or ecological, exposure. Moreover, a typical SSF CWS is smaller than FWS CWS and thus is common in areas where land availability is limited, such as Europe, whereas FWS CWS are more common in North America and Australia (Cabrera et al., 2023; Ziajahromi et al., 2020).

9.4.1.1 HSSF CWS Integrated with Carrícola's WWTP

Cabrera et al. (2023) conducted field study at Carrícola's WWTP in Valencia, Spain to gain knowledge on MPs behavior and its fate in a full-scale CWS. The plant is designed to serve a community of 115 population equivalents (PE). One PE represents the organic pollutant load produced by one person per day, expressed in grams of BOD per day. The primary treatment is done by grids after which the wastewater is transferred to two Emscher settlers (18 m^3). Following this, the wastewater is transferred through pipes to two horizontal subsurface flow CWs (HSSF CWS1 each measuring 200 m^2 each), which are arranged in parallel fashion.

The connecting pipe (0.3 m in diameter) opens into the water distribution chamber where a vertical pipe is present, 1 m above the ground. This elevation results in settling of solids present in the pipes connecting the Emscher tanks with CWS1. This pipe is drained monthly to CWS2. Two types of soil are filled in the CWSs, 40 mm coarse gravels and 10–25 mm fine gravels and arrangement in which they are filled is shown in Figure 9.2. The basins are planted with common reed plants (*Phragmites*). A second CWS (HSSF CWS2) with similar configuration is present in series to CWS1 for extended exposure time of water. As per earlier studies, the inflow is recorded to be 16 m^3per day, with 98% of total solids removed and COD removal of 94% in the final effluent (Hernandez-Crespo et al., 2022). Four sampling sites are chosen, namely: 1.

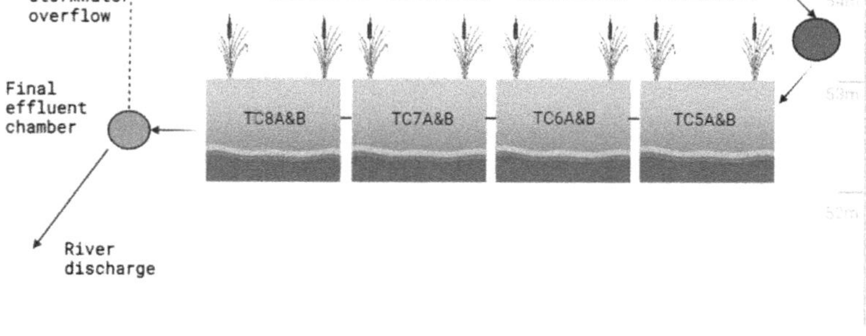

FIGURE 9.2 Cromhall's ICWS scheme showing 12 FWSCW connected in series. The system serves as tertiary wastewater treatment. (Bydalek et al., 2023.)

influent, 2. effluent of primary treatment, 3. Outlet of CWS1, and 4. Final effluent of the treatment plant. The sampling was done twice. On average,17.4% MPs were removed in the Emscher settlers and around 98.6% from CWS1. It was observed that MP concentration slightly increased upon subsequent sampling at outlets of CWS1 and CWS2, which was attributed to monthly sludge drainage operation (Cabrera et al., 2023). Three groups of MPs were found, namely fibers, films, and particles, of which the predominant form of MPs were fiber, followed by films. Around 97.5% of fibers were removed and almost complete removal of particles was observed. This was likely due to the filtering effect of the CWS.

9.4.1.2 Cromhall Integrated Constructed Wetland System

Bydalek et al. (2023) conducted a field study at Cromhall integrated constructed wetland system (ICWS), which is a wastewater improving FWS CWS (Figure 9.2). The wastewater is first treated in an oxidation ditch located in South West Gloucestershire, UK. This sewage treatment plant is part of a combined sewage system which collects sewage and runoffs from the rural catchment of population equivalent to 2100. The ICWS was commissioned in March 2020 and since then has been receiving wastewater of 1.4 ML per day. It has an active surface area of approx. 8000 m^2 which has been arranged in a sectional design arrangement. It consists of 12 segregated treatment cells (TC1- TC8). TC 1 functions as a settling pond with very less vegetation coverage (10%). TC 2- TC 4 has a large vegetation coverage (around 90%). TC 5-12 has open water surface zones with moderate vegetation coverage (30–80%). The vegetation primarily comprises emerging macrophytes such as *Schoenoplectuslacustris* and *Typha angustifolia.* It was reported that ICW retained approx. 95% of MPs. Settling ponds with no macrophytes contributed 38% of removal and shallow TCs with plenty of vegetation were found capable of removing 88%. The open water TCs (TC5-TC8) downstream were not found to contribute any significant removal (3%) due to low inflow of MP concentration as well as possible preferential flow paths that are generated between patches of vegetation.

9.4.2 Floating Treatment Wetlands

Floating treatment wetlands (FTWs) are a variation of traditional constructed wetlands. The variation lies in the fact that the plants are grown on a matrix which is placed on a floating frame and held by use of ropes. The plants grown here are emergent and floating with roots submerged in the water. This essentially creates small artificial islands floating on the surface of water (Fletcher et al., 2020). Some examples include *Typha latifolia, Canna indica,* and water hyacinth. The FTW are typically constructed using PVC pipes, foam, wire mesh, or bamboo (Pavlineri et al., 2017). The most crucial factors to keep in mind while choosing material are buoyancy, durability, cost, and availability. Rafts or floating mats are used commonly to provide a frame for the plants to grow on. Occasionally the plants are grown on wire meshes and then placed on the rafts. Sometimes PVC pipes are also used to provide buoyancy. Few considerations are taken while designing FTWs. For example, the rafts/floating mats should be resilient and durable to withstand prolonged environmental exposure and at the same time support the load of the plant and its growth media.

The design of the raft/floating mat is such that it can cover the whole area for efficient MPs retention. Additionally, the design must ensure a proper flow distribution of water is maintained (Wei et al., 2020). Different planting methods can be used such as using seeds, planting juvenile seedlings, or by vegetative propagation. Macrophyte species such as Typha are generally planted through vegetation on the floating mat. Planting of *Phragmites* is also planted using similar techniques. Nutrient-rich growth media should be selected for better cultivation. Mostly coconut fibers with soil are used as a growth media (Shahid et al., 2018). The growth media should have optimum pH for plant growth. Additionally, the growth media must be resistant to waterlogging. It is also advisable to avoid planting tall plants, as during extreme weather conditions, the winds might cause the mats to disperse or capsize. Furthermore, dead matter accumulation should be limited.

9.4.2.1 Monasterios WWTP

Cabrera et al. (2023) also conducted a field study at Los Monasterios WWTP in Valencia, Spain. This WWTP serves a population equivalent to 1500. The pretreatment here is carried out by grit chamber and an aerated biological reactor of volume 176 m^3 is present for secondary treatment. Helophytes (*Typha* and *Phragmites*) are present, floating on top of the floating mat surface with its root system extending up to 4m in depth. The biological reactors operate on an extended aeration mode, recirculating the sludge. This is followed by a tank for sedimentation of volume 48 m^3. And lastly it has an FTW of 348 m^2 in surface area and 557 m^3 in volume for tertiary treatment. Similar to biological reactors, it has macrophytes of the same species (*Typha* and *Phragmites*) growing on a floating mat.

A Horizontal subsurface flow CWS of surface area 68 m^2 and volume 18 m^3 with gravel (2–3 cm) is present. Common reed plant is planted on HSSF CWS for remediation. It receives 100–120 m^3 of influent per day. All of the discharged water is then reused. Three different sample locations were chosen: 1. influent to the treatment plant, 2. Outlet of biological reactor, and 3. The outlet of HSSF CW. The sampling was done twice. The removal efficiency by secondary treatment alone was 53%, which improved even further to 96.2% after the water was treated in the CWS (FTW and HSSF CW combined). Fibers dominated the MPs found in all of the three sampling points (70–80%) similar to Carrícola's WWTP and the removal efficiency for fibers was nearly identical. In the case of size of the MPs, it was observed that the percentage distribution of larger MPs decreased and there was a clear percentage shift towards smaller dimensions. This trend was consistent with HSSF CWS at Carrícola's WWTP. The possible explanation given was mechanical wear-tear during HSSF-CW phase or ingestion by earthworm present (Cabrera et al. 2023; Chen et al., 2021).

9.4.3 Limitation of CWS and FTW

The CWS has many advantages over natural wetland systems in remediating wastewater from diverse sources. However, it has some disadvantages such as, inconsistency of the system to work efficiently due to changing in environment; toxic effects of prolonged exposure to MPs as well as other chemicals used for plant growth (e.g., ammonia and pesticides); mismanagement in harvesting macrophytes; and possible surges in waste water flow may reduce efficacy of CWS temporarily. Moreover, the materials used in creating FTWs (e.g., PVC pipes and polystyrene foam sheets) can also contribute to secondary MPPs (Zhang et al., 2020). It is also known that the filtered MPs settle down and accumulate in CWS. Ziajahromi et al.(2020) conducted a study by collecting sediment cores from one of the stormwaters CWS in Australia. The mean MP presence was found to be 320 MP/kg at the inlet and 595 MP/kg at the outlet. It shows that CWS can act as a potential pollutant sink. Additionally, prolonged exposure of *Chironomus tepper* with MPs has detrimental effects on its growth parameter and impedes its emergence (Lu et al., 2022).

9.4.4 Microalgae-based Bioremediation (MBR)

Microalgae-based bioremediation technique (MBR) has gained some attention recently to bioremediate MPs out of wastewater. Certain species such as *Tetraselmis* sp., and *Gloeocapsa* Sp., are able to colonize submerged MPs surfaces. Several kinds of MPs materials such as polyethylene (PE), polystyrene (PS), polyvinyl chloride (PVC), polypropylene (PP), and high-density polyethylene (HDPE), can be effectively removed by *Scenedesmus abundans* (Cheng and Wang, 2022).

Microalgae are known to secrete exopolysaccharides (EPS), which can aggregate scattered MP particles forming small clumps (Cunha et al., 2020). In addition to this, microalgae secrete ligninolytic enzymes which degrade the bond present between the constituent units of MP polymer (ethylene or propylene) into simple monomers (Reddy and Nair, 2022). The effectiveness of forming hetero-aggregates depends upon both algae species (its morphology, metabolism, and amount of EPS secreted) and the characteristics of plastics (size, density, hydrophobicity) (Cheng and Wang, 2022). A positive correlation has been reported between the surface roughness of the particle and the number of microalgae attached. It was also noted that hydrophilic MPs are more favorable for biofilm formation (Cunha et al., 2020). Study showed HDPE MPs were less likely to form hetero-aggregations with microalgae regardless of exposure time. This might have something to do with the difference in density of HDPE and microalgae. HDPE are generally less dense than microalga cells which may reduce the chances for effective interaction (Gao et al., 2023). Although many studies investigated the abilities of microalgae to form hetero-aggregates with MPs, it is not enough to quantify MP removal.

Sakhon et al. (2019) observed that even though the removal efficiency of *Tetraselmissuecica* was high, the formation of aggregates was less after it was exposed to PS MPs. This high removal can be attributed to enhanced adsorption onto the container surface by the action of secreted EPS, this observation is also corroborated by Long et al. (2017). Therefore, in order to quantify the removal efficiency both hetero-aggregations and surface adsorptions should be taken into account. Cheng and Wang(2022) noted that *Scenedesmus abundans* can remove MPs such as PS, polylactide, and poly (methyl methacrylate) with a removal efficiency of 84%, 87%, and 98%, respectively. Microalgae can utilize organic nutrients present in the wastewater (Wang et al., 2022). This has an added benefit, namely the production of algal biomass which can be harvested and used as feedstock for biofuel production. Other value-added products such as pigments, proteins, enzymes, and vitamins can also be obtained using the harvested microalgae (Deng et al., 2018).

9.4.4.1 Limitations to MBR Technology

The MBR technique has many advantages over conventional techniques, however there are some major challenges restricting its large-scale applications. Prolonged exposure to MP results in decreased photosynthetic efficiency. This is because MPs can adhere to microalgae cell surface and obstruct nutrient uptake, photosynthesis, and respiration (Dong et al., 2022). In some cases, MPs interaction with algal cells increased reactive oxygen species production (Zhang et al., 2017). Additionally, the MPs interact with other environmental pollutants such as heavy metals and hydrocarbons which can alter the chemical properties of the particle, thus complicating the dynamics and leads to low removal efficiency (Song et al., 2020; Gao et al., 2023). Collectively these are known as toxic effects of MPs and it is a limiting factor in up-scaling of MBR technology.

Generally, for an effective remediation of MPs present in the wastewater, long holding time is required by microalgae. This is due to the fact that MPs come in different shapes, sizes, and with different chemical properties. So, a microalga requires a long holding time to degrade MPs, but a typical biological reactor just holds MPs for 4–10 hours which is not sufficient time for effective degradation (Hamidian et al., 2021; Zhang et al., 2020).

Residual algal cells also possess an environmental risk in the treatment system. Commonly, to harvest algal cells after remediation is completed, one can use sedimentation, flocculation, and filtration. Using these methods one can recover somewhere between 10 and 90% of algal cells (Liu et al., 2023). But even with such efficient recovery, a small quantity of algal cells can still remain in the system after harvesting. This has the potential to negatively impact the aquatic system and cause environmental toxicity (Gao et al., 2023).

Future investigation should focus on overcoming the above challenges. The size, shape, and chemical characteristics are important parameters determining the MPs level of toxicity and its retention time in the system. Knowing these parameters will be helpful in developing new strategies to tackle the drawbacks. Some strategies developed may aim to understand the microalgal species-plastic particles dynamics and choose appropriate species to minimize MPs toxicity. Some strategies can augment the existing harvesting protocol to reduce environmental risks. While some other strategies may focus on integrating MBR technology with other conventional phytoremediation methods, which will lead to novel treatment solutions (Gao et al., 2023).

9.5 SUMMARY AND CONCLUSION

From above studies it is clear that macrophytes possess the ability to remediate MPs out of the wastewater using mechanisms such as adsorption, aggregation, and translocation. This has certainly opened up an avenue for phytoextraction, rhizofiltration, and rhizostabilization of MPs. Macrophytes planted in CWs have shown great removal efficiency, and when these systems are integrated with existing wastewater remediation technologies, close to 100% of removal efficiency was achieved. Still, much is left to be desired from macrophytes. Like their terrestrial counterparts, macrophytes do not seem to possess any mechanism to degrade those accumulated MPs. This might be due to a lack of development of any metabolic pathway to utilize the carbon present in the MP. Additionally, the wetlands (both natural and constructed) harbor great diversity of marine and terrestrial wildlife that rely on aquatic vegetation. So, without any degradation, it is inevitable that the accumulated particles in the aquatic system will find their way to higher organisms. MBR technology does show a great potential to degrade MPs, and integrating with macrophytes' remediation capability, we can address accumulation of MPs in higher order organisms. But first we have to address the drawbacks of using MBR technology, namely toxicity to prolonged MP exposure. Genetic modification and transgenics can be used to augment adhesion capability of macrophyte and increase algal cell resistance to MP toxicity.

In conclusion, more studies are required to holistically integrate the existing and emerging phytoremediation strategies with our current WWTP infrastructure and water cleaning technologies. A special emphasis on the problem of MP accumulation in plant body and better optimization of MBR technology is needed. Additionally, we have to consider other biotic and abiotic factors before introducing macrophytes species. With these we can come closer to a future of sustainable, economical, and environmentally friendly solutions to treat our industrial wastewater.

REFERENCES

Acarer, S. (2023). Microplastics in wastewater treatment plants: sources, properties, removal efficiency, removal mechanisms, and interactions with pollutants. *Water Science & Technology*, 87(3), 685–710.

Bajt, O. (2021). From plastics to microplastics and organisms. *FEBS Open bio*, 11(4), 954–966.

Bydalek, F., Ifayemi, D., Reynolds, L., Barden, R., Kasprzyk-Hordern, B., & Wenk, J. (2023). Microplastic dynamics in a free water surface constructed wetland. *Science of the Total Environment*, 858, 160113.

Cabrera, D. C., Wang, Q., Martín, M., Rajadel, N. O., Rousseau, D. P., & Hernández-Crespo, C. (2023). Microplastics occurrence and fate in full-scale treatment wetlands. *Water Research*, 240, 120106.

Capozzi, F., Carotenuto, R., Giordano, S., & Spagnuolo, V. (2018). Evidence on the effectiveness of mosses for biomonitoring of microplastics in freshwater environment. *Chemosphere*, 205, 1–7.

Chen, Y., Li, T., Hu, H., Ao, H., Xiong, X., Shi, H., & Wu, C. (2021). Transport and fate of microplastics in constructed wetlands: A microcosm study. *Journal of Hazardous Materials*, 415, 125615.

Cheng, Y. R., &Wang, H. Y. (2022). Highly effective removal of microplastics by microalgae Scenedesmus abundans. *Chemical Engineering Journal*, 435, 135079.

Conley, K., Clum, A., Deepe, J., Lane, H., & Beckingham, B. (2019). Wastewater treatment plants as a source of microplastics to an urban estuary: Removal efficiencies and loading per capita over one year. *Water Research X*, 3, 100030.

Cunha, C., Silva, L., Paulo, J., Faria, M., Nogueira, N., & Cordeiro, N. (2020). Microalgal-based biopolymer for nano-and microplastic removal: a possible biosolution for wastewater treatment. *Environmental Pollution*, 263, 114385.

Deng, X., Gao, K., Addy, M., Chen, P., Li, D., Zhang, R., ... & Ruan, R. (2018). Growing Chlorella vulgaris on mixed wastewaters for biodiesel feedstock production and nutrient removal. *Journal of Chemical Technology & Biotechnology*, 93(9), 2748–2757.

Dong, J., Li, L., Liu, Q., Yang, M., Gao, Z., Qian, P., ...& Deng, X. (2022). Interactive effects of polymethyl methacrylate (PMMA) microplastics and salinity variation on a marine diatom Phaeodactylumtricornutum. *Chemosphere*, 289, 133240.

Duan, J., Han, J., Cheung, S. G., Chong, R. K. Y., Lo, C. M., Lee, F. W. F., ...& Zhou, H. C. (2021). How mangrove plants affect microplastic distribution in sediments of coastal wetlands: Case study in Shenzhen Bay, South China. *Science of the Total Environment*, 767, 144695.

Esiukova, E., Zobkov, M., & Chubarenko, I. (2020). Data on microplastic contamination of the Baltic Sea bottom sediment samples in 2015–2016. *Data in brief*, 28, 104887.

Fletcher, J., Willby, N., Oliver, D. M., & Quilliam, R. S. (2020). Phytoremediation using aquatic plants. *Phytoremediation: In-situ Applications*, 205–260.

Franco, A. A., Arellano, J. M., Albendín, G., Rodríguez-Barroso, R., Quiroga, J. M., & Coello, M. D. (2021). Microplastic pollution in wastewater treatment plants in the city of Cádiz: Abundance, removal efficiency and presence in receiving water body. *Science of the Total Environment*, 776, 145795.

Gao, N., Ning, R., & Deng, X. (2023). Feasibility, challenges, and future prospects of microalgae-based bioremediation technique for removing microplastics from wastewater. *Frontiers in Bioengineering and Biotechnology*, 11, 1288439.

Giorgetti, L., Spanò, C., Muccifora, S., Bottega, S., Barbieri, F., Bellani, L., & Castiglione, M. R. (2020). Exploring the interaction between polystyrene nanoplastics and Allium cepa during germination: Internalization in root cells, induction of toxicity and oxidative stress. *Plant Physiology and Biochemistry*, 149, 170–177.

Hamidian, A. H., Ozumchelouei, E. J., Feizi, F., Wu, C., Zhang, Y., & Yang, M. (2021). A review on the characteristics of microplastics in wastewater treatment plants: A source for toxic chemicals. *Journal of Cleaner Production*, 295, 126480.

Hernández-Crespo, C., Oliver, N., Peña, M., Añó, M., & Martín, M. (2022). Valorisation of drinking water treatment sludge as substrate in subsurface flow constructed wetlands for upgrading treated wastewater. *Process Safety and Environmental Protection*, 158, 486–494.

Helcoski, R., Yonkos, L. T., Sanchez, A., & Baldwin, A. H. (2020). Wetland soil microplastics are negatively related to vegetation cover and stem density. *Environmental Pollution*, 256, 113391.

Kalčíková, G., Gotvajn, A. Ž., Kladnik, A., & Jemec, A. (2017). Impact of polyethylene microbeads on the floating freshwater plant duckweed Lemna minor. *Environmental Pollution*, 230, 1108–1115.

Li, L., Luo, Y., Peijnenburg, W. J., Li, R., Yang, J., & Zhou, Q. (2020). Confocal measurement of microplastics uptake by plants. *MethodsX*,7, 100750.

Liu, K., Wang, X., Fang, T., Xu, P., Zhu, L., & Li, D. (2019). Source and potential risk assessment of suspended atmospheric microplastics in Shanghai. *Science of the Total Environment*, 675, 462–471.

Liu, Z., Hao, N., Hou, Y., Wang, Q., Liu, Q., Yan, S., ... & Zhao, L. (2023). Technologies for harvesting the microalgae for industrial applications: Current trends and perspectives. *Bioresource Technology*, 129631.

Long, M., Paul-Pont, I., Hegaret, H., Moriceau, B., Lambert, C., Huvet, A., & Soudant, P. (2017). Interactions between polystyrene microplastics and marine phytoplankton lead to species-specific hetero-aggregation. *Environmental Pollution*, 228, 454–463.

Lu, H. C., Ziajahromi, S., Locke, A., Neale, P. A., & Leusch, F. D. (2022). Microplastics profile in constructed wetlands: Distribution, retention and implications. *Environmental Pollution*, 313, 120079.

Lwanga, E. H., Thapa, B., Yang, X., Gertsen, H., Salánki, T., Geissen, V., & Garbeva, P. (2018). Decay of low-density polyethylene by bacteria extracted from earthworm's guts: A potential for soil restoration. *Science of the Total Environment*, 624, 753–757.

Manjate, E., Ramos, S., & Almeida, C. M. R. (2020). Potential interferences of microplastics in the phytoremediation of Cd and Cu by the salt marsh plant Phragmites australis. *Journal of Environmental Chemical Engineering*, 8(2), 103658.

Manual Constructed wetlands treatment of municipal wastewaters. (2000). United States Environmental Protection Agency, EPA/625/R-99/010. https://nepis.epa.gov/Exe/ZyPURL.cgi?Dockey=30004TBD.txt. (Accessed March 9, 2024.)

Mateos-Cárdenas, A., Scott, D. T., Seitmaganbetova, G., van Pelt Frank, N. A. M., & AK, J. M. (2019). Polyethylene microplastics adhere to Lemna minor (L.), yet have no effects on plant growth or feeding by Gammarus duebeni (Lillj.). *Science of the Total Environment*, 689, 413–421.

Mateos-Cárdenas, A., van Pelt, F. N., O'Halloran, J., & Jansen, M. A. (2021). Adsorption, uptake and toxicity of micro-and nanoplastics: Effects on terrestrial plants and aquatic macrophytes. *Environmental Pollution*, 284, 117183.

Murphy, F., Ewins, C., Carbonnier, F., & Quinn, B. (2016). Wastewater treatment works (WwTW) as a source of microplastics in the aquatic environment. *Environmental Science & Technology*, 50(11), 5800–5808.

Newete, S. W., & Byrne, M. J. (2016). The capacity of aquatic macrophytes for phytoremediation and their disposal with specific reference to water hyacinth. *Environmental Science and Pollution Research International*, 23(11), 10630–10643.

Ou, H., Liu, R., Liao, Z., & Zeng, E. Y. (2024). Occurrence and fate of microplastics in urban water management systems. In *Microplastic Contamination in Aquatic Environments* (pp. 181–228). Elsevier.

Pavlineri, N., Skoulikidis, N. T., & Tsihrintzis, V. A. (2017). Constructed floating wetlands: a review of research, design, operation and management aspects, and data meta-analysis. *Chemical Engineering Journal*, 308, 1120–1132.

Perdana, M.C., Sutanto, H.B. and Prihatmo, G. (2018). Vertical Subsurface Flow (VSSF) constructed wetland for domestic wastewater treatment. In IOP Conference Series: Earth and Environmental Science. IOP Publishing, 148, 012025.

Prata, Joana Correia (2018). Microplastics in wastewater: State of the knowledge on sources, fate and solutions. *Marine Pollution Bulletin*, 129(1), 262–265.

Reddy, A. S., & Nair, A. T. (2022). The fate of microplastics in wastewater treatment plants: An overview of source and remediation technologies. *Environmental Technology & Innovation*, 28, 102815.

Rozman, U., Kokalj, A. J., Dolar, A., Drobne, D., &, G. (2022). Long-term interactions between microplastics and floating macrophyte Lemna minor: The potential for phytoremediation of microplastics in the aquatic environment. *Science of the Total Environment*, 831, 154866.

Sadia, M., Mahmood, A., Ibrahim, M., Irshad, M. K., Quddusi, A. H. A., Bokhari, A., ... & Show, P. L. (2022). Microplastics pollution from wastewater treatment plants: A critical review on challenges, detection, sustainable removal techniques and circular economy. *Environmental Technology & Innovation*, 102946.

Sakhon, E. G., Mukhanov, V. S., & Khanaychenko, A. N. (2019). Phytoplankton exopolymers enhance adhesion of microplastic particles to submersed surfaces. *Ecologica Montenegrina*, 23, 60–69.

Shahid, M. J., Arslan, M., Ali, S., Siddique, M., & Afzal, M. (2018). Floating wetlands: a sustainable tool for wastewater treatment. *Clean–Soil, Air, Water*, 46(10), 1800120.

Sharma, M., Gupta, V., & Gupta, D. (2024). Advancement in technologies of water management in rural areas. In *Water Resources Management for Rural Development* (pp. 117–126). Elsevier.

Song, C., Liu, Z., Wang, C., Li, S., & Kitamura, Y. (2020). Different interaction performance between microplastics and microalgae: The bio-elimination potential of Chlorella sp. L38 and Phaeodactylum tricornutum MASCC-0025. *Science of the Total Environment*, 723, 138146.

Tamburri, M. N., Soon, Z. Y., Scianni, C., Øpstad, C. L., Oxtoby, N. S., Doran, S., & Drake, L. A. (2022). Understanding the potential release of microplastics from coatings used on commercial ships. *Frontiers in Marine Science*, 9, 1074654.

Turan, N. B., Erkan, H. S., & Engin, G. O. (2021). Microplastics in wastewater treatment plants: Occurrence, fate and identification. *Process Safety and Environmental Protection*, 146, 77–84.

Van Do, M., Le, T. X. T., Vu, N. D., & Dang, T. T. (2022). Distribution and occurrence of microplastics in wastewater treatment plants. *Environmental Technology & Innovation*, 26, 102286.

Wang, Q., Hernández-Crespo, C., Santoni, M., Van Hulle, S., & Rousseau, D. P. (2020). Horizontal subsurface flow constructed wetlands as tertiary treatment: can they be an efficient barrier for microplastics pollution? *Science of the total environment*, 721, 137785.

Wang, S., Shi, W., Huang, Z., Zhou, N., Xie, Y., Tang, Y., ...& Zheng, H. (2022). Complete digestion/biodegradation of polystyrene microplastics by greater wax moth (Galleria mellonella) larvae: direct in vivo evidence, gut microbiota independence, and potential metabolic pathways. *Journal of Hazardous Materials*, 423, 127213.

Wei, F., Shahid, M. J., Alnusairi, G. S., Afzal, M., Khan, A., El-Esawi, M. A., ...& Ali, S. (2020). Implementation of floating treatment wetlands for textile wastewater management: A review. Sustainability, 12(14), 5801.

Zhang, C., Chen, X., Wang, J., & Tan, L. (2017). Toxic effects of microplastic on marine microalgae Skeletonema costatum: interactions between microplastic and algae. Environmental pollution, 220, 1282–1288.

Zhang, J., Gao, D., Li, Q., Zhao, Y., Li, L., Lin, H., ...& Zhao, Y. (2020). Biodegradation of polyethylene microplastic particles by the fungus Aspergillus flavus from the guts of wax moth Galleria mellonella *Science of the Total Environment*, 704, 135931.

Zhou, X., Zhao, Y., Pang, G., Jia, X., Song, Y., Guo, A., ...& Ji, M. (2022). Microplastic abundance, characteristics and removal in large-scale multi-stage constructed wetlands for effluent polishing in northern China. *Chemical Engineering Journal*, 430, 132752.

Ziajahromi, S., Drapper, D., Hornbuckle, A., Rintoul, L., & Leusch, F. D. (2020). Microplastic pollution in a stormwater floating treatment wetland: Detection of tyre particles in sediment. *Science of the Total Environment*, 713, 136356.

Ziajahromi, S., Neale, P. A., Rintoul, L., & Leusch, F. D. (2017). Wastewater treatment plants as a pathway for microplastics: development of a new approach to sample wastewater-based microplastics. *Water Research*, 112, 93–99.

10 Elimination of Microplastics from Textile Industry Wastewater Using Various Treatment Technologies

Muskan Sharma, Varsha Gupta, Mandeep Prakash Parhi, A. A. Narmada Thathsarani Perera, and Izharul Haq

10.1 INTRODUCTION

Plastic has been in extensive use in many different industries. Over 400.3 million tons of plastics were produced in 2022 worldwide (Plastics Europe, 2023). Despite being a widely used polymeric material and playing a significant role in the economy, they are known to be a potentially dangerous source of environmental pollution. The United Nations Environment Program (UNEP, 2017) reports that plastic makes up more than 80% of marine trash. Furthermore, it is projected that by 2050, the oceans may contain more plastic by weight than fish if the pace of plastic trash release into the oceans does not decline (Auta et al., 2017).

The textile industry, which has been around forever and is huge, is a major polluter mainly because of all the chemicals it uses. About 70% of the pollution, it creates comes from chemical processing. India's fabric industry is crucial for the nation's economic development. The expansion of the textile sector not only affects a country's earnings but also plays a role in the global economy and how nations exchange their technological advancements. The debris of fiber pieces is the most frequent and durable change to the surface area of our planet. The demand for synthetic polymers and plastics has grown quickly since plastic was first produced in large numbers in the 1950s; in 2021, 391 million metric tons (MMT) of these materials were produced worldwide (Statista, 2021). Moreover, it is projected that by 2050, this amount would rise to 589 MMT (Satista, 2021). Tiny and big pieces of plastic,

DOI: 10.1201/9781003379386-10

like microplastics and larger plastic items, are being found all over the place now – in the air around Earth, even on Mount Everest according to Napper et al. in 2020, on coastlines, remote islands, and even in the deepest parts of the ocean (Ferreira et al., 2020) due to their massive production and consumption rates.

The textile industry also churns out loads of waste, like fibers, sludge, and water polluted with chemicals. When this contaminated water mixes with natural resources and ecosystems, like soil and waterways, it messes up their quality. The issue of solid and liquid waste has resulted in significant environmental contamination issues for the textile industry (Haq, Singh, and Kalamdhad, 2022b). As a result, scientists and the textile sector are concentrating on lowering the amount of wastewater generated by the textile industry and developing effective replacement methods that don't harm the environment.

Using techniques including screening, mechanical flocculation, sedimentation, neutralization, and chemical coagulation, the textile sector treats its effluent to remove suspended materials and lower its quality (Khandegar et al., 2013). The primary goals of the secondary treatment are to eliminate colloidal and dissolved organic compounds, as well as any dyes or color that may be present in the waste water. These tasks are typically accomplished by trickling filtration, anaerobic digestion, and aerated lagoons (Haq, Kalamdhad, and Panday, 2022a). The tertiary treatment, on the other hand, aims to improve the efficiency of the little contaminated water left over from the previous process so that it can be released into waterbodies without causing any issues. This water is obtained using a variety of techniques, including oxidation, electrolytic precipitation, and membrane technologies (Haq, Roy, and Kalamdhad, 2022c).

This chapter will focus on the fibrous microplastic that are released from the textile industries and will present techniques that are used and can be used for the elimination of these fibrous microplastics.

10.2 ELIMINATION OF MICROPLASTICS RELEASED FROM TEXTILE INDUSTRY

Microfibers (MFs) released from textiles have been seen in a variety of settings, suggesting that human activity has an effect on natural systems. According to reports, the main cause of microplastic contamination is synthetic textile microfibers, a subgroup of microplastic fibers (MPFs) (Haq, Raj, and Markandenya, 2018). It is anticipated that as textile production increases, the issue of MF pollution will exacerbate and become more difficult to handle.

10.2.1 TEXTILE MICROFIBERS

In terms of size, microfibers and microplastics are comparable; they both have a diameter of less than 5 mm. It should be noted that their makeup is not limited to plastic alone. Natural fibers like cotton, wool, silk, and hemp, as well as plastic microfibers made of synthetic materials like polyacrylonitrile (PAN), polyester, polypropylene, polyamide (PA), and polyethylene, can seriously harm an organism's internal organs when consumed. Naturally derived microfibers usually biodegrade in the atmosphere. Nevertheless, the process of biodegradation may be slowed down by the functionalization of textiles, and aquatic creatures are adversely affected by these materials. The textile industry uses large amounts of several chemicals during its production process (Luongo, 2015). Endocrine disruptors constitute the majority of polymeric compounds discovered in coastal settings (Periyasamy, 2021). A multitude of additives are added during the fiber-making process with the aim of improving the fibers' functionality and processability. Phthalates often get mixed into PVC coatings for textiles to boost flexibility and strength. In textile making, flame-retardant fabrics are crafted using hexafluorotitanate salts, TiO2 NPs, and brominated and phthalate chemicals (Cheng et al., 2017). These materials have been demonstrated to be hazardous to reproduction and development. Formaldehyde-based resins are frequently added to cellulose-based fabrics to enhance their crease healing capabilities, while polyfluoroalkyl compounds are applied to textiles to enhance their water repellency (Lassen et al., 2015). Textile microfibers have been found in marine sediments and creatures, and it is the textile industry that is to blame for this unseen pollution (Periyasamy and Tehrani-Bagha, 2022).

10.2.1.1 Microfiber Toxicity: The Role of the Textile Industry

One of the crucial and valuable steps in the chain of manufacturing for textiles is the coloring and finishing process. Usually, textile materials are functionalized with synthetic dyes and chemicals to enhance certain aspects of the textiles (Periyasamy and Militky, 2020). To create dark and vivid hues, metal complex dyes as well as chrome or mordant dyes are used. Lead (Pb) (Liu et al., 2019), copper (Cu),mercury (Hg) (Michielssen et al., 2016), arsenic (As) (Mintenig et al., 2017), chromium (Cr), cadmium (Cd), nickel (Ni), and zinc (Zn) are among the heavy metals found in most dyes (Oginawati et al., 2022). If the density of a heavy metal is greater than 5 mg/cm3, it is generally considered to have a high density.

Because of their complex molecular makeup, these heavy metals are not biodegradable and are challenging to remove from the environment (Murphy et al., 2017). Because of this, toxic heavy metals, which are harmful, cancer-causing, and do not break down naturally, end up in the tiny fibers released by these textiles, causing significant environmental problems (Periyasamy and Periyasami, 2023). To create conductive textiles for electromagnetic shielding purposes, the metals are also coated on the fibers' surface (Hu et al., 2021). These heavy metals are widely known for their toxicity, negative effects on human health, and harm to the environment. In addition, organic and inorganic chemicals, as well as leftovers of metals, a monomer, and other substances that can be discharged, are sources of pollution that are more dangerous to the health of humans than the microfibers themselves (Liu et al., 2020).

10.3 FIBROUS MICROPLASTICS

Both of the secondary and primary microplastics can be found in a broad range of environmental forms, including as beads, films, fragments, and fibers. FMPs (fibrous microplastics) have become increasingly intriguing lately because of their unique features like shape, exact measurements, where they might come from, and how common they are in nature. Dimension and composition are two of the many factors that need to be taken into account when defining MPs since the values of these factors might fluctuate depending on the kind of MP. According to the ECHA (2020), fibrous MPs are defined as MPs with a length to diameter ratio more than three and 15 mm as the maximum length. Previous research has mostly focused on FMPs up to 5000 μm in length. Moreover, according to the European Chemicals Agency (ECHA), to categorize and describe the shape of microplastics (MP), it is necessary to consider at least two dimensions, typically length and diameter (Haq, Mazumder, and Kalamdhad, 2020). It is suggested that for non-circular cross-sections, such as those of fragmented microplastics (FMPs), the third dimension should be expressed, along with providing the first two dimensions (Nakajima et al., 2019).

10.3.1 Composition of the FMPs

Artificial fibers may be divided into two categories: petrochemical-derived synthetic fibers (like PES, PA, and PP) and natural fibers (like cellulose acetate) that have been chemically altered. Biopolymer-based or biodegradable fibers, like PLA, must also be taken into account as a source of FMPs (Haq et al., 2016a). They are not anticipated to make a significant contribution to the current FMPs in the environment because they are a new and niche material category, but this might alter in the future (Cesa et al., 2017). According to this definition, particles released from textiles made of cellulose and regenerated cellulose don't meet the criteria to be labeled as microplastics (MPs). However, particles released from synthetic textiles containing materials like PES, cellulose acetate, and PA can be classified as microplastics (Haq et al., 2016b).

According to Horton, Walton, and Spurgeon (2017), FMPs lost from synthetic fabrics are frequently handled as secondary MPs. Secondary MPs are, in this case, smaller FMPs that have broken down or separated from the initial shed fibers.

Several publications have referred to fiber-shaped MPs as "microfibres" (Hartline, Bruce, and Karba, 2016). Within the textile business, this word is frequently employed to designate a particular type of fiber with a fineness of less than 1 dtex, precisely a linear density of 1 g/10 km. Therefore, this phrase somewhat describes the diameter of a fiber rather than its length. For example, a diameter of 5–10 μm is often associated with microfibers with a circular cross-section and fineness ranging from 0.3 to 1.0 dtex. Nevertheless, it is evident that these microfibers cannot be regarded as MPs as they have the ability to develop as kilometers-long, very thin threads. Although certain fiber products, such as pima cotton fibers measuring 3–4 μm in diameter and New Zealand merino wool fibers measuring 11 μm in diameter, fall within the microfiber size range, they cannot be considered FMPs (synthetic microplastics) because they are not synthetic materials (Sundt, Schultze, and Syversen, 2014).

10.3.2 The Environment's Microplastic Fibers Presence and their Impact

FMPs (fibrous microplastic) have been discovered in large quantities in a variety of urban settings, including wastewater treatment facilities (WWTPs) and in-house settings. They have also been discovered in lakes, oceans, soils, and air, as well as in isolated ice and polar areas (Lares et al., 2018). FMPs have also been shown to be the predominant MP type in a large number of field samples, including those taken from sediments, coastal regions, and surface waters of rivers and the sea. The broad variety of detection circumstances suggests that FMPs are widely distributed throughout all global ecosystems (Statista, 2021). The concentrations of FMP differ greatly depending on the occurrence in the environment. In sediments and soils, they might number in the hundreds or thousands of pieces per kilogram, or per square meter in air fallout, but in surface waters, they are usually much lower, often less than one item per liter (Haq et al., 2017).

FMP consumption has been shown in a variety of aquatic and terrestrial taxa, including the snail *Achatina fulica* and marine fish species like *Daphnia magna* and copepod *Calanus finmarchicus* (Song et al., 2019). Adverse consequences from consuming fibers include loss of appetite, aberrant development, neurotoxicity, oxidative stress or damage, tissue damage, and death. Moreover, the existence of airborne FMPs and their identification in lung biopsies of humans underscore the possible risk associated with fiber ingestion via inhalation. MPs often serve as carriers of diseases and chemicals. FMPs may potentially display additive-associated toxicities as a result of chemical leaching, however conclusive research is needed to determine the consequences. The possible effects cannot be overlooked, especially in light of the fact that synthetic fibers in the textile industries are frequently functionalized by surface coating and dyeing (Tan et al., 2019).

10.3.3 Fibrous Microplastic Source

There are several possible origins for FMPs that are found in the environment and have an influence on it. To lessen FMP pollution, it is crucial to locate these sources and measure the ensuing emissions.

10.3.3.1 Fiber Production

According to the reports of DNFI (2023), the fiber production globally has reached over 115 million tons in 2023. It consists of 6% of man-made cellulosic fibers, 27% natural fibers, and 67% synthetic fibers (Figure 10.1). In the sector of synthetic fibers, 67% is occupied by the synthetic filament and the remaining portion of synthetic fibers in 2023 was occupied by cellulosic fibers and synthetic staple that consisted of 9% and 24% respectively. By 2025, it is projected that the textile industry would require more than 60 million tons of PES overall. The market for synthetic fibers is anticipated to increase fast going forward due to factors such as the rising demand for diverse applications, the expanding global population, and rising living standards. In this specific context, it is estimated that the production of synthetic fibers will increase by more than 9 million tons annually, resulting in a total output exceeding 80 million tons by the year 2025.

10.3.3.2 Synthetic Textiles

Synthetic textiles are among the top consumers of plastic, coming in second out of seven categories, and they make up about 12% of the world's total plastic consumption, which amounts to 300 million tons. According to the International Union for Conservation of Nature (IUCN) in 2017, synthetic textiles are considered one of the major contributors to microplastic pollution in the oceans (Boucher and Friot, 2017).

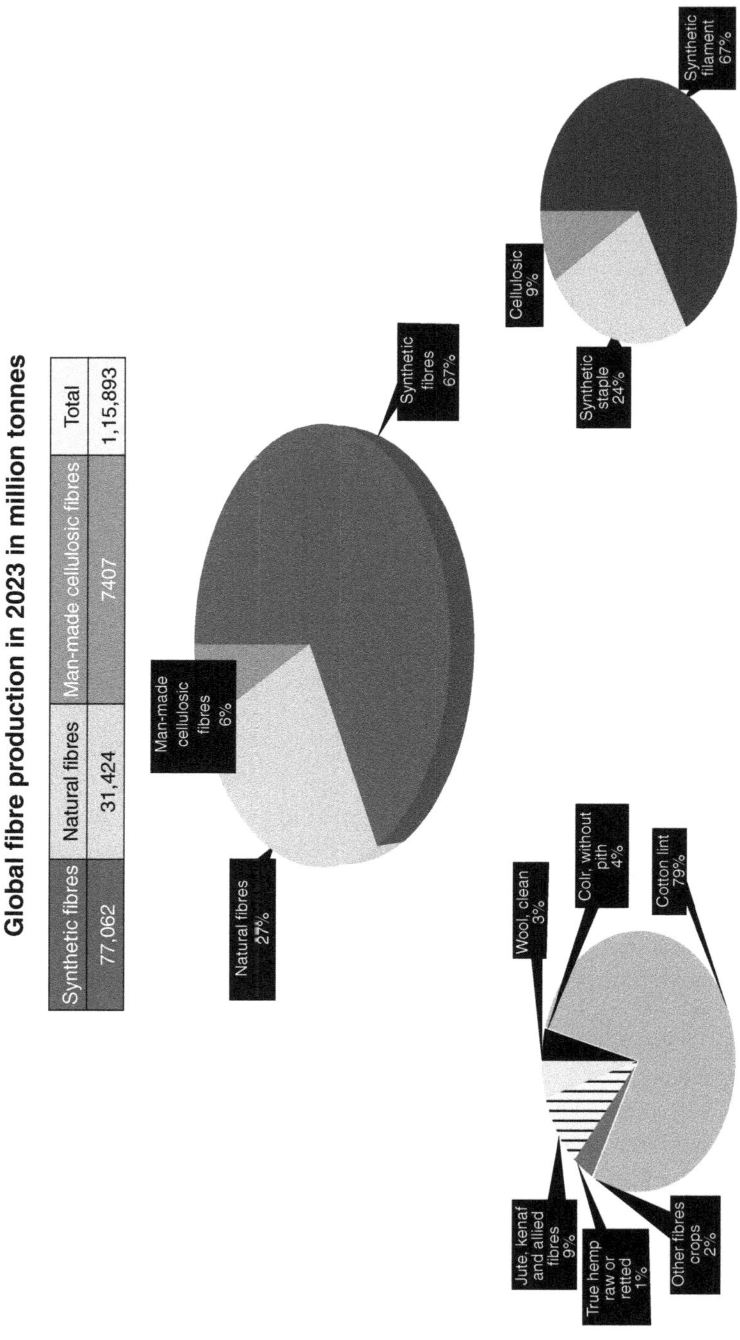

FIGURE 10.1 Global output of fibers.

It is anticipated that FMP emissions would occur at every stage of the synthetic textile life cycle, including manufacture (in a variety of goods), washing, usage (such as wear and weathering in applications), and disposal (O'Brien et al., 2020). Based on the IUCN proposal, synthetic textiles release FMPs (fibrous microplastics) ranging from 0.74% w/w to 5% w/w throughout their life cycle, with a median of 2%.

10.4 TECHNOLOGICAL TREATMENT OF WASTEWATER RELEASED FROM TEXTILE INDUSTRIES

Since the discovery of microplastic in water in 2004, research has been done on plastic deterioration and remediation techniques. Plastic use and pollution will naturally decline if plastic manufacturing declines. Many ways are being utilized to remove microplastic that has already been found in water; these strategies may be of two types: technical methods and biological methods.

To resolve the problem of microplastics from textile factories polluting water, scientists are studying different ways to clean up wastewater. Many researchers are testing special filters that can catch tiny bits of plastic. Researchers are also using chemicals to make the plastic bits stick together so they can be removed more easily. Another method involves using a type of carbon that sucks up the plastic. Some techniques use powerful reactions to break down the plastic into harmless pieces. Others rely on tiny living organisms to eat away at the plastic. These methods offer hope for keeping our water cleaner and safer.

10.4.1 Density-based Approaches

Density is one factor used in the process of separating microplastics from water. Microplastics will float if the density of the solution is more than that of the microplastics; conversely, if the density of the microplastics is greater than that of the solution, they will sink into it (Rodriguez et al., 2021). If the density of the microplastic matches that of the solution, some of the microplastic will float partially within the solution. Due to microplastic's low density, a number of techniques have been presented; the majority of these techniques were put out by Thompson et al. in 2004. He separated the microplastics using a NaCl solution. Following that, changes were made, and a variety of salt solutions were employed to separate microplastics based on their density. The salt solution employed in the density separation process provides the microplastic with buoyancy. The primary considerations for choosing a salt solution to be used for density separation are the salt solution's environmental effect, the expense of the necessary processing, and the capacity to recover microplastics. Ethanol and other solvents have the ability to harm certain tiny microplastics. Flotation is an effective method for separating microplastics since it can separate particles down to millimeter sizes. Moreover, due to the challenge of distinguishing small microplastics from the interface between air and liquid, this flotation method is most effective for separating microplastics, which are difficult to separate by hand. However, because there would be little buoyant force at work and a potential for surface fouling, it is challenging to remove microplastic in small pieces from this flotation. It is also concerning because occasionally the bubbles may catch the non-plastic stuff that has more density and bring it to the top.

In 2019, Nakajima and colleagues developed a small glass device inspired by a Utermöhl chamber, which doesn't have a shut-off valve. This creative tool enables the easy and fast separation of microplastics from sediments in just one goes. Additionally, it is portable and easy to tidy up. One benefit of utilizing the JAMSS unit is that it allows the user to rinse the whole headspace several times without having to resuspend the accumulated sediment. This minimizes the requirement for repeated processing and lowers the risk of microplastic loss and external contamination. There is no possibility of a blockage because it does not have a shut-off valve. In contrast to plastic-based devices, the glass-made JAMSS unit separates microplastics from sediments without contaminating them chemically, allowing for additional study of the adsorption of pollutants and additives onto microplastics.

10.4.2 Removal of Microplastic in Wastewater Treatment Plant (WWTPs)

Microfibers (MFs) released from textiles have been seen in a variety of settings, suggesting that human activity has an effect on natural systems. According to reports, the main cause of microplastic contamination is synthetic textile microfibers, a subgroup of microplastic fibers (MPFs). It is anticipated that as textile production increases, the issue of MF pollution will exacerbate and become more difficult to handle. The majority of microplastics (MPs) that are released from textile industries and found in wastewater treatment facilities come from activities in urban areas, such as tire wear and tear, paint chipping, and industries dealing with plastic and textiles, as well as from everyday household items like cosmetics, exfoliants, scrubbers, and textiles (Akdogan and Guven, 2019). According to Alimba and Faggio (2019), the frequent usage of microfibers in garments may potentially lead to their release into the environment, making them a major source of MPs.

Microplastics (MPs) that are in the air can come back down to land, rivers, and seas either when it rains or through dry sedimentation. Rainfall plays a big role in moving these microplastics around, especially if land is degrading, which can bring more plastic debris into waterways (Benson and Fred-Ahmadu, 2020). When considering the routes through which microplastics enter aquatic ecosystems, it is crucial to tackle unmanaged plastic waste. Research indicates that garbage could be a primary source rather than just an accidental release of marine microplastics. Organic fertilizers, plastic films, and soil enhancers are known as the main sources of soil contamination. Microplastics are conveyed from wastewater treatment plants during the skimming and settling processes, often being mistaken for organic fertilizers (Mason et al., 2016). Microplastics (MPs) can undergo different treatment processes, with wastewater treatment plants (WWTPs) being the most widespread method. WWTPs utilize various treatment techniques such as primary settling, A2O, grease and grit removal, biofilters, bioreactors, biologically active filters, disc filters, and rapid sand filters (RSFs).

10.4.2.1 Preliminary Treatment or Primary Treatment

The majority of MPs in wastewater may be effectively removed by preliminary and primary treatment. The study found that following the first treatment, between 50% and 98% of the MPs may have been segregated, and between 35% and 59% may have been eradicated (Liu et al., 2021). Before, the main ways to get rid of microplastics involved filtering out the lighter ones during grit removal and separating the heavier ones through grease skimming or surface skimming in clarifiers. The initial treatment had the most significant effect on where different sizes of microplastics ended up because it was better at getting rid of the larger ones.

As per Dris et al. (2015), there was a notable decrease in the percentage of large molecules (ranging from 1000 mm to 5000 mm), dropping from 45% to 7% following the initial intervention. Research on microplastic morphologies indicates that pre-treatment is more effective than fragmentation in removing fibers from wastewater; moreover, the presence of fibers is reduced after pre-treatment (Ziajahromi et al., 2017). The ease with which fibers can be collected in emulsification particles and separated by deposition might offer an explanation for this phenomenon. According to Michielssen et al. in 2016, they found no microbeads in the effluent from wastewater treatment facilities.

A survey conducted in NY, USA found that microbeads were prevalent in every four out of five WWTPs. This was attributed to differences in technology used, sources of influent, and removal efficiency of the WWTPs.

10.4.2.1.1 Sedimentation of Microfibers

The grit chamber, which serves as the second stage of the main treatment process, is situated before the primary sedimentation tank and after the screen. Its goal is to successfully remove grit and sand particles from wastewater. A crucial component of any city's sewage treatment plant is the grit chamber, which is not always a standard feature in industrial wastewater treatment setups. These chambers are designed to effectively filter out inorganic particles with a certain weight and size, typically around 0.20 mm in diameter and a specific gravity of approximately 2.65. In fact, more than 41% of microplastics are removed at this stage according to Liu et al. in 2019. In their study, the incoming sewage had a microplastic concentration of 47.4 MP/L, which increased to 79.9 MP/L after treatment. Murphy and colleagues in 2017 studied the effectiveness of the initial phase at a public wastewater treatment plant in Glasgow, Scotland. After this step, the average concentration of microplastics (MPs) decreased from 15.7 MP/L to 3.4 MP/L, showing a removal rate of nearly 78%. Similarly, in Spain, there was a 74% reduction in MPs during the primary stage of an urban wastewater treatment plant. However, a large wastewater treatment facility in Canada achieved a high elimination rate of MPs (92%) during its main stage. The majority of the MPs were in fibrous form. Lares et al. (2018) reported that their study's first phase removed 99% of MPs, particularly those in fiber form, with an initial input of 57.6 MP/L. The fact that over 96% of MPs are fibrous may have contributed to the study's exceptional efficiency. During this pretreatment step, most MPs were effectively removed; any remaining MPs can be eliminated in the final stage.

10.4.2.2 Secondary Treatment

The time of MPs' contact with effluent in the treatment line is another crucial component in the removal of microplastics from secondary discharges. Longer contact durations for MPs increase the likelihood of surface biofilm coating (Carr et al., 2016). As per Rummel and colleagues (2017), these biological coatings might serve as supplements, changing the surface properties or amounts of microplastics. Because particles that are neutrally buoyant are harder to settle or skim off, these alterations could noticeably impact how well microplastics are removed. Given this, it could be beneficial to investigate the ways in which wastewater nutrient concentrations and contact time impact the removal of microplastic surface fouling.

According to Talvitie et al. (2017), the investigations demonstrated that while the relative abundance of fibers increased, the corresponding high number of microplastic pieces decreased after the subsequent treatment. A plausible hypothesis is that the fibers that were readily settled or skimmed were mostly eliminated during the pre-treatment, leaving behind fibers that could have had certain innate properties (such as neutral buoyancy) that prevented further separation. Large microplastic particles might be removed even more during secondary treatment, resulting in secondary pollutants with low MP concentrations.

Studies have shown that there were almost no microplastics greater than 500 mm in the secondary effluent (Mintenig et al., 2017). Talvitie et al. (2017) found that only 8% of the total microparticles are larger than 300 mm after secondary treatment. After subsequent treatment, Dris et al. found that MPs with diameters between 500 and 1000 mm accounted for 43 percent of the total. There was no recognized reason for the higher percentage. Dris et al. (2015) suggested that upcoming studies ought to concentrate on how different secondary treatment methods perform in different situations when it comes to removing microplastics.

10.4.2.3 Tertiary Treatment

In the process of getting rid of MPs, the tertiary treatment could provide a substantial additional polish. Following tertiary treatment, the proportion of MPs in the effluent was lowered from the wastewater influent to between 0.2% and 2%. Membrane-based technologies that yield the greatest outcomes are those that employ treatment approaches that determine the removal efficiency of MPs. In 2017, Talvitie and his team looked into how well different tertiary treatment techniques worked in getting rid of pollutants. They tested methods like membrane bioreactor (MBR) for dealing with initial effluent, dissolved air flotation (DAF) for secondary treatment, and disc filter (DF) for quick sand filtering. They discovered that rapid sand filtering (RSF) and dissolved air flotation (DAF) removed around 97% and 95% of the pollutants respectively, while membrane bioreactor (MBR) came out on top with a removal rate of 99.9%.

The effectiveness in removing dissolved organic matter ranged from 40% to 98.5%. Likewise, when evaluating wastewater treatment plants (WWTPs) in New York, two plants with membrane systems didn't release microbeads. However, the other four plants, which used various filtration methods including two unspecified types, a quicksand filter, and a continuous backwash filter, did release microbeads. The particular requirements fractions were discovered to be the most common following the tertiary therapy (Bilgin et al., 2020). The enormous amount of fiber in the final effluent may occasionally be higher than in the secondary effluent. This might be because filters or membranes pass through fibers more easily along their longitudinal axes. As a result, steps for the last stage are necessary to remove certain microscopic and fiber-like MPs from the effluent.

10.4.3 Approaches based on Hydrophobicity

According to this theory, hydrophobic contact is the basis for separation. Here, the froth floatation-produced bubble surface attracts the hydrophobic particles, which are then transported to the air–liquid interface. However, the significant particle loss and unpredictable bubbles of this froth floatation make it unsuitable for analytical plastic separation. While only around 55% of the tiny microplastic (<1 mm) was recovered, the recovery rate for big microplastic (1–5 mm) was 100%. The oleophilic connection between fuel and microplastic allowed for a greater rate of microplastic recovery in the Nguyen et al. (2019) experiment. Moreover, ethanol is used in this process to eliminate oil residue, which frequently dissolves the tiny microplastic.

10.5 BIOTECHNOLOGICAL TREATMENT OF WASTEWATER RELEASED FROM TEXTILE INDUSTRIES

Biotechnological treatment of wastewater originating from textile industries presents a sustainable and effective approach to mitigating environmental pollution. This approach utilizes the inherent abilities of living organisms like bacteria, fungi, or algae to clean up pollutants found in wastewater from textile production. These organisms possess enzymes or metabolic pathways that enable them to degrade or metabolize pollutants, including dyes, solvents, and other chemicals commonly used in textile manufacturing processes. Biotechnological treatment systems typically involve the use of specialized reactors, such as bioreactors or biofilters, where these microorganisms are cultured and encouraged to thrive in a controlled environment. In bioreactors, wastewater is circulated through a chamber where microbial communities break down pollutants, while biofilters utilize microbial biofilms attached to solid support media to treat wastewater as it flows through.

The advantages of biotechnological wastewater treatment include its eco-friendly nature, as it avoids the use of harsh chemicals, and its ability to target a wide range of pollutants (Napper et al., 2020). Additionally, bioremediation processes can be tailored and optimized to address specific contaminants present in textile wastewater, offering a versatile and adaptable solution to water pollution challenges. By employing biotechnological approaches, textile industries can play a pivotal role in promoting sustainable water management practices and reducing their environmental footprint.

10.5.1 Bioremediation

Bioremediation is a naturally occurring technique that uses microbes—mostly fungus or bacteria—to clean the environment (Haq et al., 2016a b; 2017, 2018, 2020, 2022a, b, c; Haq and Kalamdhad, 2021; Haq and Kalamdhad, 2022, 2023; Haq et al., 2024; Katare et al., 2021). A variety of environmental variables cause plastic materials to degrade into microplastic and macroplastic. Bioremediation is a bioengineered technique that uses micro-organisms, to break down microplastics biologically (Moon et al., 2022). Furthermore, these microbial degradations are coupled with physico-chemical processes to attain greater efficiency. Degradation is a method in this case that results in various modifications to the physical or chemical characteristics of polymers. And environmental elements, molecular conditions, or bacterial activity are the causes of these physical and chemical changes. Because oligomers are soluble, organic intermediates such as aldehydes, acids, ketones, alcohols, etc. can develop following the main breakdown of the polymer. Relatively speaking, the effectiveness of removing microplastic is above 20%, however this also relies on the kind of microorganisms being used (Miloloza et al., 2022). When talking about the drawbacks of bioremediation, it might occasionally be challenging to assemble an appropriate consortium, and occasionally the microplastic is not entirely eliminated (Othman et al., 2021).

This degrading process is laborious and long; at times, it might take years to finish. There are other benefits to this approach as well, such as minimal energy requirements and cost-effectiveness. The method is safe for the environment and uses very specialized microorganisms for cleanup. Also, there is less production of hazardous byproducts (CAGR, 2025). The development of the biofilm surrounding the plastic polymer is the first stage of biodeterioration. After that, assimilation takes place once these polymers are transformed into oligomers, dimers, or monomers. In these microbes, biofragmentation occurs, where specific enzymes are produced outside the cells. These enzymes work on the polymer, transforming it into different forms.

They assemble on the microorganisms, after which simple or assisted diffusion takes place to absorb them by the microbe's cells. Table 10.1 discusses the various fungi and bacteria stain effectiveness to decompose microplastic.

10.5.2 Photocatalytic Degradation

Under the category of AOPs (Advance Oxidative Processes), photocatalysis is a technology that is thought to be energy-efficient for degrading microplastic in an aqueous environment. When light hits the semiconductor, it creates electrons in the holes. These holes then interact with water (H_2O) or hydroxide (OH^-) to produce hydroxyl radicals (OH) and oxygen (O_2). Consequently, microplastics are broken down, leading to processes like rupturing, branching, crosslinking, and even mineralization into water (H_2O) and carbon dioxide (CO_2). In nature, this process, known as photolysis, affects the carbon-carbon backbone of plastic.

Three phases comprise this whole process: commencement, propagation, and termination. After 15 days of photo-irradiation, microplastics (MPs) go through a sequence of reactions that lead to the creation of peroxyl radicals. These radicals play

TABLE 10.1
Biodegradation time of different types microplastic through bacterial and fungal strains

Bacteria/ Fungus	Types/strain	Types of Microplastic	Time span of Degradation	References
Fungi	*Penicillium Pinophilum, Aspergillus Niger*	Low Density Polyethylene	It takes these fungi 31 months to degrade Polyethylene	Paco et al. (2017)
Fungus	*Aspergillus Oryzae*	Polyethylene	It takes this fungus 16 days to degrade Polyethylene	Muhonja et al. (2018)
Bacteria	*Ideonellasakaiensis*	Polyethylene terephthalate	Six weeks is all it takes this bacterium to degrade microplastic.	Yoshida et al. (2016)
Bacteria	*Sporosarc-inaglobispora*	Polypropylene	It degrades the microplastic in 40 days	Auta et al. (2018)
Fungus	*Phanerochae-techrysosporium*	Polyvinyl chloride	In 90 days, this fungus degrades the microplastic	Ali et al. (2014)
Bacteria	*Bacillus cereus*	Polyethylene terephthalate	It degrades the microplastic in 40 days	Auta et al., (2017)

a vital role in processes such as photodegradation addition, chemical chain breaking, addition reactions with oxygen and water, and the development of environmentally persistent free radicals (EPFRs) (Zhu et al., 2019). Reactive oxygen species (such as OH) are produced as a result of EPFRs; once these species are created, they will target microplastic and break the chain, which will lead to a decrease in molecular weight and the production of tiny molecular products.

In liquid-phase photocatalysis, plastic present in water is utilized to decompose polyvinylidene chloride copolymer (PVLC), consisting of 5% PVLC and 95% PVC. ZnO dechlorinates at a higher rate than TiO_2, although during photo-oxidation, the process is blocked by the presence of multiple catalysts. The stratosphere emits damaging radiation, like ultraviolet-A (about 315–400 nm) and ultraviolet-B terrestrial radiation (around 295–315 nm), due to photolysis, which starts a process called photooxidation. This process is known as photodegradation. The visible spectrum of sunlight (400–760 nm) speeds up the breakdown of polymers, whereas infrared radiation (760–2500 nm) speeds up heat oxidation (Shahnawaz et al., 2019). Degradation results in a moderate beginning rate, but rapid propagation. As long as high-energy radiations are not employed, the photodegradation process is environmentally benign.

Although there is acceptance of this deterioration, the method is quite expensive (Ebrahimbabaie et al., 2022).

10.5.3 Thermo-oxidative Degradation

Oxygen and temperature are the primary elements of the microplastic breakdown process in the context of thermal oxidative degradation. When the microplastic polymer is exposed to high temperatures, the constituents of a lengthy chain of microplastic separate, engage in component reactions, and alter the microplastic polymer's characteristics (Sharma, Gupta, and Gupta, 2024). This phenomenon is often referred to as molecular degradation due to warming. Compared to prior qualities, the new physical or visual properties cause chemical reactions. When thermal degradation occurs, it leads to changes in properties like reduced ductility, increased brittleness, altered appearance, and cracking, as well as an overall decline in physical characteristics due to changes in molecular mass and distribution of molecular weight (Zhang et al., 2021).

In oxobioreduction processes like photodegradation (UV) and oxidation, two processes are required for biodegradation. The microplastic polymer is degraded by UV radiation, and the oxidation process requires high temperature and time to degrade the microplastic (Sharma et al., 2021). These techniques cause the molecular weight to decrease, allowing the polymer to break down. Even though the breakdown happens fairly fast and needs higher temperatures than the surrounding environment, this method isn't environmentally friendly. The degradation of microplastics takes time (Wang et al., 2017).

10.6 TEXTILE SUSTAINABILITY IN INDIA

India would not be able to achieve the Sustainable Development Goals (SDG) by 2030 unless it addresses the problem of microfibers. Tiny fragments shred when we wash our garments in washing machines. Microfibers are these invisible, thread-like particles that are typically smaller than 5 mm. They originate from natural, semi-synthetic, and synthetic clothing. Although synthetic microfibers—also referred to as microplastics—have received a lot of attention, there are other areas in which they should be concerned. Microfibers, both natural and semi-synthetic, have lately come to light as equally concerning pollutants because of their propensity to leak a multitude of chemicals and colors.

Microfibers enter aquatic systems and eventually the marine environment directly in developing nations like India, where wastewater treatment facilities appear to be insufficient. Despite the fact that wastewater treatment plants are able to trap some of these microfibers, they nonetheless find their way into the environment through sludge applications or landfills. This pollutant and its importance have not yet been recognized by India's textile and plastic waste management regulations. It is still present and poses an unseen menace. Considering the size of India's population, microfiber pollution is greatly increased by even a tiny portion of households using washing machines. In growing economies such as India, the fast fashion culture and the expanding textile sector are tightly gripping the younger generation and unstable consumerist society. Because of this use-and-throw mentality, which prioritizes quantity over quality, textile designs are less expensive, fabric quality is reduced, and excess manufacturing occurs (Alex, 2023).

Unsustainability is a result of expanding gross domestic product, rapid urbanization, rising purchasing power, and greater brand penetration in the Indian market. The manufacturing, the import, and export of both synthetic and natural textiles are not entirely under the jurisdiction of Indian textile regulations, despite this business being highly significant and operating in an uncontrolled manner. It has a substantial portion of the worldwide market. Long-term, sustainable solutions are required to address microfiber contamination, such as permanently limiting the discharge of microfibers from textiles. Strategies for cutting emissions include utilizing liquid detergents, front-loading washers, full loads at cold temperatures, and regular drum speeds, which greatly decrease shedding by stressing out the clothing.

10.7 SUMMARY AND CONCLUSION

The elimination of microplastics from wastewater discharged by the textile industry is a critical environmental concern addressed through various treatment technologies. Microplastics, tiny plastic particles shed during textile production and laundering, pose significant toxicity risks to aquatic ecosystems and human health. Understanding the composition of these microfibers, which consist predominantly of synthetic polymers like polyester and nylon, is crucial for developing effective treatment strategies. Technological treatment methods for textile industry wastewater encompass primary, secondary, and tertiary stages. Primary treatment involves physical processes such as screening and sedimentation to remove larger solid particles, including microplastics. Secondary treatment employs biological processes like activated sludge treatment to further degrade organic pollutants, while tertiary treatment utilizes advanced filtration and chemical treatments to achieve higher levels of purification. Biotechnological approaches offer promising avenues for microplastic remediation. Bioremediation utilizes microorganisms to break down microplastics into non-toxic byproducts, while photocatalytic degradation harnesses light-activated catalysts to degrade plastic pollutants. Thermo-oxidative degradation involves subjecting microplastics to high temperatures in the presence of oxygen to promote decomposition.

The pursuit of textile sustainability is integral to addressing microplastic pollution, with India emerging as a significant player in this realm. Efforts towards sustainable textile production encompass eco-friendly materials, water-saving technologies, and waste reduction measures. Initiatives promoting circular economy principles, such as textile recycling and upcycling, contribute to minimizing microplastic release into wastewater streams.

In conclusion, addressing the pervasive issue of microplastic contamination in wastewater from the textile industry demands a concerted effort and the implementation of intelligent solutions. Various technological interventions such as advanced filtration systems, chemical treatments facilitating agglomeration, biological degradation employing microorganisms, and sophisticated degradation techniques hold promise in mitigating this environmental challenge effectively. Concurrently, ongoing research endeavors focus on refining materials for filtration and enhancing the efficacy of biodegradation processes. However, the efficacy of these measures is contingent upon collaborative engagement from stakeholders across sectors, including industry, governance, and individual consumers. By fostering a collective commitment to reducing plastic usage, increasing recycling practices, and adopting environmentally conscious behaviors, tangible progress can be made toward safeguarding water resources. It is imperative to disseminate awareness about the significance of water quality preservation and to advocate for the adoption of sustainable practices. Through concerted action, a healthier environmental landscape can be cultivated for present and future generations.

REFERENCES

Akdogan, Z., & Guven, B. (2019). Microplastics in the environment: A critical review of current understanding and identification of future research needs. *Environmental Pollution*, 254, 113011.

Alex, K., & Riya. (2023). *Microfibre-Free Laundry: Multifaceted Approach Needed to Combat These Tiny Shreds of Pollutants*. Down to Earth. Retrieved from www.downtoearth.org.in/blog/pollution/microfibre-free-laundry-multifaceted-approach-needed-to-combat-these-tiny-shreds-of-pollutants-90564.

Ali, M. I., Ahmed, S., Javed, I., Ali, N., Atiq, N., & Hameed, A. (2014). Biodegradation of starch blended polyvinyl chloride films by isolated Phanerochaetechrysosporium PV1. *International Journal of Environmental Science and Technology*, 11, 339–48.

Alimba, C. G., & Faggio, C. (2019). Microplastics in the marine environment: Current trends in environmental pollution and mechanisms of toxicological profile. *Environmental Toxicology and Pharmacology*, 68, 61–74.

Auta, H. S., Emenike, C. U., & Fauziah, S. H. (2017). Screening of Bacillus strains isolated from mangrove ecosystems in Peninsular Malaysia for microplastic degradation. *Environmental Pollution*, 231, 1552–9.

Auta, H. S., Emenike, C. U., Jayanthi, B., & Fauziah, S. H. (2018). Growth kinetics and biodeterioration of polypropylene microplastics by Bacillus sp. and Rhodococcus sp. isolated from mangrove sediment. *Marine Pollution Bulletin*, 127, 15–21.

Benson, N. U., & Fred-Ahmadu, O. H. (2020). Occurrence and distribution of microplastics-sorbed phthalic acid esters (PAEs) in coastal psammitic sediments of tropical Atlantic Ocean, Gulf of Guinea. *Science of The Total Environment*, 730, 139013.

Bilgin, M., Yurtsever, M., & Karadagli, F. (2020). Microplastic removal by aerated grit chambers versus settling tanks of a municipal wastewater treatment plant. *Journal of Water Process Engineering*, 38, 101604.

Boucher, J., & Friot, D. (2017). Primary microplastics in the oceans: a global evaluation of sources (Vol. 10). Gland. Switzerland: IUCN.

Brien, S., Okoffo, E. D., O'Brien, J. W., Ribeiro, F., Wang, X., Wright, S. L., ...& Thomas, K. V. (2020). Airborne emissions of microplastic fibres from domestic laundry dryers. *Science of The Total Environment*, 747, 141175.

Carr, S. A., Liu, J., & Tesoro, A. G. (2016). Transport and fate of microplastic particles in wastewater treatment plants. *Water Research*, 91, 174–182.

Cesa, F. S., Turra, A., & Baruque-Ramos, J. (2017). Identification of microplastic in effluents of wastewater treatment plants using focal plane array-based micro-Fourier-transform infrared imaging. *Science of The Total Environment*, 598, 1116–1129. https://doi.org/10.1016/j.scitotenv.2017.03.047

Cheng, X. W., Guan, J. P., Yang, X. H., & Tang, R. C. (2017). Improvement of flame retardancy of silk fabric by bio-based phytic acid, nano-TiO2, and polycarboxylic acid. *Progress in Organic Coatings*, 112, 18–26. https://doi.org/10.1016/j.porgcoat.2017.07.012

Discover Natural Fibres Initiative. (2023). Global fiber production reaches over 115 million tons. (accessed on: 11 March, 2024). Retrieved from www.dnfi.org

Dris, R., Gasperi, J., Rocher, V., Saad, M., Renault, N., & Tassin, B. (2015). Microplastic contamination in an urban area: a case study in Greater Paris. *Environmental Chemistry*, 12(5), 592–599.

Ebrahimbabaie, P., Yousefi, K., & Pichtel, J. (2022). Photocatalytic and biological technologies for elimination of microplastics in water: Current status. *Science of The Total Environment*, 806, 150603.

European Chemicals Agency (ECHA), Committee for Risk Assessment (RAC), & Committee for Socio-economic Analysis (SEAC). (2020). Background document. Helsinki, Finland: ECHA.

Ferreira, M., Thompson, J., Paris, A., Rohindra, D., & Rico, C. (2020). Presence of microplastics in water, sediments and fish species in an urban coastal environment of Fiji, a Pacific Small Island Developing State. *Marine Pollution Bulletin*, 153, 110991. https://doi.org/10.1016/j.marpolbul.2020.110991

Hartline, N. L., Bruce, N. J., Karba, S. N., Ruff, E. O., Sonar, S. U., & Holden, P. A. (2016). Microfiber masses recovered from conventional machine washing of new or aged garments. *Environmental Science & Technology*, 50(21), 11532–11538.

Haq, I., Kumar, S., Kumar, V., Singh, S.K., & Raj, A. (2016a). Evaluation of bioremediation potentiality of ligninolytic Serratia liquefaciens for detoxification of pulp and paper mill effluent. *Journal of Hazardous Materials* 305:190–199.

Haq, I., Kumari, V., Kumar, S., Raj, A., Lohani, M., & Bhargava, R.N. (2016b). Evaluation of the phytotoxic and genotoxic potential of pulp and paper mill effluent using Vigna radiata and Allium cepa. *Advanced Biology*: 8065736.

Haq, I., Kumar, S., Raj, A., Lohani, M., & Satyanarayana, G.N.V. (2017). Genotoxicity assessment of pulp and paper mill effluent before and after bacterial degradation using Allium cepa test. *Chemosphere 169*: 642–650.

Haq, I., Raj, A., &Markandeya (2018). Biodegradation of Azure-B dye by Serratia liquefaciens and its validation by phytotoxicity, genotoxicity and cytotoxicity studies. *Chemosphere* 196:58–68.

Haq, I., Mazumder, P., & Kalamdhad, A.S. (2020). Recent advances in removal of lignin from paper industry wastewater and its industrial applications – *A review. Bioresource Technology* 312:123636.

Haq, I., & Kalamdhad, A.S. (2021). Phytotoxicity and cyto-genotoxicity evaluation of organic and inorganic pollutants containing petroleum refinery wastewater using plant bioassay. *Environmental Technology & Innovation* 23:101651.

Haq, I., Kalamdhad, A.S., & Pandey, A. (2022a). Genotoxicity evaluation of paper industry wastewater prior and post-treatment with laccase producing Pseudomonas putida MTCC 7525. *Journal of Cleaner Production* 342:130981.

Haq, I., Singh, A., & Kalamdhad, AS (2022b). Integrated technologies for the remediation of paper industry waste in a circular bioeconomy. In Editor(s): Sunita Varjani, Ashok Pandey, Mohammad J. Taherzadeh, Huu Hao Ngo, R.D. Tyagi, *Biomass, Biofuels, Biochemicals*, Elsevier, Pages 351–363, ISBN 9780323885119.

Haq, I., Roy, S., & Kalamdhad, A.S. (2022c). Characterization of Pulp and Paper Mill Wastewater and Its Toxicity Analysis Using Vigna Radiata. In *Environmental Degradation: Monitoring, Assessment and Treatment Technologies*; Springer International Publishing: Cham, Switzerland, pp. 261–269. ISBN 9783030941475.

Haq, I., & Kalamdhad, A.S. (2022). Advanced and Ecofriendly Technologies for the Treatment of Industrial Wastewater to Constrain Environmental Pollution. In Arora, S., Kumar, A., Ogita, S., Yau, Y.Y. (eds) *Innovations in Environmental Biotechnology*. Springer, Singapore. https://doi.org/10.1007/978-981-16-4445-0_22.

Haq, I., & Kalamdhad, A.S. (2023). Enhanced biodegradation of toxic pollutants from paper industry wastewater using Pseudomonas sp. immobilized in composite biocarriers and its toxicity evaluation. *Bioresources Technology Reports* 24:101674.

Haq, I., Kalamdhad, A.S., & Malik, A. (2024) Bioremediation of petroleum refinery wastewater using Bacillus subtilis IH-1 and assessment of its toxicity. *Archives of Microbiology*, 206, 296.

Horton, A. A., Walton, A., Spurgeon, D. J., Lahive, E., & Svendsen, C. (2017). Microplastics in freshwater and terrestrial environments: Evaluating the current understanding to identify the knowledge gaps and future research priorities. *Science of the Total Environment*, 586, 127–141.

Hu, S., Kremenakova, D., Militký, J., & Periyasamy, A. P. (2021). Copper-coated textiles for viruses dodging. In *Textiles and their use in microbial protection*, 1st ed.; Series: Textile Institute Professional Publications; CRC Press: Boca Raton, FL, USA, pp. 235–250.

Katare, P. Y., Sankhla, M. S., Singhal, M., Ekta, B., Jadhav, K. P., & Bhagyashri, T. N. (2021). Microplastics in aquatic environments: Sources, ecotoxicity, detection & remediation. *Biointerface Research in Applied Chemistry*, 12, 3407–28.

Khandegar, V., & Saroha, A. K. (2013). Electrocoagulation for the treatment of textile industry effluent – a review. *Journal of Environmental Management*, 128, 949–963.

Lares, M., Ncibi, M. C., Sillanpää, M., & Sillanpää, M. (2018). Occurrence, identification and removal of microplastic particles and fibers in conventional activated sludge process and advanced MBR technology. *Water Research*, 133, 236–246.

Lassen, C. (2015). *Survey of PFOS, PFOA and Other Perfluoroalkyl and Polyfluoroalkyl Substances*. Copenhagen K, Denmark: The Danish Environmental Protection Agency.

Liu, P., Zhan, X., Wu, X., Li, J.. Wang, H., & Gao, S. (2020). Effect of weathering on environmental behavior of microplastics: Properties, sorption and potential risks. *Chemosphere*, 242, 125193.

Liu, W., Zhang, J., Liu, H., Guo, X., Zhang, X., Yao, X., ...& Zhang, T. (2021). A review of the removal of microplastics in global wastewater treatment plants: Characteristics and mechanisms. *Environment International*, 146, 106277.

Liu, X., Yuan, W., Di, M., Li, Z., & Wang, J. (2019). Transfer and fate of microplastics during the conventional activated sludge process in one wastewater treatment plant of China. *Chemical Engineering Journal*, 362, 176–182.

Luongo, G. (2015). *Chemicals in Textiles: A Potential Source for Human Exposure and Environmental Pollution*. Gothenburg, Sweden: Stockholm University, Faculty of Science.

Mason, S. A., Garneau, D., Sutton, R., Chu, Y., Ehmann, K., Barnes, J., ...& Rogers, D. L. (2016). Microplastic pollution is widely detected in US municipal wastewater treatment plant effluent. *Environmental Pollution*, 218, 1045–1054.

Michielssen, M. R., Michielssen, E. R., Ni, J., & Duhaime, M. B. (2016). Fate of microplastics and other small anthropogenic litter (SAL) in wastewater treatment plants depends on unit processes employed. *Environmental Science: Water Research & Technology*, 2(6), 1064–1073.

Miloloža, M., Cvetnić, M., KučićGrgić, D., OcelićBulatović, V., Ukić, Š., Rogošić, M., ... & Bolanča, T. (2022). Biotreatment strategies for the removal of microplastics from freshwater systems: A review. *Environmental Chemistry Letters*, 6, 1–26.

Mintenig, S. M., Int-Veen, I., Loder, M. G. J., Primpke, S., & Gerdts, G. (2017). Microfibers in oceanic surface waters: A global characterization. *Science Advances*, 6(23), eaay8493. https://doi.org/10.1126/sciadv.aay8493

Mintenig, S. M., Int-Veen, I., Löder, M. G., Primpke, S., & Gerdts, G. (2017). Identification of microplastic in effluents of wastewater treatment plants using focal plane array-based micro-Fourier-transform infrared imaging. *Water Research*, 108, 365–372.

Moon, U. M., Mondal, C., Saha, N. C., &Hossain, A. (2022). Microplastics in freshwater riverine systems: Brief profile, trophic-level transfer and probable remediation. In *River Health and Ecology in South Asia Springer* (pp. 103–126).

Muhonja, C. N., Makonde, H., Magoma, G., & Imbuga, M. (2018). Biodegradability of polyethylene by bacteria and fungi. *Plos One*, 13, e0198446.

Murphy, F., Russell, M., Ewins, C., & Quinn, B. (2017). The uptake of macroplastic & microplastic by demersal & pelagic fish in the Northeast Atlantic around Scotland. *Marine Pollution Bulletin*, 122(1–2), 353–359.

Nakajima, R., Tsuchiya, M., Lindsay, D. J., Kitahashi, T., Fujikura, K., & Fukushima, T. (2019). A new small device made of glass for separating microplastics from marine and freshwater sediments. *PeerJ*, 7, e7915. https://doi.org/10.7717/peerj.7915

Napper, I. E., Davies, B. F. R., Clifford, H., Elvin, S., Koldewey, H. J., Mayewski, P. A., ... & Gajurel, A. P. (2020). Reaching new heights in plastic pollution—Preliminary findings of microplastics on Mount Everest. *One Earth*, 3, 621–630. https://doi.org/10.1016/j.oneear.2020.09.011

Nguyen, B., Claveau-Mallet, D., Hernandez, L. M., Xu, E. G., Farner, J. M., & Tufenkji, N. (2019). Separation and analysis of microplastics and nanoplastics in complex environmental samples. *Accounts of Chemical Research*, 52, 858–66.

Oginawati, K., Suharyanto, Susetyo, S. H., Sulung, G., Muhayatun, Chazanah, N., ...& Fahimah, N. (2022). Investigation of dermal exposure to heavy metals (Cu, Zn, Ni, Al, Fe and Pb) in traditional Batik Industry workers. *Heliyon*, 8, e08914. https://doi.org/10.1016/j.heliyon.2022.e08914.

Othman, A. R., Hasan, H. A., Muhamad, M. H., Ismail, N. I., & Abdullah, S. R. (2021). Microbial degradation of microplastics by enzymatic processes: A review. *Environmental Chemistry Letters*, 19, 3057–73.

Paço, A., Duarte, K., da Costa, J. P., Santos, P. S., Pereira, R., Pereira, M. E., ...& Rocha-Santos, T. A. (2017). Biodegradation of polyethylene microplastics by the marine fungus Zalerionmaritimum. *Science of the Total Environment*, 586, 10–5.

Periyasamy, A. P. (2021). Evaluation of microfiber release from jeans: The impact of different washing conditions. *Environmental Science and Pollution Research*, 28, 58570–58582. https://doi.org/10.1007/s11356-021-15236-4

Periyasamy, A. P., & Militky, J. (2020). Sustainability in Textile Dyeing: Recent Developments. In J. Militky (Ed.), *Sustainability in the Textile and Apparel Industries* (pp. 37–79). Cham, Switzerland: Springer. https://doi.org/10.1007/978-3-030-50515-5_2

Periyasamy, A. P., & Tehrani-Bagha, A. (2022). A review on microplastic emission from textile materials and its reduction techniques. *Polymer Degradation and Stability*, *199*, 109901.

Periyasamy, A. P., & Periyasami, S. (2023). Critical review on sustainability in denim: A step toward sustainable production and consumption of denim. *ACS Omega*, 8, 4472–4490.

Plastics Europe. (2023). Plastics – the Fast Facts 2023. Retrieved from plasticseurope.org/knowledge-hub/plastics-the-fast-facts-2023/.

Rodriguez-Narvaez, O. M., Goonetilleke, A., Perez, L., & Bandala, E. R. (2021). Engineered technologies for the separation and degradation of microplastics in water: A review. *Chemical Engineering Journal*, 414, 128692. https://doi.org/10.1016/j.cej.2021.128692

Rummel, C. D., Jahnke, A., Gorokhova, E., Kühnel, D., & Schmitt-Jansen, M. (2017). Impacts of biofilm formation on the fate and potential effects of microplastic in the aquatic environment. *Environmental Science & Technology Letters*, 4(7), 258–267.

Shahnawaz, M., Sangale, M. K., & Ade, A. B. (2019). Plastic waste disposal and reuse of plastic waste. In *Bioremediation Technology for Plastic Waste* Springer (pp. 21–30).

Sharma, M., Gupta, V., & Gupta, D. (2024). Advancement in technologies of water management in rural areas. In *Water Resources Management for Rural Development* (pp. 117–126). Elsevier.

Sharma, M., Gupta, D., Parhi, M. P., Gurung, S., Parasar, H., & Kumar, A. (2021). Phosphate solubilizing microbes: A sustainable tool for healthy agriculture. *Ecology Environment & Conservation*, *27*(August Suppl. Issue), S303–S310.

Song, Y., Cao, C., Qiu, R., Hu, J., Liu, M., Lu, S., ...& He, D. (2019). Uptake and adverse effects of polyethylene terephthalate microplastics fibers on terrestrial snails (Achatina fulica) after soil exposure. *Environmental Pollution*, 250, 447–455.

Statista. (2021). *Annual Production of Plastics Worldwide from 1950 to 2021*. Hamburg, Germany: Statista.

Statista. (2021). *Production Forecast of Thermoplastics Worldwide from 2025 to 2050*. Hamburg, Germany: Statista.

Sundt, P., Schulze, P. E., & Syversen, F. (2014). Sources of microplastic-pollution to the marine environment. *Mepex for the Norwegian Environment Agency*, 86, 20.

Synthetic fibers market size worth $88 billion by 2025: CAGR 6.3%, www.grandviewresearch.com/ press-release/global-synthetic-fibers-market, (accessed 10 March 2024).

Talvitie, J., Mikola, A., Koistinen, A., & Setälä, O. (2017). Solutions to microplastic pollution – Removal of microplastics from wastewater effluent with advanced wastewater treatment technologies. *Water Research*, 123, 401–407.

Tan, X., Yu, X., Cai, L., Wang, J., & Peng, J. (2019). Microplastics and associated PAHs in surface water from the Feilaixia Reservoir in the Beijiang River, China. *Chemosphere*, 221, 834.

United Nations Environment Programme. (2017). *Marine plastics litter and microplastics: Global lessons and research to inspire action and guide policy change*. (accessed on: 21-12-2024) https://wesr.unep.org/media/docs/early_warning/marine_plastics_litter_and_microplastics.pdf

Wang, J. C., & Wang, H. (2017). Fenton treatment for flotation separation of polyvinyl chloride from plastic mixtures. *Separation and Purification Technology*, 187, 415–25.

Yoshida, S., Hiraga, K., Takehana, T., Taniguchi, I., Yamaji, H., Maeda, Y., ...& Oda, K. (2016). A bacterium that degrades and assimilates poly (ethylene terephthalate). *Science*, 351, 1196–9.

Zhang, K., Hamidian, A. H., Tubić, A., Zhang, Y., Fang, J. K., Wu, C., & Lam, P. K. J. (2021). Understanding plastic degradation and microplastic formation in the environment: A review. *Environmental Pollution*, 274, 116554.

Zhu, K., Jia, H., Zhao, S., Xia, T., Guo, X., Wang, T., &Zhu, L. (2019). Formation of environmentally persistent free radicals on microplastics under light irradiation. *Environmental Science & Technology*, 53, 8177–86.

Ziajahromi, S., Neale, P. A., Rintoul, L., & Leusch, F. D. (2017). Wastewater treatment plants as a pathway for microplastics: development of a new approach to sample wastewater-based microplastics. *Water Research, 112*, 93–99.

11 Treatment of Microplastics from Pharmaceutical Industrial Wastewater

A. A. Narmada Thathsarani Perera, Varsha Gupta, Muskan Sharma, Mandeep Prakash Parhi, and Izharul Haq

11.1 INTRODUCTION

Microplastics (MP) are widely acknowledged, sparking worries about their possible negative impacts on nature and humans. The direct disposal including fishing gears, granules used in the manufacturing of larger products, and cargo ships is one of the main sources and to aquatic environment (Haq et al., 2016a). Additionally, the mechanical breakdown of sizable plastic waste existing in the environment stands out as the second most significant source. Contributions to microplastics also arise from discharges from sewage and water treatment plants, along with ineffective practices in urban waste separation and disposal. The effects of microplastic particles are alarmingly increasing due to pharmaceutical industry wastewater, disturbing aquatic ecosystems and health. In the current scenario, the escalating discharge of pharmaceutical industry wastewater has resulted in the impurity of surface waters, groundwater, lagoons, arctic freshwaters, coastal shorelines, and oceans with microplastics (Fischer et al., 2016; Peng et al., 2018; Jeyasanta et al., 2020; Kye et al., 2023). Throughout the pandemic, the rapid increase in microplastic contamination occurred due to the heightened consumption of personal care products for COVID-19 prevention and certain pharmaceuticals (Pashaei et al., 2022). The utilization of disposable plastics, encompassing post-used personal protective equipment (PPEs), vaccination residues, testing materials, and hospital waste, has directly led to the production of significant amounts of waste (de Sousa, 2021; Lee & Kim, 2022). Due to the highest population density, Asia and Africa experience a significant concentration of microplastics. Elevated levels of plastic use and industrialization could lead to a higher content of microplastics in Europe (Pashaei et al., 2022; Lusher et al., 2013).

Microplastics, with a core size varying from 50–500 μm, may exist as primary or secondary particulates based on their formulation pathway resulting from the weathering or degradation of initial plastic components (Andrady, 2011; Pashaei

DOI: 10.1201/9781003379386-11

et al., 2022). The influence of MPs is determined by their dimensions, shape, concentration, polymer type, and pigmentation (Pashaei et al., 2022). Plastic beads, pellets, fibers, nurdles, and powders produced synthetically for commercial use are identified by primary microplastics. MP particles of secondary origin are formed through the breakdown and disintegration of large plastic fragments (macroplastic) and medium-sized plastic debris (mesoplastic) within the environment. Primary microplastics find extensive applications as raw materials for plastic production or in the textile fibers, industrial abrasives, manufacturing of appliances, and exfoliants used in cosmetic materials (Duis & Coors, 2016; Atugoda et al., 2021). Organic pollutants accumulate due to the hydrophobic properties and high surface area-to-volume ratio of microplastics. These contaminants encompass pharmaceuticals and personal care products, polyaromatic hydrocarbons (PAHs), polybrominated diethers (PBDs), polychlorinated biphenyls (PCBs), perfluorinated alkyl substances, and trace metal contaminants such as Co, Ag, Hg, Cr, Ni, Cu, Zn, Pb, and Cd on the microplastics' surfaces (Guzzetti et al., 2018).

An investigation carried out in the Southeastern Black Sea throughout the years of 2009–2020 revealed that the microplastics concentration increased by over threefold during that period. Thirteen different polymer types were identified, with 25.3% of polyethylene terephthalate and 44.9% of polyethylene being the most prevalent. The sizes of microplastic particles ranged from 118–4998µm, with 50% measuring less than 2000 µm. The predominant form of microplastics was found to be fragments, making up 56.3% of the total. In 2020, during the pandemic period, the highest particle concentration (0.944 particlesm^{-3}) was observed (Eryaşar et al., 2021). The surge in the usage of masks, gloves, and various single-use plastic items significantly impacted pharmaceutical industry wastewater. Additionally, the task of identifying and classifying procedures becomes more challenging when polymers are incorporated into medication formulations, packaging materials, and equipment, as they collectively give rise to a complex mixture of microplastics (Pashaei et al., 2022). MPs offer a substantial solid surface, particularly in aquatic settings, and the quantities of organic pollutants deposited on the outer layer of the plastic can be significantly higher—often by several orders of magnitude—than in the nearby waters.

Industrial wastewater from pharmaceuticals is mostly produced during the manufacturing of raw materials, cosmetics, and a range of medications. This process results in the production of effluent that contains possibly hazardous components. While the pharmaceutical sector has been a major driver of economic expansion, it has also been a major source of environmental degradation (Rana et al., 2014). Pharmaceutical wastewater has a complex composition that includes a high level of organic matter, high salinity, microbial toxicity, and difficulty in biodegradation (Guo et al., 2017). The negative impact of microplastics in pharmaceutical wastewater on aquatic ecosystems and overall human health has raised concerns about finding suitable treatment processes. Contaminants in pharmaceutical wastewater have access to almost all aquatic environments, resulting in various health hazards. Hence, the removal of these contaminants, including MPs, is highly important. Conventional techniques of wastewater treatment show success to some extent but not the total removal of MPs. Chemical and biological strategies are utilized in wastewater treatment. However, removal of MPs has to be done with specific techniques as most of the traditional methods prove ineffective in eliminating all particles. Hence, the primary aim of this chapter is to point out the cutting-edge methods used in removal of microplastics along with the existing techniques.

11.2 OCCURRENCE OF MICROPLASTICS FROM THE PHARMACEUTICAL INDUSTRY

Pharmaceuticals and personal care products (PPCPs) are one of the growing sources of pollutants in the pharmaceutical industry. Excess amounts of production and discarding of these materials are raising increasing questions about their possible negative impacts on public health and ecosystems. Microplastics can function as carriers for hydrophobic organic contaminants (HOCs) such as PPCPs (Atugoda et al., 2021; Martinho et al., 2022). HOCs exhibit a strong inclination to adsorb onto non-polar areas, such as organic substances and sediments. Nevertheless, they frequently demonstrate a higher attraction for plastic surfaces (Carbery et al., 2018; Atugoda et al., 2021). The increased population in a particular country is directly influencing the elevated production of pharmaceutical products, encompassing drugs such as antibiotics, β-blockers, antidepressants, anticonvulsants, and anti-inflammatory non-steroidal medications. Hormones, on the other hand, include steroidal hormones, antimicrobial agents, and UV screening products. Sorption isotherm studies were conducted to explore their binding on microplastics (Atugoda et al., 2021). Additionally, sources of microplastic pollution in the pharmaceutical industry include hospital waste, vaccination residues, post-used PPEs, testing kits, and packaging. The mentioned sources were found to contain higher amounts of polypropylene, polyethylene terephthalate, polypropylene, and polystyrene (Pashaei et al., 2022). Therefore, it is essential to implement appropriate treatment and management measures for wastewater in the pharmaceutical industry to mitigate potential threats to humans, animals, and the environment as a whole.

Microplastics are categorized into various types, considering their source, formation pathway, size, form, and polymeric composition. Specifically, as pertheorigin (Figure 11.1), they are further categorized as primary and secondary microplastics (Duis & Coors, 2016; Bradney et al., 2019; Hamidian et al., 2021). Additionally, categorized according to their chemical properties, these encompass polyethylene, polypropylene, polystyrene, and various other materials. Most pharmaceutical and personal care products are classified as primary microplastics since they are intentionally manufactured and added to consumer goods (Osman et al., 2023). The unrestricted use of various daily care items, including toothpaste, facial cleansers, cosmetics, and resin pellets, contributes to the boost in the release of microplastic particles into the ecological systems (Fendall & Sewell, 2009; Martinho et al., 2022). The breakdown of larger plastic particle waste results in secondary microplastics and they are unintentionally formed. Plastic bottles, bags, food containers, and other items that are chemical, physical, or biological are all part of this process (Martinho et al., 2022; Osman et al., 2023).

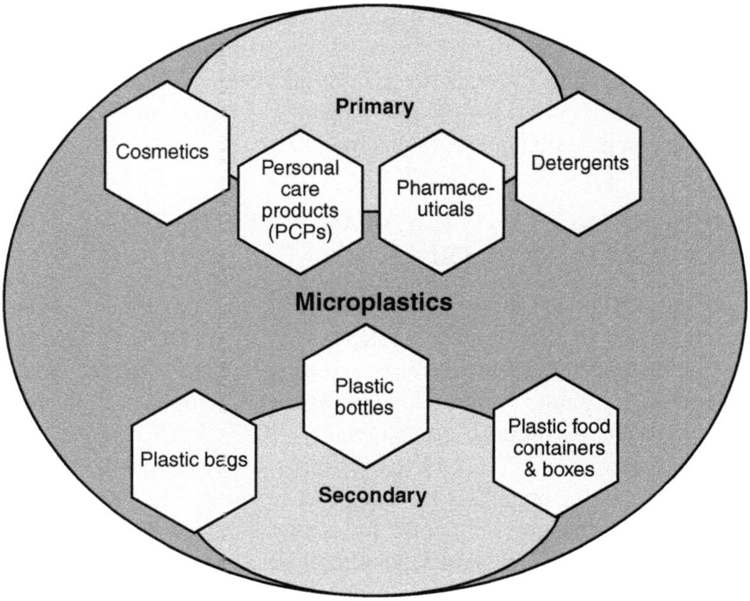

FIGURE 11.1 Types of microplastics generated from pharmaceutical waste.

TABLE 11.1
Pharmaceuticals products found in aquatic environments

Product type	Drug type	References
Antibiotics	Penicillin, tetracycline, aminoglycoside, erythromycin, enoxacin, oxytetracycline, sulfonamides, fluoroquinolone, ciprofloxacin and trimethoprim	(Fick et al., 2009; Fram & Belitz, 2011; Matongo et al., 2015; Mokh et al., 2017; Atugoda et al., 2021; Pashaei et al., 2022)
Psychiatric drugs	Antipsychotics, anxiolytics, hypnotics, antidepressants	
Cardiovascular drugs	Ca-channel blockers, β-blockers, angiotensin converting enzyme inhibitors	

The intake of pharmaceutical products has surged throughout the past ten years, and during the pandemic, the consumption of antibiotics has risen, directly contributing to the elevated presence of pharmaceutical contaminants in wastewater (Pashaei et al., 2022; Arienzo and Donadio, 2023). Further the presence of other pharma products including hormones, psychiatric drugs and cardiovascular drugs (Table 11.1) were noted in aquatic environments, suggesting that the discharge of pharmaceutical waste was not properly maintained (Pashaei et al., 2022). The prevalent antibiotics in oceans include ciprofloxacin, tetracycline, oxytetracycline, sulfamethoxazole, and trimethoprim. These substances, known as endocrine disruptors, have effects on crustaceans and fish, leading to alterations in hormone levels in the blood and the potential induction of cardiovascular collapse (Arienzo et al., 2021). Moreover, the factors affecting the transportation, degradation, and absorption of pharmaceutical contaminants depend on various parameters, including pH, water depth, temperature, ionic strength, and enzymes. Higher degradation of contaminants can be observed in summer than in winter (Nikolaou et al., 2007). The above-mentioned products contain microplastics, and they have the ability to bind with them, potentially resulting in an elevation in the amount of microplastics.

Polyethylene, polypropylene, polystyrene, polyvinyl chloride (PVC), polyamide, and polyester polyethylene terephthalate (PET) are commonly identified microplastics in water bodies (Godoy et al., 2020; Arienzo and Donadio, 2023). Within these waters bodies' microplastics have capacity to offer a substantial solid surface, particularly for hydrophobic organic contaminants. This surface area can exceed that of the nearby waters by several orders of magnitude. Origin of these non-biodegradable compounds is from plastic bottles, packaging, containers, and other products. The chart below illustrates the densities of plastic materials commonly encountered in water bodies (Duis and Coors, 2016). These substances exhibit gene-altering and disrupting the endocrine impacts on aquatic biota, posing harm to various organisms. Through the process of fragmentation, they indirectly display harm to human health (Rezania et al., 2018).

11.3 POLLUTION FROM MICROPLASTICS PRECEDING AND AMIDST THE COVID-19 PANDEMIC

During the onset of the pandemic, the World Health Organization (WHO) urged retailers worldwide to accelerate monthly manufacturing of personal protective equipment (PPE) by 40% to efficiently curb the swift transmission of SARS-CoV-2 (WHO, 2020). The use of single-use masks, testing kits, syringes, and personal protective equipment became the highest source of microplastic pollution. Biodegradable plastics are also identified as a contributor to secondary microplastic pollution, while the surge in demand for PPE represents a new major source of plastic consumption, potentially heightening health risks (Lee and Kim, 2022).

In line with research undertaken to investigate into the amount of microplastic pollution in surface water, during the COVID-19 pandemic, there was more microplastic contamination in Turkey's and Iran's surface water than there was before the outbreak (Figure 11.2). In Iran, the microplastics concentration per cubic meter was 0.000061 particles in 2019, and notably increased (43 particles/m^3) during the COVID-19 pandemic in 2021. In Turkey, it was 0.750 particles/m^3 in 2019 and increased up to 0.944 particles/m^3 after the pandemic (Pashaei et al., 2022).

11.4 IMPACT OF MICROPLASTICS ON AQUATIC ECOSYSTEMS

Large-scale plastic waste releases fragmented debris that contains microscopic particles known as microplastics into water supplies. Due to their reduced size,

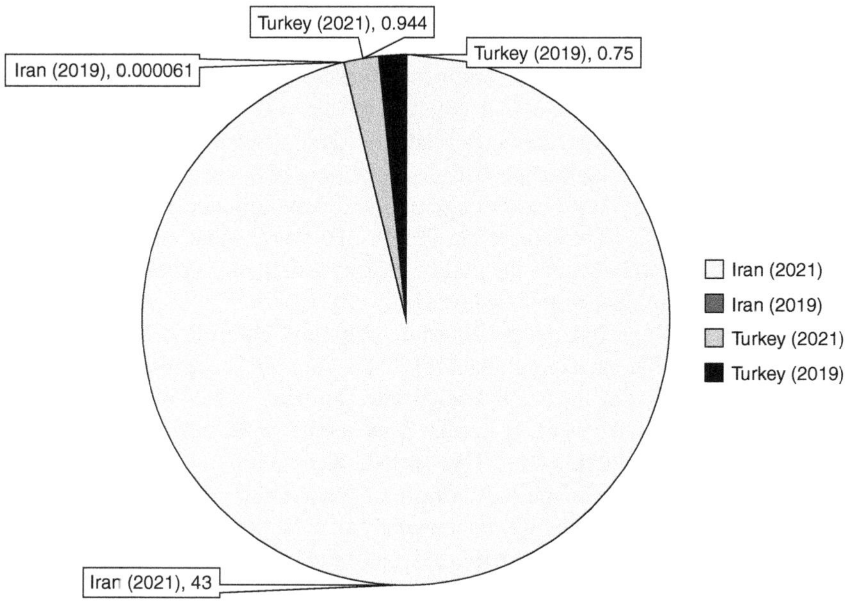

FIGURE 11.2 Comparative analysis of microplastic concentrations in pharmaceutical wastewater: Iran and Turkey (2019–2021).

marine/aquatic species may find it easier to ingest microplastics, which can cause toxic waste to build up and interfere with their physiological functions. Almost 700 aquatic species have been negatively impacted by the influx of microplastics worldwide, including different crustaceans, sea turtles, and penguins (Cincinelli et al., 2019). On the other hand, the problem caused by microplastics gets worse since most victims remain undiscovered in the wide waters (Pabortsava and Lampitt, 2020). The primary causes of plastic contamination in nature are by human actions, uncontrolled waste from sewage and water treatment facilities, and the textile industry (Sharma, Gupta, and Gupta, 2024). For the life forms in water, even one microbead is highly toxic. Following consumption, it evaluates the migration of chemicals to organisms at primary levels, which can then be passed on to upper trophic level organisms further along the food chain (Kolandhasamy et al., 2018). Several lab experiments have demonstrated that artificial additives and adsorbed toxins on the surface of plastic may seep out and enter the tissues and stomachs of marine creatures after ingestion (Sharma et al., 2021). Depending on concentration gradients, some scientists hypothesized that compounds already existing in the body would also be able to flow in the reverse way by sorbing to plastics in the intestines. As is already known, seafood contains a wide variety of potentially hazardous substances, especially in dominant predatory species like tuna and swordfish (Haq et al., 2016b).

Reactive oxygen species development and production in marine primary producers are impacted by polystyrene microbeads (Haq, Raj, and Markandeya, 2018). MPs have distinct sub-lethal effects rather than affecting survival in species at high trophic levels. For example, after being exposed to polystyrene MPs, crabs exhibit decreased fecundity, delayed offspring development, impaired food absorption and enzyme function, and altered behavior (Haq, Singh, and Kalamdhad, 2022b). Similarly, rotifers and sea urchins that consumed polystyrene microbeads of varying sizes showed decreased growth rate, fecundity, and embryotoxicity. Numerous investigations on mussels revealed the buildup of polystyrene microbeads in the gut cavity, which changes the activity of enzymes and filtration (Haq, Roy, and Kalamdhad, 2022c).

11.5 CONTEMPORARY TECHNIQUES FOR TREATING WASTEWATER FROM THE PHARMACEUTICAL INDUSTRY

Most of the conventional techniques of wastewater treatment display promising results to some extent but not the total removal of MPs. Chemical and biological strategies are utilized in wastewater treatment. This includes membrane filtration, activated carbon filtration, the Fenton process, coagulation and sedimentation, chlorination and UV-oxidation, ozonation, and biological methods (Ahmed et al., 2017; Gomes et al., 2017).

11.5.1 Membrane Filtration

The main types of membrane filtration utilized in pharmaceutical industrial waste water are microfiltration (MF), ultrafiltration (UF). MF relies on static pressure as the driving force, characterized by small pore size, where the separation process occurs through membrane sieve action, akin to traditional filtration principles (Haq, Kalamdhad, and Pandey, 2022a). Ultrafiltration (UF) operates on the principle of pressure disparity between the membranes on opposing sides, utilizing an ultrafiltration membrane as the filter agent. Larger macromolecules are maintained as water passes across the membrane surface under a particular pressure, whereas water, tiny molecules permeate and inorganic salts (Guo et al., 2017). The homogeneous pore distribution of the membranes during wastewater treatment is the benefit of membrane filtration.

According to Ma et al., (2019), the ultrafiltration membrane effectively stopped microplastics since the particles' size surpassed the threshold pore size of the UF membrane. As per Enfrin et al., (2020), as the exposed time lengthened, a greater quantity of microplastics penetrated through the membrane. Specifically, for certain membranes like polysulfone, which possessed a relatively hydrophilic nature, both microplastics and the polysulfone membranes exhibited negative charges and were hydrophobic. Consequently, the repelling electrostatic forces produced by the microplastics and membrane surface charge counterbalanced the attractive polar interactions (Enfrin et al., 2020).

11.5.2 Active Carbon Filtration/Activated Carbon Adsorption

Active carbon is regarded as among the most promising techniques for microplastic removal and can be categorized based on both physical and chemical absorption properties. Recognized for its extensive specific surface area, elevated adsorption capacity, stable chemical properties, and multi-level pore structure, it finds widespread application functioning as an adsorbent or carrier to facilitate pollutant removal via catalytic processes (Guo et al., 2017). The retention of microplastics by activated carbon materials is impacted by many things, such as the shape, chemistry, and size of the microplastic surfaces. Additionally, water properties such as saturation and ionic strength play a significant role in this process (Virpi Siipola et al., 2022).

A large-scale process developed by Alt and Mauritz (2010) utilizing powdered activated carbon, at sewage treatment plants, where activated carbon is administered following the standard treatment process. The estimated cost was 0.5 Swedish Krona (SEK) per cubic meter of water or 20 SEK per person annually, as existing equipment on site largely sufficed. By drying and gasifying the carbon-sludge mixture in this system at 900°C, all organic molecules are effectively eliminated at that temperature. The gas produced is utilized in the sludge drying procedure (Alt and Mauritz, 2010). Furthermore, an advanced technique in active carbon filtration involves the utilization of granulated active carbon biofilters, which will be further elaborated upon in this chapter.

11.5.3 Coagulation and Sedimentation

Coagulation involves the introduction of chemical compounds into wastewater, followed by quick stirring to disperse them, which subsequently transforms steady contaminants into unstable and precipitable materials. In the advanced processing of pharmaceutical wastewater, the primary focus lies in effectively extracting and eliminating bound water surrounding hydrophilic colloids. Therefore, the nature of the flocculant is crucial, as it directly influences the efficacy of coagulation (Guo et al., 2017). Soon after coagulation, sedimentation takes place, separating the pollutants according to their density.

Ma et al. (2019) reported that an aluminum origin coagulant exhibited enhanced efficacy in eliminating microplastics. This suggests the potential for enhancing the coagulation in wastewater treatment facilities (Ma et al., 2019). It has been demonstrated that coagulants, both natural and chemical thoroughly eliminate microplastics from wastewater streams. Nevertheless, research on the application of organic coagulants specifically for microplastic elimination remains scarce, with the majority of studies concentrating on reducing COD and turbidity (Reza et al., 2023).

11.5.4 Chlorination and UV-oxidation

While chlorination and UV-oxidation are commonly utilized techniques in wastewater treatment plants (WWTs), research suggests that microplastics may not be entirely resistant to chlorine exposure. Chlorine has the potential to disrupt existing bonds and create new ones (Kelkar et al., 2019). Similarly, UV-oxidation alters the chemical characteristics and surface topography of microplastics, leading to increased surface roughness and the formation of cracks and flakes. This process can result in the generation of smaller particles, including nano-scale plastics (Cooper and Corcoran, 2010; Cai et al., 2018). Hence, advanced oxidation processes were developed to meet the need of removal of microplastics.

11.5.5 Ozonation

Ozonation, a widely employed method in wastewater treatment systems, has evolved into more sophisticated advanced oxidation processes due to recent advancements (Gao et al., 2019). This method can enhance the dissolved oxygen (DO) levels in the effluent, potentially obviating the need for reaeration and augmenting the DO levels in the receiving stream. Consequently, it facilitates the elimination of a broad spectrum of pollutants (Guo et al., 2017; Gomes et al., 2017). In acidic conditions, ozone mostly operates as an oxidant; in neutral and alkaline nature, it primarily relies on free radical events. Because ozone rapidly oxidizes and decomposes organic matter in water, contaminants are successfully eliminated from the water. Although direct mineralization is challenging, an alternative approach involves the production of a readily biodegradable byproduct (Ding et al., 2021). This byproduct effectively addresses turbidity and pathogen concerns within wastewater. Ozone, when integrated with other wastewater treatment methodologies, contributes to the formation of ozone advanced oxidation technology, characterized by heightened oxidation capabilities and reduced selectivity towards reactants. Examples of such combinations include O_3-UV and O_3-H_2O_2 (Guo et al., 2017). Research indicates that ozonation, advanced oxidation processes, and other conventional techniques are inadequate on their own for completely eliminating all microplastic residues from water. Ozonation, for example, alters polymer structures but does not enhance microparticle removal (Cherniak et al., 2022; Pashaei et al., 2022).

11.5.6 Fenton Process

The conventional Fenton process operates on the principle that when hydrogen peroxide interacts with soluble iron (II) salts, hydroxyl radicals are formed. These radicals target the organic pollutant and either partially or totally oxidize it (Dos Santos et al., 2021). Iron has been studied for AOP in wastewater treatment as it is non-toxic and abundant in nature (Ahmed et al., 2021). Nonetheless as multiple studies have demonstrated, MPs are hardly impacted by the convention Fenton process (Tagg et al., 2017). Some studies have used Fenton process to "weather" MPs for degradation (Liu et al., 2019).

Liu et al. (2022) tried to age and degrade polystyrene (PS) and nylon (PA6) MPs on four Fenton treatment cycle with each cycle lasting for five hours. The experimental conditions were set as follows: pH 3, 0.5% hydrogen peroxide and 0.2 M Fe^{2+}. It was noted the PVC MPs fragmented under the experimental conditions, and the mass loss was recorded approximately to be 10%. It was observed that low molecular weight products like o-xylene and ethylbenzene degraded the material. Additionally, part of PVC fragments was digested by anaerobic microorganisms, suggesting these microbes were utilizing the carbon sources from Fenton degradation compounds (Mackuľak et al., 2015).

Polystyrene (PS) degradation through Fenton was reported by (Haq et al., 2017), in which they used ferric ions and the experimental conditions were 0.42 mM concentration of Fe^{3+}, 141 mM concentration of hydrogen peroxide at pH= 2. Even at 75 °C, no signs of polymer breakdown were seen for the course of the five-hour oxidative treatment. Ferric ions are used in the process as a transition metal in the Fenton-like

Fenton method. By using the standard Fenton reaction ([Fe^{2+}] = 1 mM, [H_2O_2] = 10 mM, pH = 3,and a 360-minute reaction time, the red color found in high-density polyethylene (HDPE) microplastic was broken down by more than 90% (Luo et al., 2021). The degree of pigment degradation outweighed the amount of leaching. Aged low-density polyethylene LDPE displayed surface cracks and fragmentation following the oxidation process (Luo et al., 2021).

11.5.7 Biological Techniques

The biological degradation of plastic involves various enzymes and enzymatic reactions. The fundamental process proceeds by breaking down polymers into smaller particles, which are then broken down into dimers, oligomers, and monomers. Microbes help in the mineralization. Carbon dioxide is utilized to produce many intermediary chemicals after it has been completely mineralized (Haq, Mazumder, and Kalamdhad, 2020). These compounds are used for microbial cell miniatous and growth. Lipases, lignin peroxides, esterases, manganese peroxides, and laccases are some of the various extracellular enzymes that are essential for the breakdown of microplastics. These enzymes make microplastics more hydrophilic and transform them into alcohol remnants or carbonyl (Taniguchi et al., 2019). By accelerating chain cleavage events, hydrolase enzymes including esterase, cutinase, lipases work on plastic surfaces to break down microplastics. These enzymes act on the polymer's surface, causing cracks to form, but they are unable to diffuse within the polymer. The produced monomers eventually enter several metabolic pathways after being absorbed into the cytoplasm of microorganisms.

There are several methods to determine if biodegradation is taking place (Haq et al., 2016a, b; 2017; 2018; 2020; 2022a, b, c; 2024; Haq and Kalamdhad, 2021; Haq and Kalamdhad, 2022, 2023). This includes, to check CO_2 production from degradation of low molecular wight polymers. Morphological, chemical, thermal and structural properties of MPs can be investigated using techniques such as laser diffraction particle, differential scanning, scanning electron microscopy, X-ray diffraction, calorimetry, etc. (Huang et al., 2022). Another popular technique for determining the biodegradation of microplastics is the gravimetric weight loss method. The biodegradation of microplastics is a very slow process that depends on the test conditions and incubation period (Raddadi and Fava, 2019), therefore careful consideration should be made while using this method. Using scanning electron microscopy, one can observe the physical properties of the polymeric surface. This aids in determining the development of microbial biofilms and surface deterioration. Examining the development of holes and cracks in the polymer can help determine how well the biodegradation of the polymer is going (Raddadi and Fava, 2019).

11.6 ADVANCED TECHNIQUES USED IN TREATING MICROPLASTIC PARTICLES IN PHARMACEUTICAL INDUSTRIAL WASTEWATER

Advanced techniques for treating microplastic particles in pharmaceutical industrial wastewater offer promising solutions to mitigate pollution. These techniques include granulated active carbon (GAC) applied with membrane bioreactor (MBR), magnetic polyoxometalate-supported ionic liquid phases (magPOM-SILPs), magnetic extraction techniques and advanced oxidation processes (AOPs). They have diverse expertise areas including microplastic removal efficiency by effectively adsorbing pollutants, and utilization of magnetic nanoparticles is some of them. By comprehensive solutions for managing microplastic pollution in pharmaceutical industrial effluents, these advanced techniques represent promising strategies to safeguard aquatic ecosystems and public health from the harmful effects of microplastic contamination.

11.6.1 Granulated Active Carbon Biofilter Applied to Membrane Bioreactor

This is a combination of granulated active carbon (GAC) along with membrane bioreactor (MBR) in the waste water system to maximize microplastic removal and reduce leftover microplastics (Baresel et al., 2019).

11.6.1.1 Membrane Bioreactor (MBR)

The membrane bioreactor (MBR) is a single apparatus that combines traditional biological treatment methods with physical liquid–solid separation achieved through membrane filtration. This system combines a suspended growth bioreactor with a microfiltration or ultrafiltration membrane unit, and it is commonly utilized in both municipal and industrial wastewater treatment plants (WWTPs) (Iorhemen et al., 2016). High-quality effluent, shorter hydraulic retention times (HRT), increased volumetric loading rates, lengthened solid retention times (SRT), reduced sludge output, and the possibility of concurrent denitrification or nitrification during long SRTs are just a few benefits that MBR technology has over the activated sludge process (ASP) (Meng et al., 2009; Mutamim et al., 2013). This technology presents superior effluent quality, enables higher loading rates, and promotes effective nitrification and denitrification processes within extended solid retention times (Wang et al., 2008; Judd, 2008). Numerous WWTPs in Sweden, including the Stockholm Water and Waste Company, the nation's extensive water supply organization, are opting to upgrade from traditional ASP to MBR (Baresel et al., 2019).

11.6.1.2 Granulated Active Carbon (GAC)-Biofilter

The utilization of granular activated carbon (GAC) biofilters is extensive within wastewater treatment plants, serving as an advanced treatment method renowned for its efficacy in eliminating hydrophobic or refractory organic substances. These substances, which are often not entirely removed through conventional biological processes, are effectively targeted and eliminated through the application of GAC biofilters (Vigneswaran et al., 2007). At the outset of the process, the GAC biofilter serves to segregate sludge from polluted carbon. The digested sludge can potentially be utilized for agricultural purposes, but pre-existing compounds in the sludge must be carefully evaluated. Moreover, through the establishment of multiple columns, a higher carbon capacity is achieved, with the initial carbon-column able to handle

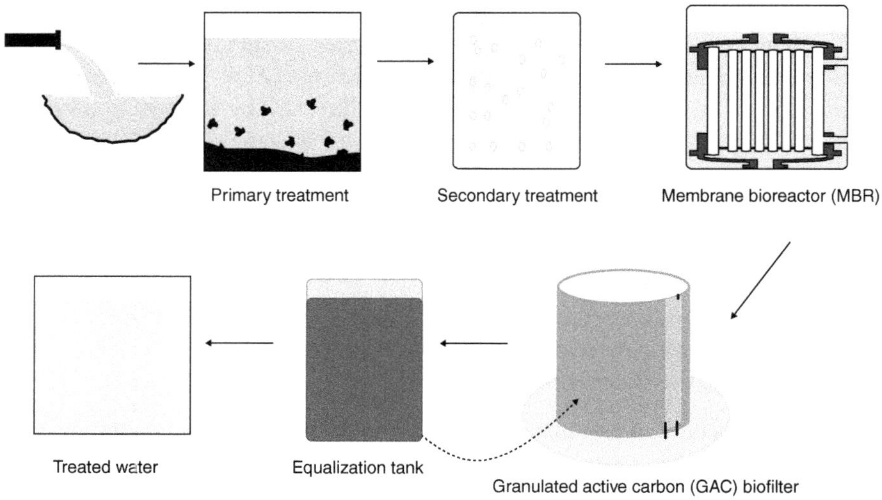

FIGURE 11.3 Tertiary wastewater treatment plant schematic diagram. (Adapted from Ahmed et al., 2017).

higher pollutant concentrations compared to subsequent ones, which are capable of handling lower concentrations. Additionally, during carbon regeneration, all organic compounds are eliminated (Ek et al., 2014).

Baresel et al. (2019) implemented a prototype configuration of MBR and GAC biofilter for an actual WWTP to study the potential functionality of removing pharmaceutical micropollutants, including microplastics (Figure 11.3). The observations of the investigation displayed that the treatment system successfully removed all of the micropollutants that were investigated, including microplastics and pharmaceutical residues, from wastewater, either to undetectable levels or extremely low quantities. This indicates that the integration of adsorption, filtration, and biodegradation offers a comprehensive and effective approach to removing micropollutants and their associated effects. Twenty-five μm-sized pores were used to filter water specimens, and the materials collected on the filters were examined using a 50x stereo microscope. Based on their shape, all microplastic particles were divided into three groups: plastic flakes, fragments, and fibers. The designation "plastic flake" applied to extremely fine particles, while bigger particles were termed "plastic fragments." All three of these groups were referred to as "microplastics" or "plastic particles." No microplastic was identified in the MBR-effluent, demonstrating a 100% effectiveness of the removal (Baresel et al., 2019).

11.6.2 Magnetic Polyoxometalate-Supported Ionic Liquid Phases (magPOM-SILPs) Technology

Polyoxometalate-supported ionic liquid phases are a composite material consisting of commercially available porous silica particles that have been modified on the surface with water-immiscible polyoxometalate ionic liquids (Herrmann et al., 2017). These ionic liquids have the ability to bind both organic and inorganic contaminants

(Herrmann et al., 2014). Microplastics can be drawn to and concentrated by persistent organic pollutants (Andrady, 2011). Additionally, magnetic molecule water purification is a viable substitute for filtration, addressing the limitations associated with filtering small water volumes, which can be energy-intensive or require significant amounts of filter materials. Magnetic water purification has the potential to treat higher water quantities and could be implemented irrespective of additional infrastructure, provided that particle removal can be achieved using a simple permanent magnet (Lu et al., 2007; Misra et al., 2019).

Misra et al. (2019) developed magnetic polyoxometalate-supported ionic liquid phases as a method for treating larger volumes of water, effectively binding inorganic, organic, microplastic, and microbial pollutants. These phases can be readily retrieved using a permanent magnet. Their findings demonstrated that the combined use of magPOM-SILPs and polystyrene beads can effectively eliminate microplastic model compounds from extensive water volumes. Furthermore, the removal of microplastics through attachment to magnetic particles presents a potential solution for treating higher water volumes that are not suitable for traditional filtration methods. Significant removal efficiencies have been documented, along with preliminary observations regarding a novel approach to removing microplastics through the surface-binding of magnetic particles (Misra et al., 2019). This technique exhibits promise for potential application in the microplastic purification of wastewater generated by the pharmaceutical industry in the future.

11.6.3 Magnetic Extraction of Microplastics

A novel method has been explored in which coated iron nanoparticles are used to magnetically charge plastics, facilitating their magnetic retrieval and isolation of microplastics (Grbic et al., 2019; Shi et al., 2022). Grbic et al. (2019) devised hydrophobic iron nanoparticles capable of adhering to plastic, thus facilitating magnetic retrieval. Their technique effectively extracted 93% of microplastics greater than 1 mm and 92% of polyethylene and PS beads ranging in size from seawater. The microplastics retrieved included PET, polyethylene, polyurethane, polypropylene, PVC. Additionally, 84% and 78% recovery rates were achieved for polyethylene, polystyrene, polyurethane, polyvinyl chloride, and polypropylene ranging from 200 μm to 1 mm in freshwater and sediments. Although not yet applied in pharmaceutical industrial wastewater, there exists significant potential for microplastics removal through this approach.

In Grbic et al. (2019) study, it has been observed that the recovery rates for the large size fraction, which ranged from 74%–105%. The average recoveries were as follows: approximately 96% for polypropylene (PP), 92% for PVC, 96% for HDPE, 105% for polyurethan, 96% for PS, and 74%for PET fibers. With the exception of PVC, it was noted that recovered PET fibers were notably lesser compared to other types. In several instances, recoveries surpassed 100% as a result of polymers breaking apart during mixing and handling. However, this experiment with higher microplastics shows successful magnetic extraction of various types of polymers. In general, the method proves effective across different sizes, types of polymers, and sample matrices, and researchers may find it suitable for incorporation into the microplastic extraction process (e.g., after density separation) or as a standalone method for obtaining cleaner samples, such as drinking water (Grbic et al., 2019). Even so, more study is required to examine the use of microplastics removal in industrial wastewater from pharmaceuticals.

In a study led by Shi et al. (2022), magnetic nano-Fe_3O_4 was employed to extract microplastics from water. Their findings revealed that the size and type of polymer of the microplastics affected the removal efficiency, and that the density of nano-Fe_3O_4 absorbed onto the microplastic surfaces strongly associated with the removal rate. The clearance rates of PE microplastic in this investigation increased from 0.5–1.3 $g \cdot L^{-1}$ nano-Fe_3O_4 in a concentration-dependent manner. They also noticed that the clearance rate of MPs increased steadily throughout the course of the magnetization treatment, stabilizing after 150 minutes. The clearance rate of microplastics peaked at a concentration of 1.3 $g \cdot L^{-1}$ nano-Fe_3O_4 and a treatment duration of 150 minutes, suggesting that these values represent the ideal concentration and duration of treatment for the established approach. Furthermore, it has been indicated the successful application of this method for eliminating microplastics from environmental water bodies such as household sewage, river, and natural seawater, achieving an elimination rate exceeding 80% (Shi et al., 2022). Although nanoencapsulation has shown encouraging results in this area, further research is required to fully understand its potential in pharmaceutical wastewater treatment.

11.6.4 Advanced Oxidation Processes (AOPs)

Advanced oxidation processes (AOPs) produce extremely reactive radicals within water, targeting the oxidation of organic contaminants with the goal of complete mineralization. AOPs facilitate the oxidation of pollutants through the generation of free radicals. These pollutants typically resist degradation by conventional oxidizing agents. Various types of AOPs exist, including supercritical water oxidation, wet air oxidation, Fenton reagent, ultrasound oxidation, photocatalytic oxidation, electrochemical oxidation, and ozonation (Guo et al., 2017). Specifically, AOPs employing ultraviolet radiation have proven effective in treating wastewater containing biologically resistant pollutants. The capacity of these processes to oxidize organic pollutants into their mineral constituents, such as CO_2 and H_2O, as opposed to shifting or condensing them into a new waste stream, like solid waste through filtration, is a significant advantage (Thiruvenkatachari et al., 2007; Deng and Zhao, 2015).

Easton et al. (2023) evaluated the elimination of polyester fiber microplastics from wastewater through a UV/H_2O_2 oxidation procedure. This study focused primarily on the effects of H_2O_2treatment and UV irradiation on the elimination of microplastics. With an estimated half-life of 45.3 hours, it was shown that a 500 mgL^{-1} H_2O_2 dosage plus 4.0mW cm-2 UVC irradiation resulted in a 52.7% mass loss after 48 hours. Additionally, the fibers degraded, causing small holes and cracks to emerge throughout their surface as well as variations in the relative number of functional groups that contain oxygen. The carbon index (CI) of treated microplastics increased by up to 10.6% when compared to untreated fibers, indicating a chemical change.

Furthermore, they observed that UVC/H_2O_2 treatment demonstrated the capability to breakdown 15.2% of microplastics and fully eliminate the COD in hospital laundry wastewater. However, the further investigation using pharmaceutical industrial waste water is required to see the state of microplastic removal.

11.7 SUMMARY AND CONCLUSION

The chapter explored the pressing problem of microplastic contamination in wastewater from the pharmaceutical industry, in terms of origin, impacts, and advanced methods to combat them. Pharmaceutical wastewater is a complex mixture that contains organic matter, high salinity, toxic microbes, and microplastics. Sources of these MPs are hospital waste, used PPE, and packaging material, and they tend to carry harmful chemicals, including PPCPs that stick more easily to plastic surfaces. Further, the chapter focused on several state-of-the-art treatment technologies to deal with these contaminants. Membrane bioreactors (MBRs) are among them, in which biological treatment is integrated with membrane filtration. The MBR system is efficient for producing high-quality effluent, for sludge reduction, and for the enhancement of the nitrogen removal process. When GAC biofilters are combined with MBR, they significantly increase the removal of pharmaceutical pollutants, including microplastics, which bring their concentration down to almost negligible quantities. Another promising method is to use Magnetic Polyoxometalate-supported Ionic Liquid Phases. These involve modified porous silica particles being used to bind and remove contaminants from large volumes of water. Finally, magnetic extraction techniques utilize coated iron nanoparticles to magnetically pull out microplastics with a high success rate across different types of plastics and water conditions. These are all robust approaches to reducing microplastic pollution in pharmaceutical wastewaters for the protection of our aquatic ecosystems and public health. This chapter dived into the integration of these techniques in light of the growing challenge of microplastic contamination and emphasized the need for effective wastewater management in the pharmaceutical sector. These innovative approaches can help us gain substantial ground in reducing the harmful effects of microplastics on the environment and human health.

In conclusion, to tackle the growing presence of microplastics in water bodies, including pharmaceutical industrial wastewater, numerous existing techniques have been developed and refined in accordance with technological advancements. Considering the pre-pandemic and post-pandemic era, there is a considerable rise of microplastics. Hence it is important to keep enhancing and continuation of research related to treatment of pharmaceutical residues. Not only that, the use and production of single-use plastics needs to reduce in the upcoming years. The utilization of pharmaceutical products containing micropollutants has surged in the modern era, exerting a direct influence on the environment and, consequently, human health. If each individual takes a proactive stance to mitigate these hazards, it can significantly contribute to creating a better world. Such actions directly support the extensive efforts of researchers striving to advance technologies aimed at reducing pollutants.

REFERENCES

Ahmed, M. B., Rahman, Md. S., Alom, J., Hasan, MD. S., Johir, M. A. H., Mondal, M. I. H., Lee, D.-Y., Park, J., Zhou, J. L., & Yoon, M.-H. (2021). Microplastic particles in the aquatic environment: A systematic review. *Science of the Total Environment*, 775, 145793. https://doi.org/10.1016/j.scitotenv.2021.145793

Ahmed, M. B., Zhou, J. L., Ngo, H. H., Guo, W., Thomaidis, N. S., & Xu, J. (2017). Progress in the biological and chemical treatment technologies for emerging contaminant removal from wastewater: A critical review. *Journal of Hazardous Materials*, 323, 274–298. https://doi.org/10.1016/j.jhazmat.2016.04.045

Alt, K., & Mauritz, A. (2010). ProjektzurTeilstrombehandlungmitPulveraktivkohleimKlärwerk Mannheim (Project for sidestream treatment with powdered activated carbon in the sewage treatment plant Mannheim). *KA AbwasserAbfall*, 57(2), 161.

Andrady, A. L. (2011). Microplastics in the marine environment. *Marine Pollution Bulletin*, 62(8), 1596–1605. https://doi.org/10.1016/j.marpolbul.2011.05.030

Arienzo, M., & Donadio, C. (2023). Microplastic–pharmaceuticals interaction in water systems. *Journal of Marine Science and Engineering*, 11(7), 1437–1437. https://doi.org/10.3390/jmse11071437

Arienzo, M., Ferrara, L., & Trifuoggi, M. (2021). The dual role of microplastics in marine environment: Sink and vectors of pollutants. *Journal of Marine Science and Engineering*, 9(6), 642. https://doi.org/10.3390/jmse9060642

Atugoda, T., Vithanage, M., Wijesekara, H., Bolan, N., Sarmah, A. K., Bank, M. S., You, S., & Ok, Y. S. (2021). Interactions between microplastics, pharmaceuticals and personal care products: Implications for vector transport. *Environment International*, 149, 106367. https://doi.org/10.1016/j.envint.2020.106367

Baresel, C., Harding, M., & Fång, J. (2019). Ultrafiltration/granulated active carbon-biofilter: Efficient removal of a broad range of micropollutants. *Applied Sciences*, 9(4), 710. https://doi.org/10.3390/app9040710

Bradney, L., Wijesekara, H., Palansooriya, K. N., Obadamudalige, N., Bolan, N. S., Ok, Y. S., Rinklebe, J., Kim, K.-H., & Kirkham, M. B. (2019). Particulate plastics as a vector for toxic trace-element uptake by aquatic and terrestrial organisms and human health risk. *Environment International*, 131, 104937. https://doi.org/10.1016/j.envint.2019.104937

Cai, L., Wang, J., Peng, J., Wu, Z., & Tan, X. (2018). Observation of the degradation of three types of plastic pellets exposed to UV irradiation in three different environments. *Science of the Total Environment*, 628–629, 740–747. https://doi.org/10.1016/j.scitotenv.2018.02.079

Carbery, M., O'Connor, W., & Palanisami, T. (2018). Trophic transfer of microplastics and mixed contaminants in the marine food web and implications for human health. *Environment International*, 115, 400–409. https://doi.org/10.1016/j.envint.2018.03.007

Cherniak, S. L., Almuhtaram, H., McKie, M. J., Hermabessiere, L., Yuan, C., Rochman, C. M., & Andrews, R. C. (2022). Conventional and biological treatment for the removal of microplastics from drinking water. *Chemosphere*, 288, 132587. https://doi.org/10.1016/j.chemosphere.2021.132587

Cincinelli, A., Martellini, T., Guerranti, C., Scopetani, C., Chelazzi, D., & Giarrizzo, T. (2019). A potpourri of microplastics in the sea surface and water column of the Mediterranean Sea. *TrAC Trends in Analytical Chemistry*, 110, 321–326. https://doi.org/10.1016/j.trac.2018.10.026

Cooper, D. A., & Corcoran, P. L. (2010). Effects of mechanical and chemical processes on the degradation of plastic beach debris on the island of Kauai, Hawaii. *Marine Pollution Bulletin*, 60(5), 650–654. https://doi.org/10.1016/j.marpolbul.2009.12.026

de Sousa, F. D. B. (2021). Plastic and its consequences during the COVID-19 pandemic. *Environmental Science and Pollution Research.* https://doi.org/10.1007/s11356-021-15425-w

Deng, Y., & Zhao, R. (2015). Advanced Oxidation Processes (AOPs) in Wastewater Treatment. *Current Pollution Reports*, 1(3), 167–176. https://doi.org/10.1007/s40726-015-0015-z

Ding, H., Zhang, J., He, H., Zhu, Y., Dionysiou, D. D., Liu, Z., & Zhao, C. (2021). Do membrane filtration systems in drinking water treatment plants release nano/microplastics? *Science of the Total Environment*, 755, 142658. https://doi.org/10.1016/j.scitotenv.2020.142658

Dos Santos, N. de O., Teixeira, L. A., Zhou, Q., Burke, G., & Campos, L. C. (2021). Fenton pre-oxidation of natural organic matter in drinking water treatment through the application of iron nails. *Environmental Technology*, 43(17), 2590–2603. https://doi.org/10.1080/09593330.2021.1890838

Duis, K., & Coors, A. (2016). Microplastics in the Aquatic and Terrestrial environment: Sources (with a Specific Focus on Personal Care products), Fate and Effects. *Environmental Sciences Europe*, 28(1). https://doi.org/10.1186/s12302-015-0069-y

Easton, T., Koutsos, V., & Chatzisymeon, E. (2023). Removal of polyester fibre microplastics from wastewater using a UV/H2O2 oxidation process. *Journal of Environmental Chemical Engineering*, 11(1), 109057. https://doi.org/10.1016/j.jece.2022.109057

Ek, M., Baresel, C., Magnér, J., Bergström, R., & Harding, M. (2014). Activated carbon for the removal of pharmaceutical residues from treated wastewater. *Water Science and Technology*, 69(11), 2372–2380. https://doi.org/10.2166/wst.2014.172

Enfrin, M., Lee, J., Le-Clech, P., & Dumée, L. F. (2020). Kinetic and mechanistic aspects of ultrafiltration membrane fouling by nano- and microplastics. *Journal of Membrane Science*, 601, 117890. https://doi.org/10.1016/j.memsci.2020.117890

Eryaşar, A. R., Gedik, K., Şahin, A., Öztürk, R. Ç., & Yılmaz, F. (2021). Characteristics and temporal trends of microplastics in the coastal area in the Southern Black Sea over the past decade. *Marine Pollution Bulletin*, 173, 112993. https://doi.org/10.1016/j.marpolbul.2021.112993

Fendall, L. S., & Sewell, M. A. (2009). Contributing to marine pollution by washing your face: Microplastics in facial cleansers. *Marine Pollution Bulletin*, 58(8), 1225–1228. https://doi.org/10.1016/j.marpolbul.2009.04.025

Fick, J., Söderström, H., Lindberg, R. H., Phan, C., Tysklind, M., & Larsson, D. G. J. (2009). Contamination Of Surface, Ground, and Drinking Water from Pharmaceutical Production. *Environmental Toxicology and Chemistry*, 28(12), 2522. https://doi.org/10.1897/09-073.1

Fischer, E. K., Paglialonga, L., Czech, E., & Tamminga, M. (2016). Microplastic pollution in lakes and lake shoreline sediments – A case study on Lake Bolsena and Lake Chiusi (central Italy). *Environmental Pollution*, 213, 648–657. https://doi.org/10.1016/j.envpol.2016.03.012

Fram, M. S., & Belitz, K. (2011). Occurrence and concentrations of pharmaceutical compounds in groundwater used for public drinking-water supply in California. *Science of the Total Environment*, 409(18), 3409–3417. https://doi.org/10.1016/j.scitotenv.2011.05.053

Gao, X., Xu, Y., Ma, M., Rao, K., & Wang, Z. (2019). Simultaneous passive sampling of hydrophilic and hydrophobic emerging organic contaminants in water. *Ecotoxicology and Environmental Safety*, 178, 25–32. https://doi.org/10.1016/j.ecoenv.2019.04.014

Godoy, V., Martín-Lara, M. A., Calero, M., & Blázquez, G. (2020). The relevance of interaction of chemicals/pollutants and microplastic samples as route for transporting contaminants. *Process Safety and Environmental Protection*, 138, 312–323. https://doi.org/10.1016/j.psep.2020.03.033

Gomes, J., Costa, R., Quinta-Ferreira, R. M., & Martins, R. C. (2017). Application of ozonation for pharmaceuticals and personal care products removal from water. *Science of the Total Environment*, 586, 265–283. https://doi.org/10.1016/j.scitotenv.2017.01.216

Grbic, J., Nguyen, B., Guo, E., You, J. B., Sinton, D., & Rochman, C. M. (2019). Magnetic extraction of microplastics from environmental samples. *Environmental Science & Technology Letters*, 6(2), 68–72. https://doi.org/10.1021/acs.estlett.8b00671

Guo, Y., Qi, P. S., & Liu, Y. Z. (2017). A review on advanced treatment of pharmaceutical wastewater. *IOP Conference Series: Earth and Environmental Science*, 63, 012025. https://doi.org/10.1088/1755-1315/63/1/012025

Guzzetti, E., Sureda, A., Tejada, S., & Faggio, C. (2018). Microplastic in marine organism: Environmental and toxicological effects. *Environmental Toxicology and Pharmacology*, 64, 164–171. https://doi.org/10.1016/j.etap.2018.10.009

Hamidian, A. H., Ozumchelouei, E. J., Feizi, F., Wu, C., Zhang, Y., & Yang, M. (2021). A review on the characteristics of microplastics in wastewater treatment plants: A source for toxic chemicals. *Journal of Cleaner Production*, 295, 126480. https://doi.org/10.1016/j.jclepro.2021.126480.

Haq, I., Kumar, S., Kumar, V., Singh, S.K., & Raj A (2016a). Evaluation of bioremediation potentiality of ligninolytic Serratia liquefaciens for detoxification of pulp and paper mill effluent. *J Hazard Mater* 305:190–199

Haq, I., Kumari, V., Kumar, S., Raj, A., Lohani, M., & Bhargava, R.N. (2016b). Evaluation of the phytotoxic and genotoxic potential of pulp and paper mill effluent using Vigna radiata and Allium cepa. *Adv Biol* 8065736.

Haq, I., Kumar, S., Raj, A., Lohani, M., & Satyanarayana, G.N.V. (2017). Genotoxicity assessment of pulp and paper mill effluent before and after bacterial degradation using Allium cepa test. *Chemosphere* 169: 642–650.

Haq, I., Raj, A., & Markandeya (2018). Biodegradation of Azure-B dye by Serratia liquefaciens and its validation by phytotoxicity, genotoxicity and cytotoxicity studies. *Chemosphere* 196, 58–68.

Haq, I., Mazumder, P., Kalamdhad, A.S. (2020). Recent advances in removal of lignin from paper industry wastewater and its industrial applications – A review. *Bioresource Technology* 312:123636.

Haq, I., & Kalamdhad, A.S. (2021) Phytotoxicity and cyto-genotoxicity evaluation of organic and inorganic pollutants containing petroleum refinery wastewater using plant bioassay. *Environ. Technol. Innov* 23:101651.

Haq, I., Kalamdhad, A.S., & Pandey, A. (2022a). Genotoxicity evaluation of paper industry wastewater prior and post-treatment with laccase producing Pseudomonas putida MTCC 7525. *J. Clean. Prod* 342:130981.

Haq, I., Singh, A., & Kalamdhad, A.S. (2022b). Integrated technologies for the remediation of paper industry waste in a circular bioeconomy, In Editor(s): Sunita Varjani, Ashok Pandey, Mohammad J. Taherzadeh, Huu Hao Ngo, R.D. Tyagi, *Biomass, Biofuels, Biochemicals*, Elsevier, Pages 351–363, ISBN 9780323885119.

Haq, I., Roy, S., & Kalamdhad, A.S. (2022c). Characterization of Pulp and Paper Mill Wastewater and Its Toxicity Analysis Using Vigna Radiata. In *Environmental Degradation: Monitoring, Assessment and Treatment Technologies*; Springer International Publishing: Cham, Switzerland, pp. 261–269. ISBN 9783030941475.

Haq, I., & Kalamdhad, A.S. (2022) Advanced and Ecofriendly Technologies for the Treatment of Industrial Wastewater to Constrain Environmental Pollution. In: Arora, S., Kumar, A., Ogita, S., & Yau, Y.Y. (eds) *Innovations in Environmental Biotechnology*. Springer, Singapore. https://doi.org/10.1007/978-981-16-4445-0_22.

Haq, I., & Kalamdhad, A.S. (2023). Enhanced biodegradation of toxic pollutants from paper industry wastewater using Pseudomonas sp. immobilized in composite biocarriers and its toxicity evaluation. *Bioresour. Technol. Repor* 24:101674.

Haq, I., Kalamdhad, A.S., & Malik, A. (2024) Bioremediation of petroleum refinery wastewater using Bacillus subtilis IH-1 and assessment of its toxicity. *Arch. Microbiol*, 206, 296.

Herrmann, S., Andrey Seliverstov, & Streb, C. (2014). Polyoxometalate-Ionic Liquids (Pom-Ils) – the ultimate soft polyoxometalates? A critical perspective. *Journal of Molecular and Engineering Materials*, 02(01), 1440001–1440001. https://doi.org/10.1142/s2251237314400012

Herrmann, S., De Matteis, L., de la Fuente, J. M., Mitchell, S. G., & Streb, C. (2017). Removal of multiple contaminants from water by polyoxometalate supported ionic liquid phases (POM-SILPs). *AngewandteChemie International Edition*, 56(6), 1667–1670. https://doi.org/10.1002/anie.201611072

Huang, Z., Hu, B., & Wang, H. (2022). Analytical methods for microplastics in the environment: a review. *Environmental Chemistry Letters*, 21(1). https://doi.org/10.1007/s10311-022-01525-7

Iorhemen, O., Hamza, R., & Tay, J. (2016). Membrane bioreactor (MBR) technology for wastewater treatment and reclamation: Membrane fouling. *Membranes*, 6(2), 33. https://doi.org/10.3390/membranes6020033

Jeyasanta, K. I., Sathish, N., Patterson, J., & Edward, J. K. P. (2020). Macro-, meso- and microplastic debris in the beaches of Tuticorin district, Southeast coast of India. *Marine Pollution Bulletin*, 154, 111055. https://doi.org/10.1016/j.marpolbul.2020.111055

Judd, S. (2008). The status of membrane bioreactor technology. *Trends in Biotechnology*, 26(2), 109–116. https://doi.org/10.1016/j.tibtech.2007.11.005

Kelkar, V. P., Rolsky, C. B., Pant, A., Green, M. D., Tongay, S., & Halden, R. U. (2019). Chemical and physical changes of microplastics during sterilization by chlorination. *Water Research*, 163, 114871. https://doi.org/10.1016/j.watres.2019.114871

Kolandhasamy, P., Su, L., Li, J., Qu, X., Jabeen, K., & Shi, H. (2018). Adherence of microplastics to soft tissue of mussels: a novel way to uptake microplastics beyond ingestion. *Science of the Total Environment*, 610, 635–640. https://doi.org/10.1016/j.scitotenv.2017.08.053

Kye, H., Kim, J., Ju, S., Lee, J., Lim, C., & Yoon, Y. (2023). Microplastics in water systems: A review of their impacts on the environment and their potential hazards. *Heliyon*, 9(3), e14359–e14359. https://doi.org/10.1016/j.heliyon.2023.e14359

Lee, M., & Kim, H. (2022). COVID-19 pandemic and microplastic pollution. *Nanomaterials*, 12(5), 851. https://doi.org/10.3390/nano12050851

Liu, B., Jiang, Q., Qiu, Z., Liu, L., Wei, R., Zhang, X., & Xu, H. (2022). Process analysis of microplastic degradation using activated PMS and Fenton reagents. *Chemosphere*, 298, 134220. https://doi.org/10.1016/j.chemosphere.2022.134220

Liu, P., Qian, L., Wang, H., Zhan, X., Lu, K., Gu, C., & Gao, S. (2019). New insights into the aging behavior of microplastics accelerated by advanced oxidation processes. *Environmental Science & Technology*, 53(7), 3579–3588. https://doi.org/10.1021/acs.est.9b00493

Lu, A.-H., Salabas, E., & Schüth, F. (2007). MagnetischeNanopartikel: Synthese, Stabilisierung, Funktionalisierung und Anwendung. *AngewandteChemie*, 119(8), 1242–1266. https://doi.org/10.1002/ange.200602866

Luo, H., Zeng, Y., Zhao, Y., Xiang, Y., Li, Y., & Pan, X. (2021). Effects of advanced oxidation processes on leachates and properties of microplastics. *Journal of Hazardous Materials*, 413, 125342. https://doi.org/10.1016/j.jhazmat.2021.125342

Lusher, A. L., McHugh, M., & Thompson, R. C. (2013). Occurrence of microplastics in the gastrointestinal tract of pelagic and demersal fish from the English Channel. *Marine Pollution Bulletin*, 67(1–2), 94–99. https://doi.org/10.1016/j.marpolbul.2012.11.028

Ma, B., Xue, W., Hu, C., Liu, H., Qu, J., & Li, L. (2019). Characteristics of microplastic removal via coagulation and ultrafiltration during drinking water treatment. *Chemical Engineering Journal*, 359, 159–167. https://doi.org/10.1016/j.cej.2018.11.155

Mackuľak, T., Takáčová, A., Gál, M., Marton, M., & Ryba, J. (2015). PVC degradation by Fenton reaction and biological decomposition. *Polymer Degradation and Stability*, 120, 226–231. https://doi.org/10.1016/j.polymdegradstab.2015.07.005

Martinho, S. D., Fernandes, V. C., Figueiredo, S. A., & Delerue-Matos, C. (2022). Microplastic pollution focused on sources, distribution, contaminant interactions, analytical methods, and wastewater removal strategies: A review. *International Journal of Environmental Research and Public Health*, 19(9), 5610. https://doi.org/10.3390/ijerph19095610

Matongo, S., Birungi, G., Moodley, B., & Ndungu, P. (2015). Pharmaceutical residues in water and sediment of Msunduzi River, KwaZulu-Natal, South Africa. *Chemosphere*, 134, 133–140. https://doi.org/10.1016/j.chemosphere.2015.03.093

Meng, F., Chae, S.-R., Drews, A., Kraume, M., Shin, H.-S., & Yang, F. (2009). Recent advances in membrane bioreactors (MBRs): Membrane fouling and membrane material. *Water Research*, 43(6), 1489–1512. https://doi.org/10.1016/j.watres.2008.12.044

Misra, A., Zambrzycki, C., Kloker, G., Kotyrba, A., Anjass, M. H., Franco Castillo, I., Mitchell, S. G., Güttel, R., & Streb, C. (2019). Water purification and microplastics removal using magnetic polyoxometalate-supported ionic liquid phases (magPOM-SILPs). *AngewandteChemie International Edition*, 59(4), 1601–1605. https://doi.org/10.1002/anie.201912111

Mokh, S., El Khatib, M., Koubar, M., Daher, Z., & Al Iskandarani, M. (2017). Innovative SPE-LC-MS/MS technique for the assessment of 63 pharmaceuticals and the detection of antibiotic-resistant-bacteria: A case study natural water sources in Lebanon. *Science of the Total Environment*, 609, 830–841. https://doi.org/10.1016/j.scitotenv.2017.07.230

Mutamim, N. S. A., Noor, Z. Z., Hassan, M. A. A., Yuniarto, A., & Olsson, G. (2013). Membrane bioreactor: Applications and limitations in treating high strength industrial wastewater. *Chemical Engineering Journal*, 225, 109–119. https://doi.org/10.1016/j.cej.2013.02.131

Nikolaou, A., Meric, S., & Fatta, D. (2007). Occurrence patterns of pharmaceuticals in water and wastewater environments. *Analytical and Bioanalytical Chemistry*, 387(4), 1225–1234. https://doi.org/10.1007/s00216-006-1035-8

Osman, A. I., Hosny, M., Eltaweil, A. S., Omar, S., Elgarahy, A. M., Farghali, M., Yap, P.-S., Wu, Y.-S., Nagandran, S., Batumalaie, K., Gopinath, S. C. B., John, O. D., Sekar, M., Saikia, T., Karunanithi, P., Hatta, M. H. M., & Akinyede, K. A. (2023). Microplastic sources, formation, toxicity and remediation: a review. *Environmental Chemistry Letters*. https://doi.org/10.1007/s10311-023-01593-3

Pabortsava, K., & Lampitt, R. S. (2020). High concentrations of plastic hidden beneath the surface of the Atlantic Ocean. *Nature Communications*, *11*(1). https://doi.org/10.1038/s41467-020-17932-9

Pashaei, R., Dzingelevičienė, R., Abbasi, S., Szultka-Młyńska, M., & Buszewski, B. (2022). Determination of the pharmaceuticals–nano/microplastics in aquatic systems by analytical and instrumental methods. *Environmental Monitoring and Assessment*, 194(2). https://doi.org/10.1007/s10661-022-09751-w

Pashaei, R., Dzingelevičienė, R., Bradauskaitė, A., Lajevardipour, A., Mlynska-Szultka, M., Dzingelevičius, N., Raugelė, S., Razbadauskas, A., Abbasi, S., Rees, R. M., &Buszewski, B. (2022). Pharmaceutical and microplastic pollution before and during the COVID-19 pandemic in surface water, wastewater, and groundwater. *Water*, 14(19), 3082. https://doi.org/10.3390/w14193082

Peng, G., Xu, P., Zhu, B., Bai, M., & Li, D. (2018). Microplastics in freshwater river sediments in Shanghai, China: A case study of risk assessment in mega-cities. *Environmental Pollution*, 234, 448–456. https://doi.org/10.1016/j.envpol.2017.11.034

Raddadi, N., & Fava, F. (2019). Biodegradation of oil-based plastics in the environment: Existing knowledge and needs of research and innovation. *Science of the Total Environment*, 679, 148–158. https://doi.org/10.1016/j.scitotenv.2019.04.419

Rana, R. S., Singh, P., Kandari, V., Singh, R., Dobhal, R., & Gupta, S. (2014). A review on characterization and bioremediation of pharmaceutical industries' wastewater: an Indian perspective. *Applied Water Science*, 7(1), 1–12. https://doi.org/10.1007/s13201-014-0225-3

Reza, T., Huda, Z., Sheikh, R., Hasan, H. A., Ismail, N. A., & Othman, A. R. (2023). Microplastic removal in wastewater treatment plants (WWTPs) by natural coagulation: A literature review. *Toxics*, 12(1), 12. https://doi.org/10.3390/toxics12010012

Rezania, S., Park, J., Md Din, M. F., Mat Taib, S., Talaiekhozani, A., Kumar Yadav, K., & Kamyab, H. (2018). Microplastics pollution in different aquatic environments and biota: A review of recent studies. *Marine Pollution Bulletin*, 133, 191–208. https://doi.org/10.1016/j.marpolbul.2018.05.022

Sharma, M., Gupta, V., & Gupta, D. (2024). Advancement in technologies of water management in rural areas. *In Water Resources Management for Rural Development* (pp. 117–126). Elsevier.

Sharma, M., Gupta, D., Parhi, M. P., Gurung, S., Parasar, H., & Kumar, A. (2021). Phosphate solubilizing microbes: A sustainable tool for healthy agriculture. *Ecology Environment & Conservation*, 27(August Suppl. Issue), S303–S310.

Shi, X., Zhang, X., Gao, W., Zhang, Y., & He, D. (2022). Removal of microplastics from water by magnetic nano-Fe3O4. *Science of the Total Environment*, 802, 149838. https://doi.org/10.1016/j.scitotenv.2021.149838

Tagg, A. S., Harrison, J. P., Ju-Nam, Y., Sapp, M., Bradley, E. L., Sinclair, C. J., & Ojeda, J. J. (2017). Fenton's reagent for the rapid and efficient isolation of microplastics from wastewater. *Chemical Communications*, 53(2), 372–375. https://doi.org/10.1039/c6cc08798a

Taniguchi, I., Yoshida, S., Hiraga, K., Miyamoto, K., Kimura, Y., & Oda, K. (2019). Biodegradation of PET: Current status and application aspects. *ACS Catalysis*, 9(5), 4089–4105. https://doi.org/10.1021/acscatal.8b05171

Thiruvenkatachari, R., Kwon, T. O., Jun, J. C., Balaji, S., Matheswaran, M., & Moon, I. S. (2007). Application of several advanced oxidation processes for the destruction of terephthalic acid (TPA). *Journal of Hazardous Materials*, 142(1), 308–314. https://doi.org/10.1016/j.jhazmat.2006.08.023

Vigneswaran, S., Shon, H. K., Kandasamy, J., & Shim, W. G. (2007). Performance of Granular Activated Carbon (GAC) adsorption and biofiltration in the treatment of biologically treated sewage effluent. *Separation Science and Technology*, 42(14), 3101–3116. https://doi.org/10.1080/01496390701588992

Virpi Siipola, Romar, H., & Lassi, U. (2022). Microplastic removal from water and wastewater by carbon-supported materials. Elsevier EBooks, pp. 361–393. https://doi.org/10.1016/b978-0-12-822225-6.00007-5

Wang, Z., Wu, Z., Mai, S., Yang, C., Wang, X., An, Y., & Zhou, Z. (2008). Research and applications of membrane bioreactors in China: Progress and prospect. *Separation and Purification Technology*, 62(2), 249–263. https://doi.org/10.1016/j.seppur.2007.12.014

WHO. (2020, March 3). Shortage of personal protective equipment endangering health workers worldwide. World Health Organization. www.who.int/news/item/03-03-2020-shortage-of-personal-protective-equipment-endangering-health-workers-worldwide.

12 Biodegradation and Removal of Phthalate Esters from Wastewater
An In-depth Analysis

Sonia Sethi

12.1 INTRODUCTION

Contaminants of Emerging Concern (CECs), alternatively referred to as emerging contaminants or pollutants, constitute a diverse array of compounds that are not subject to routine monitoring or robust regulation. This heterogeneous group encompasses substances from various sources, including personal care products (e.g., detergents, cosmetics), plastic-derived compounds (such as phthalates and bisphenols), pharmaceuticals (both prescribed and over-the-counter), pesticides, and chemical additives (e.g., preservatives). Their ingress into aquatic systems occurs via multiple pathways, including wastewater treatment plants, sewage, urban runoff, agricultural runoff, and septic systems (Pironti et al., 2021). The prominence of emerging contaminants has surged in recent times, precipitating concerns due to their pervasive presence and associated health risks for both human and animal populations. This escalating prevalence underscores the ongoing evolution and production of novel compounds, spanning chemical pollutants (e.g., pharmaceutical derivatives, phthalates, and bisphenol by-products) and biological agents (e.g., the SARS-CoV-2 virus) (Pastorino and Ginebreda, 2021).

Of particular concern is the capacity of certain CECs to act as endocrine disruptors, perturbing hormonal balance in organisms and precipitating reproductive disorders. Additionally, these contaminants are implicated in a spectrum of health afflictions, including respiratory ailments, neurological dysfunctions, and thyroid irregularities. The intricate interplay between emerging contaminants and physiological systems underscores the multifaceted nature of the challenges posed by these substances.

Phthalates (PAEs), integral additives within the plastics industry, stand as quintessential synthetic organic compounds. Employed for decades as plasticizers and enhancers in plastics production, their versatility extends across various sectors including cosmetics, flooring, pharmaceuticals, and medical devices, among others. This widespread utilization not only augments product performance but also enhances human livelihoods and living standards (Zhu et al., 2018).The distinctive feature of PAEs lies in their non-covalent attachment to substrate materials like plastics,

DOI: 10.1201/9781003379386-12

facilitating their facile release into environmental matrices such as water, soil, sediment, and air. Possessing estrogenic properties, PAEs disrupt normal endocrine functions upon entry into organisms, thereby impeding vital physiological processes. Moreover, their propensity to bind with hemoglobin molecules in the human body raises concerns regarding erythrocytic functionality (Barbagallo et al., 2022).

Ubiquitous in aquatic environments, notable PAEs including dimethyl phthalate (DMP), diethyl phthalate (DEP), dibutyl phthalate (DBP), and di-(2-ethylhexyl) phthalate (DEHP) feature prominently among detected compounds. Their reproductive toxicity towards aquatic organisms underscores potential risks to human health, prompting their classification as priority pollutants warranting stringent regulatory measures. Given the imperative to mitigate environmental PAE contamination, concerted efforts are directed towards devising efficacious removal strategies and curtailing discharge into water bodies. This endeavor assumes paramount importance in safeguarding ecological integrity and public health.

12.2 OCCURRENCE OF PHTHALATES

Phthalates, ubiquitous synthetic compounds with diverse industrial applications, pervade various environmental matrices owing to their widespread use. Their occurrence spans a plethora of ecosystems, including aquatic environments, terrestrial systems, and atmospheric compartments. In aquatic environments, phthalates infiltrate surface waters, sediments, and even groundwater reservoirs through multiple pathways such as industrial discharges, urban runoff, and agricultural runoff. Terrestrial environments also serve as reservoirs for phthalates, with soil acting as a sink for these compounds. Atmospheric dispersion constitutes another route for phthalate dissemination, with volatile phthalate species undergoing atmospheric transport over regional and even global scales. Their atmospheric presence results from emissions during manufacturing processes, vehicular exhaust, and volatilization from consumer products. The omnipresence of phthalates underscores the need for comprehensive monitoring efforts across diverse environmental compartments to assess their spatial distribution, fate, and potential impacts on ecosystems and human health (Figure 12.1).

Phthalates, particularly prevalent near industrial estates, pose a significant atmospheric pollution concern due to their lipophilic nature, facilitating absorption into human blood and fluids. Upon absorption, these compounds undergo rapid transformation into primary and secondary metabolites, culminating in adverse health effects. Distinguishing between long-chain and short-chain phthalates reveals their varied applications and potential risks. Long-chain phthalates, such as di(2-ethylhexyl) phthalate (DEHP), di(2-propylheptyl) phthalate (DPHP), di-iso-decyl phthalate (DiDP), and di-iso-nonyl phthalate (DiNP), find utility in PVC polymer production. In contrast, short-chain phthalates like dimethyl phthalate (DMP), butyl benzyl phthalate (BBzP), diethyl phthalate (DEP), di-iso-butyl phthalate (DnBP), and di-n-butyl phthalate (DnBP) are common in non-PVC applications (Benjamin et al., 2017).

The US EPA has identified several PAEs as priority toxic contaminants, notably dimethyl phthalate (DMP), di-2-ethylhexyl phthalate (DEHP), dibutyl phthalate (DBP), and diethyl phthalate (DEP). Among these, DMP stands out for its extensive

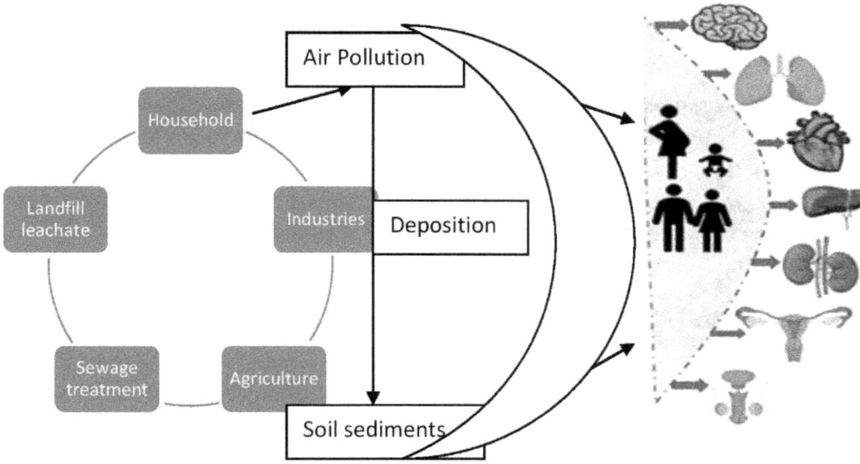

FIGURE 12.1 Occurrence of Phthalates in environment.

industrial use across various sectors owing to its high saturation solubility, rendering it suitable for applications ranging from insect repellents to plastics production, solid rocket propellants, pesticides, and protective coatings (Ahmadi et al., 2016).

DBP, DMP, DEP, and DEHP are extensively employed as plasticizers across various industrial sectors, contributing to product flexibility and durability. However, their pervasive presence raises concerns due to their classification as endocrine disruptors, capable of adversely affecting human health and the environment even at minute concentrations (Wang et al., 2019a). The decomposition of DMP holds significance not only from a scientific perspective but also for industrial and medical purposes. DEP, prominently utilized in a plethora of industrial applications ranging from tools to cosmetics, stands out as a widely recognized plasticizer. Despite its widespread usage, DEP has garnered attention for its association with various diseases and its capacity to impact neuroendocrine functions, male reproductive health, childhood obesity, lipid metabolism, and fecundity. Consequently, the imperative to remove DEP from aquatic ecosystems arises.

Similarly, DBP finds extensive application, particularly in personal care products such as perfume solvents and hair spray fixatives (Li et al., 2015a). DEHP, present in non-polymeric products like adhesives, cosmetics, inks, and perfumes, carries notable environmental ramifications, even at minimal concentrations. The management of wastewater containing high levels of DEHP necessitates stringent measures, including treatment at the source before discharge into urban sewage networks. Furthermore, DEHP's documented endocrine-disrupting properties and potential carcinogenic effects underscore the urgency for effective regulatory interventions.

Solvents imbued within adhesives, cosmetics, inks, perfumes, munitions, paints, insect repellents, hairspray additives, and lubricating oils harbor di(2-ethylhexyl) phthalate (DEHP), a pervasive constituent with multifarious industrial applications. Despite its widespread usage, DEHP exhibits detrimental environmental ramifications, even at minute concentrations. Consequently, wastewater laden with elevated DEHP levels necessitates stringent regulatory measures, including preemptive treatment modalities at the source, prior to discharge into urban sewage networks. Extensive research underscores DEHP's propensity for endocrine disruption and carcinogenicity, underscoring the imperative of mitigating its environmental dissemination. The variability in phthalate concentrations within wastewater across different nations mirrors disparities in production volumes and regulatory frameworks. Notably, these phthalates emerge as priority environmental contaminants, precipitating urgent attention and regulatory scrutiny despite evolving regulatory paradigms.

Solvents in adhesives, cosmetics, inks, perfumes, munitions, paints, insect repellents, hairspray additives, and lubricating oil are non-polymeric products that contain DEHP. DEHP has been shown in several studies to have harmful environmental consequences, even at low concentrations. Wastewater with a high concentration of DEHP must be subjected to a restrictive condition. Treatment at the source, for example, can also be used until effluent is discharged into sewage urban networks. DEHP has been shown in several studies to have endocrine disrupting properties and may cause cancer.

The amount of phthalates in wastewater in a given country varies depending on the quantity of production and restrictions in that country. So these PAEs are now the environmental priority contaminants and one of the most important environmental issues, despite increasingly strict regulations. To safeguard humans and the environment, these dangerous substances should be eliminated from wastewater. Several methods for removing PAEs from water and wastewater have been established due to their widespread production and high toxicity.

12.3 DEGRADATION OF PHTHALATES

Phthalates, notorious for their recalcitrance in conventional wastewater treatment, persist in aquatic environments, accumulating in lakes and rivers and permeating through the intricate ecosystems. This insidious infiltration into the food web engenders deleterious effects on both human populations and diverse aquatic biota, underscoring the urgency of remedial action (Kasonga et al., 2021). Traditional treatment modalities such as adsorption, flocculation, and biodegradation confront significant limitations in the remediation of phthalate esters. Challenges include the disposal of adsorbates, copious sludge generation, and the recalcitrance of non-biodegradable compounds, exacerbating the efficacy of these approaches. Consequently, the imperative arises for the exploration of alternative remediation methodologies.

Advanced oxidation processes (AOPs) emerge as promising strategies for phthalate degradation, offering a suite of techniques encompassing photocatalysis, photolysis, Fenton processes, ozonation, UV-based methods, sonolysis, electrochemical methods, radical-based approaches, and coupled AOPs. Noteworthy examples include the integration of photo-Fenton processes, sono-Fenton systems, O_3/UV processes, O_3/H_2O_2 systems, and peroxymonosulfate (PMS)/O_3 treatments, demonstrating enhanced efficacy in phthalate ester degradation. The synergistic application of AOPs with secondary wastewater treatment methodologies holds particular promise, heralding a paradigm shift towards comprehensive remediation strategies. By harnessing the synergies between AOPs and conventional treatment processes, the environmental burden posed by phthalate contamination can be effectively mitigated, fostering ecological resilience and safeguarding public health.

12.3.1 Conventional Treatment Methods of Phthalates

Phthalates exhibit a propensity for diverse fate pathways, including photolysis, volatilization, hydrolysis, adsorption, and biotransformation. Leveraging these intrinsic properties, the concurrent recovery and elimination of contaminants from water matrices underscore the appeal of "reverse osmosis & ion exchange" as a pre-eminent remedial procedure. In the realm of wastewater treatment, the adsorption technique stands as a stalwart method for the removal of both organic and inorganic contaminants. This approach capitalizes on the affinity of pollutants for sorbents, facilitating their sequestration from aqueous environments. The pursuit of cost-effective pollutant removal sorbents has witnessed exponential growth, fueled by a burgeoning demand for sustainable solutions. Researchers have increasingly turned towards locally available agricultural wastes, natural materials, and industrial by-products as viable sources of sorbent materials (Chandarana et al., 2021). These materials are judiciously harnessed for the production of activated carbon, which serves as a cornerstone in wastewater treatment endeavors.

The schematic depiction (Figure 12.2) delineates the multifaceted approaches encompassing physical, chemical, and biological remediation strategies for phthalate esters (PAEs). By amalgamating diverse remediation techniques, this holistic framework embodies the collective effort towards mitigating the environmental impact of phthalate contamination and fostering sustainable water management practices.

12.3.2 Physical/Chemical Treatments

Physical and chemical treatment methods, such as coagulation/flocculation and adsorption, represent pivotal approaches for mitigating the burden of suspended solids, ions, colloidal particles, floating materials, oil and grease, colorants, heavy metals, and toxic compounds in wastewater. Despite their widespread application,

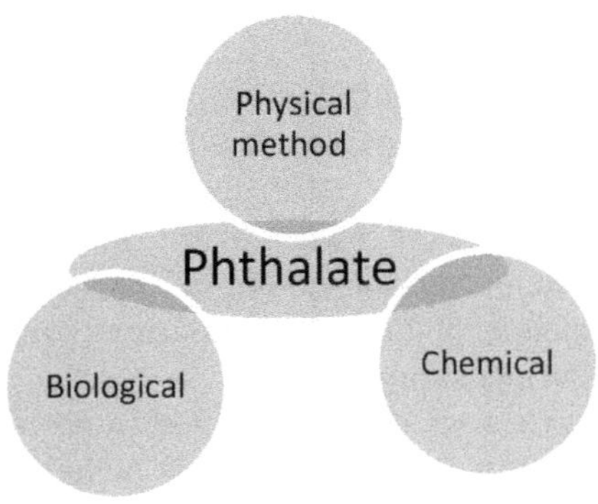

FIGURE 12.2 Methods for remediation of Phthalate compounds.

these methods are encumbered by inherent limitations, including the generation of sludge, the release of metal ions, challenges in regenerating adsorbents, the scarcity of cost-effective sorbents, and the incomplete mineralization of toxic organic compounds into benign by-products (Kumar et al., 2017).

The utilization of activated carbon, characterized by a substantial surface area ranging from 500 m^2/g for diethyl phthalate (DEP) sorption to 1224 m^2/g for DEP removal, underscores its efficacy in adsorbing phthalate esters. Notably, the efficiency of activated carbon exhibits a positive correlation with temperature, with optimal removal observed at elevated temperatures (25–55 °C). Furthermore, modified activated carbon, incorporating elements such as copper and tetrabutylammonium iodide (TBA), demonstrates enhanced di-isopropyl phthalate removal from industrial effluents. In anticipation of employing biological/tertiary treatment techniques, precipitation and/or coagulation serve as pre- or post-treatment measures. A notable example entails the facile chemical co-precipitation technique employing Fe and Ag co-doped ZnO nanorods, efficacious in phthalate removal and the reduction of dibutyl phthalate (DBP) under visible irradiation. These multifaceted approaches underscore the synergistic integration of physical and chemical treatment modalities, bolstering the arsenal of wastewater treatment strategies aimed at ameliorating the environmental impact of phthalate contamination.

In the quest to treat phthalates effectively, traditional inorganic coagulants like alum, ferric chloride, ferrous sulfate, poly aluminum sulfate, poly aluminum chloride, poly alkylene, and polyamine have long been employed. However, the extensive and sustained use of these chemicals is fraught with drawbacks. To circumvent these limitations, there's a burgeoning interest in natural coagulants, prized for their efficiency, renewability, and non-toxic properties. Moreover, the utilization of natural coagulants tends to result in reduced sludge production, further enhancing their appeal.

Natural coagulants possess a plethora of surface charges, augmenting the efficacy of the coagulation process. This intrinsic characteristic contributes to their superior performance compared to their synthetic counterparts (Naruka et al., 2021). Among the diverse array of natural coagulants, plant-based options hold particular promise. Species such as Moringa oleifera, guar gum, mustard seeds, nirmali seeds, Cereus peruvianus, Ocimumbasilicum, various species of the Genus Cassia, Plantago species, fava bean seeds, tannin, cactus, chestnut, and acorn have demonstrated efficacy in phthalate treatment.

For instance, Dillenia indica L. (Dilleniaceae) seed extract has emerged as a noteworthy natural coagulant for phthalate remediation. Under optimal conditions, a substantial reduction in dibutyl phthalate (DBP) concentration was achieved, with DBP levels decreasing by 95% at a pH of 3 when treated with 150 mg/L of photocatalyst (Akbari et al., 2018). Such findings underscore the potential of natural coagulants in the sustainable management of phthalate contamination, paving the way for environmentally benign treatment methodologies.

Advanced oxidation processes (AOPs) represent a potent arsenal in combating the scourge of phthalate esters (PAEs) and organic pollutants, encompassing methodologies such as sonolysis, photocatalytic oxidation, UV irradiation, ozonation, and Fenton systems. Semiconductors like titanium dioxide (TiO2) and zinc oxide

(ZnO), augmented by redox agents, catalyze the breakdown of organic constituents, heralding a paradigm shift in remediation strategies (Moradeeya et al., 2022).

A hallmark of AOPs lies in the generation of hydroxyl radicals (OH), renowned for their potent oxidizing capabilities, alongside other free radicals that synergistically enhance the mineralization process. In an electro-Fenton system, operating under optimal conditions, significant removal efficiencies of di-isobutyl phthalate (DIBP) reaching 93% were achieved, underscoring the efficacy of this approach. High-frequency ultrasound has emerged as a frontrunner among AOPs, revered for its cleanliness, minimal chemical dosing requirements, and environmental benignity (Moradeeya et al., 2022). Notably, catalyst-free UV irradiation coupled with high-frequency ultrasonication demonstrated remarkable efficacy in degrading dimethyl phthalate (DMP). While UV irradiation alone degraded 16.6% of DMP within 120 minutes using six lamps, the incorporation of ultrasonication led to a substantial increase in breakdown efficiency, with DMP levels plummeting by 98.4% within the same timeframe. These findings underscore the synergistic potential of AOPs in addressing phthalate contamination, underscoring their pivotal role in advancing sustainable wastewater treatment methodologies.

The integration of ultrasonic and electrochemical techniques offers several advantages, including enhanced mass transfer, increased surface area for reactions, accelerated pollutant degradation kinetics, and reduced energy consumption. Moreover, the synergistic effects observed in combined processes often result in improved pollutant removal efficiencies compared to individual treatment methods.

12.3.3 Biological Treatments

Microbial degradation stands out as a pivotal route for the environmental breakdown of phthalate esters (PAEs), ubiquitous pollutants with a proclivity for persistence. Unlike abiotic degradation, which progresses sluggishly over years, microbial degradation offers a rapid, cost-effective, and eco-friendly mechanism for PAE removal from contaminated sites. Bacteria inhabiting wastewater ecosystems exhibit remarkable versatility, capable of degrading various PAEs under aerobic, anaerobic, and facultative conditions. Extensive research has elucidated the efficacy of microbial breakdown in pure cultures under aerobic conditions, representing a cornerstone in PAE elimination from the environment. Furthermore, researchers have explored avenues to enhance PAE biodegradation by manipulating environmental variables such as temperature, pH, co-substrate availability, surfactant presence, and other factors.

Noteworthy advancements include the utilization of yeast extract as a co-metabolic substrate in conjunction with Bacillus subtilis, resulting in enhanced PAE biodegradation. However, under anaerobic conditions, the rate of PAE biodegradation is markedly reduced compared to aerobic conditions. Despite this, select fungal and algal species have demonstrated proficiency in PAE biodegradation, expanding the repertoire of microbial agents capable of remediation. Fungal species such as Phanerochaetechryososporium, Trametes versicolor, and Daldiniaconcentrica have emerged as notable contributors to PAE degradation, underscoring the multifaceted nature of microbial remediation strategies in combating PAE pollution.

Gao and Chi (2015) delved into the biodegradation dynamics of diethyl phthalate (DEP) and dibutyl phthalate (DBP) using three marine microalgae species: Dunaliella salina, Chaetoceros muelleri (both planktonic), and Cylindrothecaclosterium (a benthic microalgae). Their findings underscored the capacity of microalgae to effectively degrade phthalate esters (PAEs) in aquatic environments. Contrastingly, the biodegradation of DBP necessitates the concerted action of both external and intracellular enzymes. In natural settings, the complete breakdown of complex organic compounds like PAEs typically involves a collaborative effort among diverse bacterial consortia.

A pivotal step in the biodegradation pathway of PAEs involves the sequential hydrolysis of ester linkages, leading to the formation of phthalate monoesters and subsequently phthalic acid. The de-esterification process yields alcohols from alkyl or aryl chains. Given the complexity of this metabolic pathway, microbial degradation of PAEs necessitates the orchestration of numerous metabolic genes and enzymes. This implies that a single organism may not possess the capacity for the complete mineralization of PAEs, necessitating a synergistic interplay among microbial communities for effective remediation (Gani et al., 2017).

12.3.3.1 Bacterial Degradation of Phthalates

The complete mineralization of phthalates (PAEs) can be achieved through the concerted action of bacterial consortia possessing specialized metabolic capabilities. For instance, one consortium comprises Pseudomonas fluorescens, Pseudomonas aureofaciens, and Sphingomonaspaucimobilis, while another includes Xanthomonas maltophilia and Sphingomonaspaucimobilis. Recent studies have also identified a binary combination of Variovorax sp. BS1 and Achromobacterdenitricans as proficient degraders of dimethyl phthalate (DMP). This binary mixture exhibited rapid degradation kinetics, surpassing individual strains in efficacy. The degradation of PAEs by microbial consortia involves a cascade of metabolic reactions, commencing with the hydrolysis of ester linkages by esterases. Notably, esterases purified from Micrococcus sp. strain YGJ1 and Bacillus sp. have demonstrated the ability to hydrolyze medium-chain (C3–C5) mono/di-alkyl phthalate esters, underscoring the diversity of enzymatic pathways involved in PAE degradation.

Araki et al. (2019) employed Pseudomonas sp. strain PTH10 for the growth with phthalic acid (PA), utilizing enzymes such as phthalate-2,3-dioxygenase and adecarboxylating 2,3-dihydro-cis-2,3-dihydroxy-PA dehydrogenase. Notably, the 3,4-dioxygenase enzyme directly cleaved the ring system of the cis-2,3-dihydroxybenzoate intermediate formed during PA degradation. In another study, Wang et al.(2018) demonstrated that optimal conditions for the biodegradation of diethyl phthalate (DEP) by Sphingobiumyanoikuyae SHJ were observed at a temperature of 30°C, pH 7.0, and agitation at 150 rpm. This highlights the importance of environmental parameters in influencing microbial activity and degradation efficiency. Specifically, factors such as pH and temperature play crucial roles in shaping microbial growth and activity, ultimately affecting the rate of biodegradation.

In a recent investigation, the versatile capabilities of Nocardia spp. in mitigating rubber contamination and degrading phthalate esters (PAEs) have been elucidated. Notably, the degradation efficiency of PAEs by Nocardia strain LMB-7 exhibited a positive correlation with the length of alkyl chains. Remarkably, the degradation prowess of Nocardia strain LMB-7 in targeting di(2-ethylhexyl) phthalate (DEHP) was particularly noteworthy. Under optimized conditions of pH 8.0 and 30°C, DEHP disappearance reached an impressive 98.9% within a mere 24-hour timeframe. This acceleration in degradation kinetics far exceeds previous reports for other microbial strains(Yue et al., 2017).While previously isolated strains demonstrated proficient degradation of both dibutyl phthalate (DBP) and DEHP, none achieved degradation rates surpassing 97% within the same temporal window. Amongst the array of strains previously characterized for DEHP degradation, only a select few managed to achieve degradation efficiencies exceeding 94%. Notably, the majority of these strains necessitated significantly longer degradation periods, ranging from three days to over six days (Yang et al., 2018).

Among the myriad strains previously documented for their capability in di(2-ethylhexyl) phthalate (DEHP) degradation, with the exception of two strains, namely Sphingomonas sp. DK4 and Corynebacterium sp. O18, which exhibited a weaker degradation proficiency of less than 30%, the remainder displayed remarkable efficiency, achieving degradation rates exceeding 94%. Notable examples include G. polyisoprenivorans G1, Rhodococcus sp. strain WJ4, Fusarium culmorum, Agromyces sp. MT-O, Rhodococcusruber YC-YT1, Gordonia alkanivorans YC-RL2, and Microbacterium sp. CQ0110Y. However, it is noteworthy that the swiftest two among them necessitated a timeframe of three days for complete degradation, while the others required six days or longer.

In a separate study, it was elucidated that 60 isolates exhibited the capacity to degrade various phthalate esters (PAEs), sourced from diverse habitats, with a predominant number of 44 isolates originating from sludge and soil environments (Ren et al.,2017). Biochemical analyses unveiled a spectrum of degrading capabilities among these isolates, with distinct efficiencies observed across different species. Notably, isolates belonging to the genus Rhodococcus showcased broad substrate availability and high degradation efficiencies. For instance, Rhodococcus sp. 2G demonstrated the ability to degrade seven types of PAE compounds within a span of five days, with degradation rates ranging from 70.8% to 95% (Zhao et al., 2018b).

Furthermore, certain isolates, such as Bacillus subtilis No. 66 and Delftia sp. TBKNP-05, exhibited complete metabolization of high concentrations of PAEs. Conversely, some isolates, including Paenibacillus sp. S-3 and Gordonia sp. JDC-2, were unable to mineralize PAEs into carbon dioxide and water due to the absence of genes responsible for metabolizing PAE intermediates (Zhang et al., 2018b). Additionally, the efficiency of PAE degradation varied with the molecular weight of the compounds. Low-molecular-weight PAEs, such as dimethyl phthalate (DMP), diethyl phthalate (DEP), and dibutyl phthalate (DBP), were found to be more readily metabolized by the isolates compared to high-molecular-weight PAEs, including dioctyl phthalate (DOP), benzyl butyl phthalate (BBP), and di(2-ethylhexyl) phthalate (DEHP). Notably, DBP, a frequently detected PAE in the environment, was effectively degraded by 36 isolates, with degradation rates ranging from 44% to 100%.

It's intriguing to note that a single bacterial species isolated from L. apiculatum exhibited remarkable proficiency in growing in the presence of various phthalate esters (PAEs) including diethyl phthalate (DEP), dibutyl phthalate (DBP), benzyl butyl phthalate (BBP), di(2-ethylhexyl) phthalate (DEHP), and diisononyl phthalate (DINP). This underscores the adaptability and resilience of certain bacterial species in tolerating and utilizing PAEs as carbon sources. Interestingly, the degradation of DEHP, a particularly recalcitrant PAE compound, necessitated the concerted effort of a bacterial consortium comprising multiple species. This consortium comprised Serratia ficaria, Enterobacter asburiae, Pantoeaagglomerans, Acinetobacter johnsonii, Serratia rubidaea, Serratia marcescens, Staphylococcus warneri, Microbacteriumhy drocarbonoxydans, Pseudomonas graminis, Rhodococcus, and Bacillus simplex. Such collaborative degradation strategies highlight the complexity and synergistic interactions inherent in microbial degradation processes, especially when targeting challenging compounds like DEHP.

The elucidation of metabolites in the di(2-ethylhexyl) phthalate (DEHP) biodegradation process has unveiled the multifaceted metabolic pathways employed by the L. apiculatum bacterial consortium. Notably, this consortium has demonstrated proficiency in degrading DEHP through two distinct pathways: de-esterification and β-oxidation. However, an analysis of metabolite presence throughout the degradation process indicates that the primary metabolic route employed by the consortium is de-esterification. This finding aligns with previous studies, which have consistently reported that certain bacterial strains, such as those belonging to the Gordonia and Acinetobacter genera, exhibit the capacity to degrade phthalic acid esters predominantly through the β-oxidation degradation pathway (Xia et al., 2017). The prevalence of de-esterification as the principal metabolic pathway in DEHP degradation by the L. apiculatum consortium underscores the diversity of metabolic strategies employed by microbial communities in tackling complex environmental contaminants.

In another comprehensive investigation, a novel bacterium, identified as Pseudomonas sp. YJB6, emerged as a potent agent for degrading di-n-butyl phthalate (DBP) within contaminated soil environments. Remarkably, under-optimized conditions, including a temperature of 31.4°C, initial inoculation size of 0.6 (OD600), and pH 7.6 in basic mineral salts liquid medium (MSM), YJB6 achieved complete degradation of DBP at a concentration of 200 mg/L within a mere three-day period. Intriguingly, the analysis of degradation intermediates revealed the emergence of butyl-ethyl phthalate (BEP), mono-butyl phthalate (MBP), phthalic acid (PA), and benzoic acid (BA), suggesting the delineation of a novel and intricate biodegradation pathway orchestrated by YJB6.Further the immobilization of strain YJB6 using sodium alginate (SA), polyvinyl alcohol (PVA), and SA-PVA matrices significantly enhanced cellular stability and adaptability, resulting in elevated volumetric DBP-degrading rates compared to freely suspended cells. Moreover, these immobilized cells demonstrated impressive reusability across multiple cycles while retaining substantial DBP-degrading activity(Nai-Xian Feng et al., 2021).

12.3.3.2 Fungal Degradation of Phthalates

Thermophilic fungi exhibit a remarkable capacity to engage in diverse metabolic and catabolic activities even under elevated temperatures, making them valuable reservoirs of enzymatic prowess. In a recent endeavor, eleven wood-decay thermophilic fungal strains were meticulously isolated from decaying organic materials

(DOM) sourced from arid locales in Khulais, Saudi Arabia. Among these strains, six displayed the remarkable ability to produce laccase enzymes at a thriving temperature of 50°C.Laccase-positive (Lac+) isolates, particularly Chaetomium brasiliense (G3), Canariomyces notabilis (KW1), and Paecilomycesformosus (KW3), were identified as promising candidates for combating select endocrine-disrupting chemicals (EDCs) hailing from disparate classes, including 17α-ethinyl estradiol (EE2), a synthetic steroid hormone, and 4-tert-butylphenol (4-t-BP), an alkylphenol.

Subsequently, Chaetomium sp. was singled out for its pronounced efficacy against target EDCs, where its laccases were harnessed for the biocatalytic degradation of treated municipal sewage wastewaters (TMWW) laden with 4-t-BP and EE2. Notably, the catalytic prowess of laccase was demonstrated by the efficient degradation of 4-t-BP, with approximately 60 ± 4.8% degradation achieved within a mere two-hour timeframe at 50°C. However, the oxidation of EE2 proved to be less efficient, attaining a removal rate of 35 ± 4.1%.Further exploration into the catalytic system revealed the impact of redox mediators such as 1-hydroxybenzotriazole (HBT) and syringaldehyde, which significantly augmented the transformation rates of EE2 to approximately 80 ± 2.4%. Notably, near-complete removal (90 ± 7.2%) of 4-t-BP was achieved within two hours with the aid of TEMPO (Daâssi & Alharbi, 2023).

Among the microbial contenders, certain fungi have emerged as adept degraders of phthalate esters, particularly DBP. A notable example is A. flavus SDBP4, isolated from a hazardous sanitary landfill site, which has been scrutinized for its DBP degradation capacity. This isolate showcases robust growth attributes, including biomass production and total soluble protein content, when cultivated in a DBP-amended medium, underscoring its ability to utilize the pollutant as a carbon source. Moreover, A. flavus SDBP4 exhibits proficiency in degrading DBP by leveraging esterases, elucidating a comprehensive biodegradation pathway for DBP. This groundbreaking pathway delineates the intricate breakdown of DBP into various metabolites, including DIBP and BBP, alongside commonly reported byproducts such as DMP and PA. Such detailed characterization not only sheds light on the degradation process but also underscores the potential environmental applications of strain SDBP4. With its adeptness in DBP biodegradation, this strain holds promise for remediation efforts in contaminated agricultural soils, landfills, and other matrices, offering an eco-friendly alternative for addressing phthalate pollution (Zhu et al., 2022).

The efficacy of the white-rot fungus Peniophoralycii LE-BIN 2142 in degrading a range of phthalic acid esters (PAEs), including diethyl phthalate (DEP), dibutyl phthalate (DBP), di(2-ethylhexyl) phthalate (DEHP), diisobutyl phthalate (DiBP), and n-butylbenzyl phthalate (BBP), was investigated. Notably, DEHP emerged as the most efficiently biodegraded compound by the fungus, with over 98% degradation observed by day 6 of cultivation. However, the degradation of DBP and DiBP in the fungal culture liquid reached only around 17–18% by the end of the 10-day cultivation period. BBP, on the other hand, proved to be the most recalcitrant compound, with approximately 40% residual content remaining on day 10 of P. lycii cultivation.

DEP exhibited resistance to fungal biodegradation and even demonstrated a toxic effect at a concentration of 1.5 g/L. This adverse impact was evidenced by a threefold decrease in the rate of radial growth of the fungus on agar compared to the control, along with a reduction in fungal biomass during liquid-phase submerged cultivation by about 1.5 times. Interestingly, cultivation of P. lycii on media containing PAEs led to a notable increase in esterase activity, approximately doubling compared to the control medium without phthalates. Conversely, oxidase activity exhibited a significant decline, decreasing by 2–4 times under the influence of PAEs. These findings underscore the complex interplay between fungal metabolism and PAE degradation, shedding light on the enzymatic mechanisms underlying the biodegradation process (Savinova et al., 2023).

In a rigorous investigation, the prowess of Fusarium culmorum in metabolizing a substantial concentration (3 g/L) of di(2-ethylhexyl) phthalate (DEHP) as the exclusive carbon and energy source in solid-state fermentation (SSF) was meticulously examined. For comparative analysis, cultures propagated on glucose served as controls. Astoundingly, the biodegradation of DEHP by F. culmorum surged to an impressive 96.9% within a span of 312 hours. Notably, this fungal entity exhibited a remarkable surge, reaching threefold higher levels of esterase activity in DEHP-supplemented cultures compared to control counterparts (1288.9 vs. 443.2 U/L, respectively).Furthermore, zymographic scrutiny unveiled a spectrum of nine discernible bands showcasing esterase activity (ranging from 24.6 to 214.5 kDa) in DEHP-amended cultures, distinctly divergent from those observed in control cultures and from prior observations in submerged fermentation settings.

González-Márquez et al., (2019a) delved into the metabolic dynamics of F. culmorum when confronted with di(2-ethylhexyl) phthalate (DEHP) as the exclusive carbon source in Submerged Fermentation (SmF). Astonishingly, this investigation unveiled F. culmorum's prowess in metabolizing 92% of DEHP within a mere 36 hours, with a staggering 99.6% degradation achieved within the extended timeframe of 144 hours. Similarly, Ahuactzin-Pérez et al. (2016) embarked on a parallel exploration, orchestrating F. culmorum's growth in the ambience of DEHP (1 g/L), complemented by glucose as a cosubstrate in SmF. The findings echoed a remarkable degradation trajectory, wherein 95% of DEHP succumbed to metabolic transformation within a concise window of 60 hours, culminating in complete metabolization by the culmination of 156 hours. Furthermore, Puranik et al. (2023) contributed insights into the metabolic prowess of Aspergillus flavus, showcasing its remarkable aptitude in mineralizing 99.34% of dibutyl phthalate, a structurally less complex phthalate analog, at a concentration of 0.1 g/L within the expanse of 360 hours of incubation in SmF. In a related endeavor, Savinova et al. (2023) demonstrated the efficacy of white-rot fungi, specifically Peniophoralycii and Trameteshirsuta, in dismantling DEHP (1 g/L) within the confinements of SmF. Notably, both fungal species exhibited commendable degradation efficiencies, achieving 99% and 99.2% degradation of DEHP, respectively, within a compact timeframe of 240 hours.

12.3.3.3 Algal Degradation of Phthalates

In addition to bacterial agents, microalgae emerge as formidable contenders in the biodegradation of Phthalic Acid Esters (PAEs). The distinctive metabolic repertoire of microalgae distinguishes them as efficient bioremediators of organic contaminants. Unlike bacteria, microalgae possess the unique ability to conduct photosynthesis, a process that not only fuels their own growth but also fosters the proliferation of heterotrophic bacteria, thereby facilitating the degradation of organic pollutants. Moreover, microalgae exhibit direct enzymatic prowess in the degradation of organic contaminants. Notably, cyanobacteria species such as Cyanothece sp. PCC7822, Syenchocystis sp. PCC6803, and Synechococcus sp. PCC7942 have been documented to metabolize Diethyl Phthalate (DMP), underscoring their intrinsic capacity for biotransformation (Zhang et al., 2016). Similarly, green microalgae exemplified by Chlorella pyrenoidosa and the single-celled dinoflagellate Karenia brevis have demonstrated the capability to degrade PAEs, further consolidating the potential of microalgae-based bioremediation strategies in combating environmental pollution (Sun et al., 2019).

It has been documented that certain aquatic algae exhibit an intriguing ability to colonize the surfaces of plastic substrates, utilizing them as platforms for growth and development. This colonization induces alterations in the physical characteristics of the plastic material. Notably, species such as Scenedesmus dimorphus, Anabaena spiroides, and Naviculapupula adhere to plastic surfaces by secreting extracellular polysaccharides, leading to the formation of surface fissures and a reduction in mass. Additionally, these algae have been observed to accumulate phthalate esters (PAEs) within their cellular structures, wherein they undergo conversion into monoester compounds. Subsequently, these compounds undergo further biochemical transformations, including hydroxylation, oxidation, glycosylation, and malonylation, yielding conjugated metabolites (Mondal et al., 2022).

In comparison to single microorganism systems, the co-cultivation of algae with bacteria has emerged as a highly promising approach in wastewater treatment endeavors. The intricate interplay between microalgae and bacteria, particularly pertaining to biodegradation and biosorption mechanisms, has garnered considerable scientific interest. Within the bioremediation process, microalgae play a pivotal role by supplying oxygen molecules via photosynthesis, thereby fostering the growth of bacteria. Simultaneously, they utilize carbon dioxide, a byproduct of bacterial metabolism, for photosynthetic activities, while actively removing organic pollutants through biosorption or bioaccumulation mechanisms. Notably, algae act as hosts for cohabitating bacteria, facilitating their adherence onto algal surfaces through the secretion of extracellular polymers (EPS), thereby forming biofilm-like microenvironments. Additionally, bacteria serve as protective shields for symbiotic algae, effectively mitigating the toxic effects of pollutants (Ji et al., 2019).

In an experimental investigation aimed at assessing the efficacy of various processes—such as hydrolysis, photolysis, adsorption, and biodegradation—in the removal of diisobutyl phthalate (DiBP), six different species of microalgae were scrutinized. The study delineated the average removal efficiency of DiBP, at initial concentrations of 5, 0.5, and 0.05 mg L−1, across all six microalgae (initial cell density of 1 × 106 cells mL−1). Notably, the removal hierarchy was observed as Scenedesmus obliquus (95.39 %) >Chlorella vulgaris (94.78 %) >Chroococcus sp. (91.16 %) >Cyclotella sp. (89.32 %) >Nitzschia sp. (88.38 %) >Nostoc sp. (84.33 %).

The outcomes of hydrolysis and photolysis experiments unveiled minimal impacts on DiBP removal, with respective efficiencies of merely 0.89% and 1.82%. Conversely, the adsorption efficiency of all six microalgae exhibited a notable decline with escalating initial DiBP concentrations, while biodegradation efficiency experienced an increment. Remarkably, Chlorella vulgaris and Chroococcus sp. showcased superior adsorption and biodegradation efficiencies among the tested microalgae. Subsequently, Scenedesmus obliquus was singled out for the assessment of DiBP degradation products owing to its exceptional removal efficacy. This analysis yielded insightful findings, identifying monoisobutyl phthalate (MiBP), phthalic acid (PA), and salicylic acid (SA) as potential degradation products of DiBP. The postulated degradation pathways primarily encompassed dealkylation, hydroxyl group addition, and decarboxylation mechanisms (Xiaoyue al., 2024).

A novel system merging visible-light photocatalysis with biological treatment was devised for the continuous elimination of phthalate esters (PAEs) from both synthetic and authentic aquaculture wastewater. This system's efficacy was scrutinized under various operational parameters, such as the presence of glucose or PAEs, on individual PAE removal within a photobiological setup (PBS). In wastewater harboring a blend of PAEs, specifically di-(2-ethylhexyl)phthalate (DEHP), dibutyl phthalate (DBP), and dimethyl phthalate (DMP), a co-immobilized bioreactor system comprising Pseudomonas putida and the microalga Chlorella vulgaris exhibited superior removal efficacy compared to immobilized P. putida alone or a coculture of immobilized P. putida and suspended C. vulgaris. Notably, the PBS deployed for the continuous treatment of authentic aquaculture wastewater containing DEHP (0.62 ± 0.05 mg/L), DBP (8.7 ± 0.9 mg/L), and DMP (17.4 ± 1.5 mg/L) achieved a minimum of 99.5% PAE removal and 99.2% mineralization efficiency under optimal conditions. Following 42 days of treatment, Pseudomonas spp. (98.12%) remained the predominant genus in the bioreactor. These findings underscore the viability of the symbiotic microalgal–bacterial system as a promising alternative to a pure P. putida immobilized bioreactor for mitigating CO_2 emissions resulting from mineralized PAEs via microalgal activity (Chung & Chen, 2023).

12.3.3.4 Enzymes Involved in Degradation

Enzymatic treatment offers numerous advantages over traditional biological waste treatment methods (Haq et al., 2016a, Haq et al., 2016b; Haq et al., 2017, Haq et al., 2018, Haq et al., 2020 ; Haq and Kalamdhad, 2021; Haq et al., 2022a, Haq et al., 2022b, Haq et al., 2022c; Haq and Kalamdhad, 2022, 2023; Haq et al., 2024). Enzymes exhibit greater resilience to toxic substances and shock loading effects, eliminating the need for acclimation. Furthermore, enzymes can function effectively across a wide range of organic compound concentrations, pH levels, temperatures, and salinity levels. When compared to conventional methods involving the inoculation or stimulation of degrading microorganisms in polluted sites with nutrient supplements, enzymatic bioremediation presents significant advantages. Inoculating and controlling microorganisms within the environment can be challenging, and adding nutrients to polluted sites may elevate the chemical oxygen demand of the water environment.

Enzymatic bioremediation addresses these challenges by isolating and applying enzymes directly to the polluted site, thereby expediting the remediation process. Key enzymes involved in the metabolism of phthalate esters (PAEs) include phthalate oxygenase, phthalate dioxygenase, phthalate dehydrogenase, and phthalate decarboxylase. Additionally, lipase and esterase enzymes derived from bacteria are employed in the initial hydrolysis of PAEs, facilitating the formation of corresponding monoesters and alcohols. This enzymatic approach holds promise for efficient and rapid remediation of polluted sites, offering a valuable alternative to conventional bioremediation methods.

12.3.3.4.1 Lipolytic Enzymes

Lipolytic enzymes, such as lipase and esterase, possess the remarkable capability to hydrolyze a diverse array of compounds, including aliphatic polyesters, fatty acid esters, and triglycerides. In the natural environment, these enzymes play a pivotal role in the degradation of aliphatic polyesters such as cutin, suberin, and esteroid. Specifically, hydrolases are instrumental in the breakdown of plastic, and this group includes esterases, lipases, and cutinases. These hydrolases are indispensable for the enzymatic cleavage of polymer structures, as they catalyze the hydrolysis of ester bonds through nucleophilic attacks on carbonyl carbon atoms, which are generated by prior oxidation reactions. This intricate enzymatic machinery underscores the crucial role of lipolytic enzymes in the natural degradation processes of various organic compounds, including synthetic polymers like plastics (Wilkes and Aristilde, 2017).

12.3.3.4.2 Esterases

Esterases exhibit the remarkable ability to catalyze hydrolysis reactions, cleaving ester bonds within molecules and leading to the formation of alcohols, phenols, and acids. This enzymatic activity is pivotal in the degradation of a wide range of substrates, irrespective of whether these substrates were generated during oxidation processes or were inherent constituents of the polymer structure. The versatile nature of esterases highlights their significance in enzymatic processes involved in the breakdown of organic compounds, contributing to various environmental and biotechnological applications (Yao et al., 2021).

12.3.3.4.3 Cutinase

Cutinase enzymes have shown promise in the degradation of phthalates, particularly in the breakdown of ester bonds present in phthalate compounds. These enzymes are part of the broader class of esterases and exhibit activity against a range of substrates, including phthalates. Cutinases are known for their ability to catalyze the hydrolysis of ester bonds found in cutin, a complex polymer present in plant cell walls, as well as similar compounds like phthalates. In the context of phthalate degradation, cutinases can cleave the ester linkages in phthalate molecules, leading to the formation of corresponding monoesters, alcohols, and phthalic acid. This enzymatic activity plays a crucial role in the biodegradation of phthalates, contributing to their breakdown into simpler and less harmful compounds.

Research has explored the potential of cutinase enzymes derived from various sources, including microorganisms and plants, for their effectiveness in degrading phthalates. By harnessing the catalytic capabilities of cutinases, scientists aim to develop enzymatic strategies for the remediation of environments contaminated with phthalate pollutants(Nakamura et al., 2017).

12.3.3.4.4 Proteolytic Enzymes

Protease enzymes, while primarily known for their role in protein degradation, have also shown potential for the degradation of phthalates. While their main function is to hydrolyze peptide bonds in proteins, some proteases exhibit promiscuous activity and can act on a wider range of substrates, including ester bonds present in phthalate compounds. Research into the use of proteases for phthalate degradation is ongoing, and scientists are exploring various sources of proteases, including microbial and plant-derived enzymes. By understanding the enzymatic mechanisms involved and optimizing conditions for protease activity, researchers aim to develop enzymatic strategies for the remediation of phthalate-contaminated environments(Mistry et al., 2022).

12.4 PHYSICO–CHEMICAL–BIOLOGICAL HYBRID TECHNIQUES

Physico–chemical–biological hybrid techniques for phthalate degradation in wastewater treatment involve combining physical, chemical, and biological methods to efficiently remove or degrade phthalate compounds from wastewater. These techniques aim to mitigate the environmental impact of phthalates, which are commonly found in wastewater and can pose risks to ecosystems and human health. Here are some examples of hybrid approaches:

12.4.1 Advanced Oxidation Processes (AOPs) Coupled with Biological Treatment

AOPs such as ozonation, photocatalysis, or Fenton oxidation are combined with biological treatment methods like activated sludge or biofilm reactors. AOPs generate highly reactive oxygen species that oxidize phthalate compounds into simpler, more biodegradable forms, which are then further degraded by microorganisms in the biological treatment stage (Hoda Amiri et al.,2022) as illustrated in Figure 12.3.

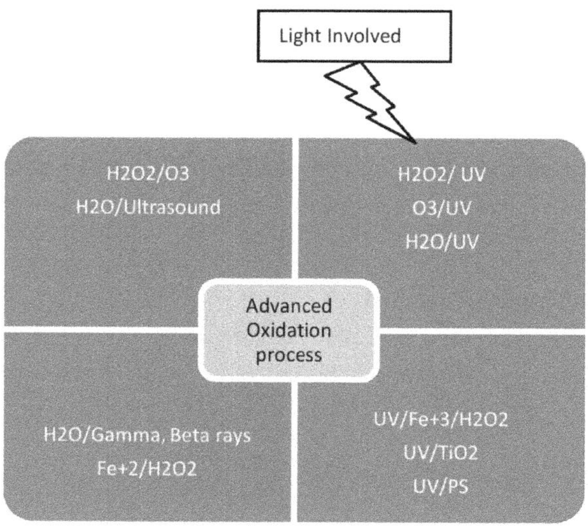

FIGURE 12.3 Advanced oxidation process.

12.4.2 Membrane Bioreactors (MBRs) with Chemical Pretreatment

MBRs integrate membrane filtration with biological treatment. In hybrid systems, chemical pretreatment methods like coagulation-flocculation or adsorption onto activated carbon are employed upstream of the MBR to remove phthalates and other contaminants before biological degradation occurs within the MBR. This combination enhances the overall removal efficiency (Khan, 2020).

12.4.3 Sequential Batch Reactors (SBRs) with Physico-chemical Pre-treatment

SBRs operate in cycles, allowing for sequential stages of physico-chemical and biological treatment. In these hybrid systems, phthalate-containing wastewater undergoes initial physico-chemical treatment processes like coagulation, adsorption, or advanced oxidation before entering the SBR for biological degradation by microorganisms.

12.4.4 Constructed Wetlands with Chemical Precipitation

Constructed wetlands utilize natural processes for wastewater treatment. In hybrid systems, chemical precipitation methods such as co-precipitation or complexation are employed upstream of the wetland to remove phthalates by forming insoluble complexes or precipitates, which are then further degraded by microbial communities within the wetland (Gorito et al., 2017).

12.4.5 Electrocoagulation combined with Microbial Consortia

Electrocoagulation involves the use of electric current to destabilize and remove contaminants from wastewater through coagulation and flocculation. In hybrid systems, electrocoagulation is followed by biological treatment using microbial consortia capable of degrading phthalate compounds. This combination enhances the overall removal efficiency and reduces the energy requirements of electrocoagulation.

12.4.6 Bioaugmentation

Bioaugmentation involves the deliberate introduction of specialized microbial communities or genetically modified microorganisms into biological treatment systems to enhance the degradation of specific pollutants like phthalates. When combined with physico-chemical pre-treatment steps such as ozonation or UV irradiation, bioaugmentation can significantly improve the efficiency of phthalate removal in wastewater treatment plants. Its primary advantages include high efficiency in reducing phthalate levels and the utilization of low-cost biosorbents. For instance, Yao et al. (2019) demonstrated the use of pepper straw biochar for removing DBP and DMP from contaminated water. This environmentally friendly approach repurposes agricultural waste, providing a sustainable solution for phthalate removal.

Other natural materials, such as beached seaweed and Sargassum siliquastrum, have also been employed for DEHP removal, exhibiting mean removal capacities of 5.68 mg g−1 and 6.54 mg g−1, respectively. The adsorption of DEHP by seaweed biomass is influenced by various factors, including temperature, with optimal conditions typically involving a specific biomass concentration, pH, temperature, and DEHP concentration.

12.4.7 Advanced Adsorption Techniques

Adsorption processes, including activated carbon adsorption or adsorption onto modified materials like graphene oxide, offer effective means to remove phthalates from wastewater. By coupling these adsorption techniques with biological treatment processes, the removal efficiency of phthalates can be significantly enhanced. Physico-chemical modifications of adsorbents, such as altering surface properties or introducing functional groups, can enhance their affinity for phthalate molecules, thereby improving adsorption capacity.For example, Li et al.(2015a) demonstrated the adsorption of DBP using vanillin-modified chitosan beads. This modification strategy enhances the adsorption performance of chitosan beads, making them more effective in capturing DBP molecules from wastewater. Subsequent biological treatment steps can then be employed to degrade the adsorbed phthalates, further enhancing the overall removal efficiency.

12.4.8 Sequential Treatment Systems

Sequential treatment systems integrate a series of physico-chemical and biological treatment steps in a sequential manner to achieve thorough removal of phthalates and other contaminants from wastewater. By combining multiple treatment processes, each targeting different aspects of pollutant removal, sequential treatment systems offer a comprehensive approach to wastewater remediation.

Such hybrid processes may be:

- Activated sludge process + membrane separation/filtration systems (reverse osmosis, ultrafiltration) + gamma radiations;
- Constructed wetlands coupled with waste stabilization ponds (removal of pharmaceuticals, beta-blockers), or biodegradation, or/and sorption, or/and volatilization, or/and hydrolysis, or/and photodegradation;
- Membrane bioreactor (MBR) + membrane separation/filtration systems (reverse osmosis, nanofiltration) (removal of pesticides, pharmaceuticals, beta-blocking drugs);
- Membrane bioreactor (MBR) + UV oxidation, or adsorption on activated carbon, or ozonation followed by ultrasounds;
- Ozonation + biological activated carbon (drugs: antibiotics, antidepressants, beta blockers, endocrine disrupting chemicals, pesticides);
- Flocculants + activated sludge + ultrafiltration (endocrine disrupting chemicals, pesticides, beta blockers);

- Ultrafiltration + activated carbon + ultrasounds (antibiotics);
- Surface flow constructed wetland (SFCW) + horizontal flow constructed wetland (HFCW) (drugs: beta blockers, stimulants).

Overall, physico-chemical-biological hybrid techniques offer promising strategies for the effective degradation of phthalates in wastewater treatment, leveraging the complementary advantages of different treatment methods to achieve efficient pollutant removal and environmental protection.

12.5 SUMMARY AND CONCLUSIONS

This chapter explores the biodegradation and removal processes of phthalate esters from wastewater. It details microbial degradation pathways, efficiency of different treatment methods, and environmental impacts. Emphasis is placed on advanced techniques for enhancing biodegradation, regulatory standards, and future research directions to improve wastewater treatment outcomes.

In conclusion, the degradation of phthalates presents a complex but vital challenge in the realm of wastewater treatment and environmental remediation. Over the years, significant progress has been made in understanding the mechanisms and developing effective strategies for phthalate degradation. From physico-chemical methods such as adsorption and ozonation to biological processes like bioaugmentation and biosorption, a diverse array of techniques have been explored to tackle phthalate pollution.

Physico-chemical methods offer rapid removal of phthalates from wastewater through adsorption onto various surfaces or oxidation processes. However, these techniques may have limitations in complete degradation and can generate potentially harmful by-products. On the other hand, biological approaches harness the enzymatic capabilities of microorganisms to degrade phthalates into simpler, less harmful compounds. Bioaugmentation, bioremediation, and biosorption are some of the promising biological strategies that have shown effectiveness in treating phthalate-contaminated water.Furthermore, hybrid techniques combining physico-chemical and biological processes have emerged as a synergistic approach to enhance phthalate degradation efficiency. Sequential treatment systems, integrating multiple steps in a systematic manner, offer a comprehensive solution for the removal of phthalates and other contaminants from wastewater.

Despite the progress made, challenges remain in optimizing the efficiency, scalability, and sustainability of phthalate degradation techniques. Future research efforts should focus on exploring novel materials, enzymes, and microbial consortia, as well as integrating advanced technologies such as nanotechnology and genetic engineering, to further improve the efficacy of phthalate degradation processes. Overall, addressing the issue of phthalate pollution requires a multidisciplinary approach, involving collaboration between researchers, engineers, policymakers, and industry stakeholders. By continuing to innovate and implement effective remediation strategies, we can mitigate the environmental impact of phthalates and safeguard water resources for future generations.

REFERENCES

Ahmadi, E., Kakavandi, B., Azari, A., Izanloo, H., Gharibi, H., Mahvi, A. H., Javid, A., Hashemi, & S. Y. (2016).The performance of mesoporous magnetite zeolite nanocomposite in removing dimethyl phthalate from aquatic environments. *Desalin Water Treat.* 57 (57); 27768–27782.

Ahuactzin-Pérez, M., Tlecuitl-Beristain, S., García-Dávila, J., González-Pérez, M., Gutiérrez-Ruíz, M.C., & Sánchez, C. (2016). Degradation of di (2-ethyl hexyl) phthalate by Fusarium culmorum: kinetics, enzymatic activities and biodegradation pathway based on quantum chemical modeling. *Sci Total Environ.* 566–567; 1186–1193.

Akbari-Adergani, B., Saghi, M.H., Eslami, A., Mohseni-Bandpei, A., & Rabbani, M. (2018). Removal of dibutyl phthalate from aqueous environments using a nanophotocatalytic Fe, Ag-ZnO/VIS-LED system: modeling and optimization. *Environ. Technol.* 39; 1566–1576.

Araki, A., Bamai, Y.A., Bastiaensen, M., Van den Eede, N., Kawai, T., Tsuboi, T., Miyashita, C., Itoh, S., Goudarzi, H., Konno, S., Covaci, A., & Kishi, R. (2019). Combined exposure to phthalate esters and phosphate flame retardants and plasticizers and their associations with wheeze and allergy symptoms among school children. *Environ. Res.* 183; 109–212.

Barbagallo, S., Baldauf, C., Orosco, E., & Roy, N. M. (2022). Di-butyl phthalate (DBP) induces defects during embryonic eye development in zebrafish. *Ecotoxicology* 31; 178–185.

Benjamin, S., Masai, E., Kamimura, N., Takahashi, K., Anderson, R. C., & Faisal P. A. (2017).Phthalates impact human health: epidemiological evidences and plausible mechanism of action. *J. Hazard. Mater.* 340; 360–383.

Chandarana, H., Subburaj, S., Kumar, P.S., Kumar, & M.A. (2021). Evaluation of phase transfer kinetics and thermodynamic equilibria of Reactive Orange 16 sorption onto chemically improved Arachis hypogaea pod powder. *Chemosphere.* 276.

Chung, Y.C., & Chen, C.Y. (2023). Coupled Photocatalysis and Microalgal–Bacterial Synergy System for Continuously Treating Aquaculture Wastewater Containing Real Phthalate Esters. *Environments 10*(12); 215.

Daâssi, D., & Alharbi, S.R. (2023). Degradation of endocrine-disrupting chemicals in wastewater by new thermophilic fungal isolates and their laccases. 3 *Biotech* 13, 26.

Gani, K. M., Tyagi, V. K., & Kazmi, A. A. (2017). Occurrence of phthalates in aquatic environment and their removal during wastewater treatment processes: a review. *Environ. Sci. Pollut. Res* 24 (21); 17267–17284.

Gao, J., & Chi, J. (2015). Biodegradation of phthalate acid esters by different marine microalgal species. *Mar. Pollut. Bull.* 99 (1–2); 70–75.

González-Márquez, A., Loera-Corral, O., Santacruz-Juárez, E., García-Dávila, J., Tlecuitl-Beristain, S., Viniegra-González, G., & Sánchez, C. (2019a). Biodegradation patterns of the endocrine disrupting pollutant di (2-ethyl hexyl) phthalate by Fusarium culmorum. *Ecotoxicol. Environ. Saf.* 170: 293–299.

Gorito, A.M., Ribeiro, A.R., Almeida, C.M.R., & Silva, A.M.T. (2017). A review on the application of constructed wetlands for the removal of priority substances and contaminants of emerging concern listed in recently launched EU legislation. *Environ. Pollut.* 227; 428–443.

Haq, I., Kumar, S., Kumar, V., Singh, S.K., & Raj, A. (2016a). Evaluation of bioremediation potentiality of ligninolytic Serratia liquefaciens for detoxification of pulp and paper mill effluent. *J Hazard Mater* 305: 190–199

Haq, I., Kumari, V., Kumar, S., Raj, A., Lohani, M., & Bhargava, R.N. (2016b). Evaluation of the phytotoxic and genotoxic potential of pulp and paper mill effluent using Vigna radiata and Allium cepa. *Adv. Biol.* 8065736.

Haq, I., Kumar, S., Raj, A., Lohani, M., & Satyanarayana, G.N.V. (2017). Genotoxicity assessment of pulp and paper mill effluent before and after bacterial degradation using Allium cepa test. *Chemosphere* 169; 642–650.

Haq, I., Raj, A., & Markandeya (2018). Biodegradation of Azure-B dye by Serratia liquefaciens and its validation by phytotoxicity, genotoxicity and cytotoxicity studies. *Chemosphere* 196; 58–68.

Haq, I., Mazumder, P., & Kalamdhad, A.S. (2020). Recent advances in removal of lignin from paper industry wastewater and its industrial applications – A review. *Bioresour. Technol* 312; 123636.

Haq, I., & Kalamdhad, A.S. (2021). Phytotoxicity and cyto-genotoxicity evaluation of organic and inorganic pollutants containing petroleum refinery wastewater using plant bioassay. *Environ. Technol. Innov* 23; 101651.

Haq, I., Kalamdhad, A.S., & Pandey, A. (2022a). Genotoxicity evaluation of paper industry wastewater prior and post-treatment with laccase producing Pseudomonas putida MTCC 7525. J. *Clean. Prod* 342;130981.

Haq, I., Singh, A., & Kalamdhad, A.S. (2022b). Integrated technologies for the remediation of paper industry waste in a circular bioeconomy, Editor(s): Sunita Varjani, Ashok Pandey, Mohammad J. Taherzadeh, Huu Ha oNgo, R.D. Tyagi, *Biomass, Biofuels, Biochemicals*, Elsevier, pp. 351–363, ISBN 9780323885119.

Haq, I., Roy, S., & Kalamdhad, A.S. (2022c). Characterization of Pulp and Paper Mill Wastewater and Its Toxicity Analysis Using Vigna Radiata. In *Environmental Degradation: Monitoring, Assessment and Treatment Technologies*; Springer International Publishing: Cham, Switzerland, pp. 261–269. ISBN 9783030941475.

Haq, I., & Kalamdhad, A.S. (2022). Advanced and Ecofriendly Technologies for the Treatment of Industrial Wastewater to Constrain Environmental Pollution. In: Arora, S., Kumar, A., Ogita, S., & Yau, Y.Y. (eds) *Innovations in Environmental Biotechnology*. Springer, Singapore. https://doi.org/10.1007/978-981-16-4445-0_22.

Haq, I., & Kalamdhad, A.S. (2023). Enhanced biodegradation of toxic pollutants from paper industry wastewater using Pseudomonas sp. immobilized in composite biocarriers and its toxicity evaluation. *Bioresour. Technol. Repor.* 24;101674.

Haq, I., Kalamdhad, A.S., & Malik, A. (2024). Bioremediation of Petroleum Refinery Wastewater Using Bacillus subtilis IH-1 and Assessment of Its Toxicity. *Arch. Microbiol*, 206; 296.

Hoda, A., Susana, S.M., Marziyeh, A. S., & Mohammad, M.S. (2022). Advanced oxidation processes for phthalateesters removal in aqueous solution: a systematic review 38(2).

Ji, X. et al. (2019). Effects of carboxyl single-walled carbon nanotubes on synthetic wastewater nutrient removal by an algal-bacterial consortium: Regulation and interaction. *Sci. Total Environ.* 1:685: 997–1005.

Kasonga, T., Coetzee, M., Kamika, I., Ngole-Jeme, V., Benteke, & Momba, M. (2021). Endocrine-disruptive chemicals as contaminants of emerging concern in wastewater and surface water: a review. *J. Environ. Manage.* 277; 111485.

Khan, N.A., Khan, S.U., Sirajuddin, K., Ahmed, S., Farooqi, I.H., Yousefi, M., Mohammadi, A.A., & Changani, F. (2020). Recent trends in disposal and treatment technologies of emerging-pollutants – A critical review. *Trends Anal. Chem.* 115744.

Kumar, M.A., Poonam, S., Kumar, V.V., Baskar, G., Seenuvasan, M., Anuradha, D., & Sivanesan, S. (2017). Mineralization of aromatic amines liberated during the degradation of a sulfonated textile colorant using Klebsiella pneumoniae strain AHM. *Process Biochem.* 57: 181–189.

Li, H., Wang, P., & Liu, W. (2015a). Removal of dibutyl phthalate (DBP) from aqueous solution by adsorption using vanillin-modified chitosan beads (CTSV). *Desalination and Water Treatment.* 56 (2); 452–462.

Mistry, A.N., Kachenchart, B., Wongthanaroj, A., Somwangthanaroj, A., & Luepromchai, E. (2022). Rapid biodegradation of high molecular weight semi-crystalline polylactic acid at ambient temperature via enzymatic and alkaline hydrolysis by a defined bacterial consortium. *Polym. Degrad. Stab.* 202; Article 110051.

Mondal, T. et al. (2022). Phthalates – A family of plasticizers, their health risks, phytotoxic effects, and microbial bioaugmentation approaches. *Environ. Res. Nov.* 214(Pt 3); 114059

Moradeeya, P.G., Sharma, A., Kumar, M.A., & Basha, S. (2022). Titanium dioxide based nanocomposites – current trends and emerging strategies for the photocatalytic degradation of ruinous environmental pollutants. *Environ. Res.* 204.

Nai-Xian Feng, Yu-Xi Feng, Qi-Feng Liang, Xin Chen, Lei Xiang, Hai-Ming Zhao, Bai-Lin Liu, Gang Cao, Yan-Wen Li, Hui Li, Quan-Ying Cai, Ce-Hui Mo, Ming Hung Wong. (2021). Complete biodegradation of di-n-butyl phthalate (DBP) by a novel Pseudomonas sp. YJB6. *Sci. Total Environ.* 761(20 March); 143208.

Nakamura, A.M., Nascimento, A.S., & Polikarpov,I. (2017). Structural diversity of carbohydrate esterases. *Biotechnol. Res. Innov.* 1; 35–51.

Naruka, A.K., Suganya, S., Kumar, P.S., Amit, C., Ankita, K., Bhatt, D., & Kumar, M.A. (2021). Kinetic modelling of high turbid water flocculation using native and surface functionalized coagulants prepared from shed-leaves of Avicennia marina plants. *Chemosphere.* 272.

Pastorino, P., & Ginebreda, A. (2021).Contaminants of emerging concern (CECs): occurrence and fate in aquatic ecosystems. *Int. J. Environ. Res. Pub. Health* 18 (24); 13401.

Pironti, C., Ricciardi, M., Proto, A., Bianco, P., Montano, L., & Motta, O. (2021). Endocrine-disrupting compounds: an overview on their occurrence in the aquatic environment and human exposure. *Water* 13 (10); 1347.

Puranik, S., Shukla, L., Kundu, A., Kamil, D., Paul, S.,Venkadasamy, G., Salim, R., Singh, S.K., Kumar, D., & Kumar, A. (2023). Exploring potent fungal isolates from sanitary landfill soil for in vitro degradation of dibutyl phthalate. *J Fungi* 9(1):125.

Ren, L., Fan, S., Wang, J., Ruth, N., Qiao, C., Jia, Y., & Yan, Y. (2017). Complete genome sequence of a phthalic acid esters degrading Mycobacterium sp. YC-RL4. *Braz. J.Microbiol.* 48; 607–609.

Savinova, O.S., Shabaev, A.V., & Fedorova, T.V. (2023). Biodegradation of Phthalic Acid Esters by the White-Rot Fungus Peniophoralycii. *Microbiology* 92; 427–433.

Sun, C., Zhang, G., Zheng, H., Liu, N., Shi, M., Luo, X., Chen, L., Li, F., & Hu, S. (2019). Fate of four phthalate esters with presence of Karenia brevis: uptake and biodegradation. *Aquat. Toxicol.* 206; 81–90.

Wang, G., Zhang, Q., Chen, Q., Ma, X., Xin, Y., Zhu, X., Ma, D., Cui, C., & Zhang, J., Xiao (2019a). Photocatalytic degradation performance and mechanism of dibutyl phthalate by graphene/TiO_2 nanotube array photoelectrodes. *Chem. Eng. J.* 358; 1083–1090.

Wang, W., Leung, A.O.W., Chu, L.H., & Wong, M.H. (2018). Phthalates contamination in China: status, trends and human exposure-with an emphasis on oral intake. *Environ. Pollut.* 238;771–782.

Wilkes, R.A., & Aristilde, L. (2017). Degradation and metabolism of synthetic plastics and associated products by Pseudomonas sp.: capabilities and challenges J. *Appl. Microbiol.* 123; 582–593.

Xia, X., Gurr, GM., Vasseur, L., Zheng, D., Zhong, H., Qin, B. (2017).Metagenomic sequencing of Diamondback moth gut microbiome unveils key holobiont adaptations for herbivory. Front Microbiol 8:1–12.

Xiaoyu, W., Jianan, W., Xinyi, Z., Qiaoshen, C., Manogaran, L., Ming, Li. (2024). Comparing the removal efficiency of diisobutyl phthalate by Bacillariophyta, Cyanophyta and Chlorophyta.Science of The Total EnvironmentVolume912, 20 February, 169507

Yang, T. et al. (2018).Biodegradation of di-(2-ethylhexyl) phthalate by Rhodococcusruber YC-YT1 in contaminated water and soil. Int. J. Environ. Res. Public Health 15; 964.

YaoS., LiX., ChengH., ZhangC., BianY., JiangX., SongY. (2019).Resource utilization of a typical vegetable waste as biochars in removing phthalate acid esters from water: a sorption case study. Bioresource Technology 293;122081.

Yao, J., Gui, L., Yin, S. (2021). A novel esterase from a soil metagenomic library displaying a broad substrate range Amb. Express,11; 1–10.

Yue, F., Zhang, L., Jing, W., Ying, Z., Bangce, Y. (2017).Biodegradation of phthalate esters by a newly isolated Acinetobacter sp. strain LMB-5 and characteristics of its esterase. Pedosphere 27; 606–615.

Zhang, K., Luo, H., Zhu, Z., Chen, W., Chen, J., Mo, Y., (2018b). Performance and microbial community structure of bioaugmentation in a sequencing batch reactor treating bis(2-ethylhexyl) phthalate wastewater at low temperature. J. Environ. Eng.144 (9); 04018085.

Zhang, X., Liu, L., Zhang, S., Pan, Y., Li, J., Pan, H., Xu., S., Luo, F. (2016). Biodegradation of dimethyl phthalate by freshwater unicellular cyanobacteria. BioMed research international.5178697

Zhao, H.M., Hu, R.W., Du, H., Xin, X.P., Li, Y.W., Li, H., Cai, Q.Y., Mo, C.H., Liu, J.S., Zhou, D.M., Wong, M.H., He, Z.L. (2018b). Functional genomic analysis of phthalate acid ester (PAE) catabolism genes in the versatile PAE-mineralisingbacteriumRhodococcus sp. 2G. Sci. Total Environ. 640–641; 646–652.

Zhu, B.Q., Gao, Z.Q., Hu, G.J., Wang, J.F., Yu, N.Y., Wei, S. (2018). Contamination levels and ecological risk assessment of phthalate esters (PAEs) in the aquatic environment of key areas of Taihu Lake. Environ. Sci. 39 (8); 3614–3621.

Zhu, Z., Rao, R., Zhao, Z., Chen, J., Jiang, W., Bi, F., Yang, Y., Zhang, X. (**2022**). Research progress on removal of phthalates pollutants from environment. J. Mol. Liq. 355;118930.

13 Use of Different Bioreactors for the Removal of Microplastics from Wastewater

Devki, Deepesh Kumar Neelam, Ruchi Vyas, and Izharul Haq

13.1 INTRODUCTION

Particles of plastic that are smaller, than 5 millimeters are commonly known as "micro plastic" (Thompson et al., 2004; Lares et al., 2018). These particles can be categorized into fibers and particles based on their production methods. They can be further classified as secondary pollutants (Sun et al., 2019; Xu et al., 2020). These microscopic plastic particles were first observed in the Sargasso Sea in the North Atlantic Ocean in 1972 by Carpenter and Smith. Since then, they have been recognized as a water contaminant known as microplastics (MPs). The presence of microplastics in bodies of water including wastewater has raised environmental concerns across the whole world (Bayo et al., 2020; Kazour et al., 2019). By various means, MPs are drained in wastewater; these include industrial discharge, drainage of rainfall water, and larger plastic objects fragments. It proves a challenge for conventional wastewater treatment technologies due to their microscopic size, resulting in their accumulation in aquatic habitats. They primarily affect the marine environment but also pose risks to surface waters, air, soils, food, drinking water, sediments, airways, and food (Ballent et al., 2016; He et al., 2018). MPs are present in wastewater treatment plants (WWTPs) on a daily basis, with variable degrees of concentration and various types of polymers (Blair et al., 2019; Gatidou et al., 2019; Magni et al., 2019; Pittura et al., 2021). These come from various sources, such as secondary microplastics, which are microscopic pieces of bigger plastic objects, and direct release of small plastic particles from personal items (Bayo et al., 2020). The detrimental impact that microplastics have on ecosystems and human health emphasizes how vital it is to create sustainable and efficient cleanup techniques.

According to Colella et al. (2021), it has been anticipated that global water usage will rise by 55% by the year 2050 with population growth and industrial development as the primary contributing factors. There is an overall increase in water usage in industries, which has led to the generation of wastewater approximately 380

DOI: 10.1201/9781003379386-13

billions cubic meters every year. Such wastewater contains a number of pollutants, including: organic waste pollutants, heavy metals and persistent contaminants according to Kishor et al.(2021). Effluent from tanneries, textiles, pharmaceuticals, and petrochemicals have posed a great degree of pollution and are discharged often without proper treatment (Haq et al., 2016a, b; 17, 18 20; Haq and Kalamdhad, 2021; Haq et al., 2022a, b, c; Haq and Kalamdhad, 2022, 2023; Haq et al., 2024; Ghumra et al., 2021). In light of the concern of depletion of freshwater resources it is of utmost importance to develop treatment facilities which effectively adhere to standards put in place for effluent discharge (Singh et al., 2022).

Bioreactors, employing biological processes and microorganisms, i.e. including bacteria and fungi, possess the enzymatic capability to cleave the chemical bonds present in plastics, so transforming them into less detrimental byproducts. It offers a potential approach for the efficient elimination of microplastics from wastewater. In this introduction, we will review the different types of bioreactors and their application in solving the problem of microplastics. In biodegradation processes, microplastic is broken down and degraded by the metabolism of the microorganisms in the bioreactors. Earlier studies had a different approach whereby they assessed microplastic removal using field samples that were collected from three treatment stages throughout the study. However, such tests were not able to result in accurate results that could provide the most effective method and techniques of microplastics eradication.

This analysis looked into treatment plant operation by using meta analysis for several technologies in WWTPs around the globe. A number of documents show what treatment technologies are used in the wastewater treatment plants—sparingly used biofilters, A2O processes, Grit and grease removal, initial settling treatment and extended bioreactor secondary treatment. Tertiary treatment includes, ubiquitously, ultraviolet radiations, ozone, chlorination, biologically active filters (BAFs), disc filters (DFs), rapid sand filters (RAF) and others. The contaminants were eradicated by mixture of advanced oxidation treatment and filtration technology (Talvitie et al., 2017).

For a long time, such biological practices as constructing wetlands, using the moving bed biofilm, trickling filters or AS processes have been employed for the purpose of wastewater treatment (Rashid et al., 2021 Saidulu et al., 2021). In their work, Rashid et al. (2021) undertook an in-depth analysis of several wastewater treatment systems comprising chemical, biological, and physical methods with different objectives. The authors provided valuable connaissances on the recent advances in the area and the practical use of these technologies in the treatment of effluents contaminated with various types of pollutants. Aside from the time-saving factor, MBR offers several advantages over traditional activated sludge (AS) systems, including improved biomass retention, higher removal efficiency, and reduced tank space requirements (McQuarrie and Boltz, 2011).

In studies, there are several techniques that have been successfully utilized to treat wastewater effluents which contain microplastics (MPs). However, some of these studies were, however, either temporally or spatially restricted in such a manner that seasonal variations in the level of MPs would not be properly registered (Lares et al.,

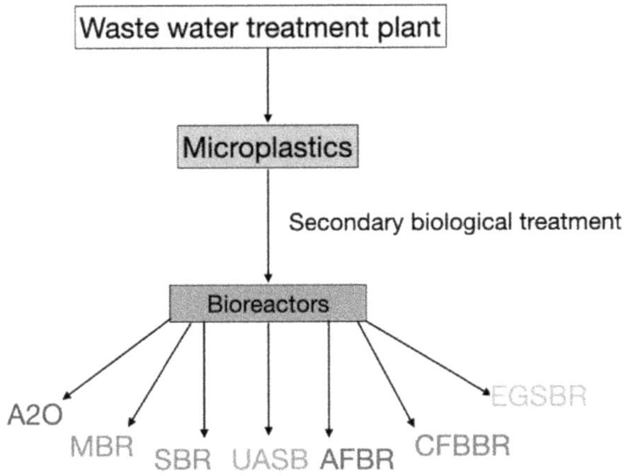

FIGURE 13.1 Bioreactors used in secondary treatment for the removal of microplastic from the wastewater treatment plant.

2018). MBR methods are well suited and can be used for treatment of industrial and municipal wastewater without the rehabilitation. It is most useful for moderately propagating microorganisms such as nitrifiers because of the enhanced concentration of suspended solids in the mixed liqueur (Dvořák et al., 2013). This method has been examined to eliminate microplastics (MP) in a smaller-scale bioreactor (Lares et al., 2018) as well as original WWTP (Talvitie et al., 2017).

13.2 TYPES OF BIOREACTORS

13.2.1 Anaerobic/anoxic/aerobic (A2O) Bioreactor

The A2O bioreactor technology is extensively utilized in wastewater treatment plants (WWTPs) (Yuan et al., 2021) . Nevertheless, its microplastics removal effectiveness was rather low due to the sludge return. Approximately 20% of the MPs that were introduced would go back to the liquid phase in the sludge. Moreover, the rate of decomposition of MPs in the A2O system was somewhat sluggish. Prior research has documented numerous microorganisms that possess the ability to degrade microplastics. The *Rhodococcus* bacterial strain exhibited a degradation rate of 6.4% on the mass of the PP polymer during a span of 40 days, as reported by Auta et al., in 2018. *Ideonellasakaiensis* successfully decomposed PET film microplastics within a span of six weeks, as demonstrated by Yoshida et al. (2016). The current 7–14 hourshydraulic retention period is inadequate for the thorough degradation of MPs solely through microbial activity in WWTPs. Consequently, the old traditional activated sludge (AS) approach is not the most effective way to get rid of MPs in WWTPs.

Numerous WWTPs employ established and commonly utilized biological nutrient removal methods, including the anaerobic/oxic (A/O) and AO processes. Phosphorus is released in the anaerobic tank, denitrification takes place at anoxic tank, and the aerobic tank is responsible for the removal of organic matter. These processes are highly cost-effective and efficient (Lai et al., 2011; Mulkerrins et al., 2000; Xu et al., 2014). The A/O (anaerobic/anoxic) technique, as demonstrated by Xu et al. (2021), can achieve a microfiber removal effectiveness of 96.5%. Alternatively, A2O technologies in WWTPs have demonstrated higher MPF removal efficiency, reaching levels of 98.3% (Lee and Kim, 2018) or nearly 100% (Leslie et al., 2017). It therefore brings with it high removal efficiencies of generally over 95%. A2O combined with advanced denitrification, ultrafiltration, and ozone treatments did remove 95.1% of pollutants, while slightly higher removal up to 96.3% was observed for A2O combined with an MBR.

13.2.2 Membrane Bioreactor (MBR)

Recently, MBR technology has gained traction among treatment methods in WWTPs. MBR there mostly are used microfiltration and ultrafiltration membranes. Microfiltration membranes have pore sizes between 0.1 and 50 μm, while ultrafiltration membranes possess a pore size ranging from 0.001 to 0.1 μm. The efficiency of the system for the removal of microplastics has reached 99.9% (Talvitie et al., 2017a;Talvitie et al., 2017b;Baresel et al., 2019; Mishra et al., 2022). A larger concentration of mixed suspended solids of 6000–10000 mg•L-1 was introduced (Dvorak et al., 2013). In theory, it integrates membrane separation with the activated sludge process, where most of microplastics were retained and stayed in the biofilm carrier side of the MBR system. This information indicates that the adsorption process is really important for the removal of microplastics in the MBR system.

Furthermore, the size of microparticles could also play into elimination. Membranes in MBR systems usually have a pore size of 0.1's μm. Thus, in theory, MBR could destroy microplastics. In a study by Lares et al. (2018), MBR was able to lower the concentration of MPs from 57600 MP/m3 to 400 MP/m3; whereas conventional activated sludge treatment reduced the concentration to 1000 MP/m3.MBR offers greater efficiency than other conventional treatment methods stated (Michielssen et al., 2016). However, much is left wanting over the advice for effluent discharge from MBR systems. Effluents mostly go into rivers and oceans. The wastewater contains MPs with a very small particle size, and smaller MPs generally have significant treatment effect when contrasted to big ones (Ogonowski et al., 2018; Zhang et al., 2020; Rubiños et al., 2023; Egea-Corbacho et al., 2023). It has been previously investigated that the accumulation of retained MPs in MBRs enhanced the secretion of organic foulants in mixed liquor (Wang et al., 2022). The influence of MPs could accelerate fouling of membrane that remains a significant challenge in the applications of MBR (Cairone et al., 2024; Kim et al., 2024; Corpuz et al., 2024).

Recent research compared the wastewater treatment processes with a membrane bioreactor system. The average concentration of 507 MPs/l was found in influent water, which increased to 1.77 MPs per kg in the reactor tank according to the results. Microplastics in the permeate decreased to 1.58 MPs per liter, a 99.69% reduction. Membrane filtration achieved 96.58% and 99.42% efficiency in clarifying and sand filtration processes, surpassing microplastic removal after biological treatment and clarification, and surpassing tertiary treatment using a sand filter (Egea-Corbacho et al., 2023).

Reverse osmosis (RO) membranes are capable of separating pollutants from wastewater that cannot be separated by MF, UF, and nanofiltration (NF) membranes. This is because RO membranes have smaller pore diameters (<1nm) and lower separation limits for molecular weight (less than 200 Daltons). Recent investigation demonstrated that uniform passing through tertiary treatment of RO membranes, wastewater can still contain substantial quantities of microplastics (MPs) (Ziajahromi et al., 2017; Sun et al., 2021; Cai et al., 2022).

In recent studies, used of primary sedimentation, microorganism treatment, MBR, and RO processes at WWTP were able to remove 93.2% and 98.0% of MPs following the MBR and RO stages, respectively. The combination of MBR and RO effectively eliminates non-fiber MPs larger than 0.5 μm from wastewater. However, MPs with a fibrous structure, particularly those smaller than 200 μm, can bypass the RO process and persist in the wastewater (Cai et al., 2022). Ziajahromi et al. (2017) have shown that PET fibers made up 88% of the total MPs present in the wastewater that was filtered using RO. Utilizing reverse RO membranes with bigger pores and higher cut-off molecular weight in WWTPs reduced the presence of MPs in the WWTP effluent within regular circumstances. Nevertheless, several studies indicate that the migration of microplastics across RO membranes may occur due to membrane faults and little gaps in pipes (Ziajahromi et al., 2017) or old membranes made up of polymers (Sun et al., 2021). In order to ascertain whether the MPs found in the final effluent of polymeric membranes used in WWTPs were derived from the membrane material itself, it was necessary to address that matter and take caution by performing further comprehensive investigations (Acarer, 2023; Nasir et al., 2024).

13.2.3 Sequencing Batch Reactor (SBR)

The bioreactor demonstrated a significant level of effectiveness in removing microplastics. SBR are an improved variant of the AS process used for WWTP. By passing it through as bubbles, oxygen is added into the combination of wastewater and biomass. The process of aeration and mixing was halted once the biological reactions have finished, allowing to settle the biomass (forming sludge), and the treated liquid above the settled sludge is extracted (Singh and Srivastava, 2011). In SBR there are four phases, i.e. fill, react, settle, and decant in one reactor (Wilderer et al., 2001). In Korea, public sewage treatment facilities having capacity of 500 m/day or more uses the SBR process in that was most adopted WWTP as in 2013 (Lee & Kim, 2018).

During the SBR process, the initial volume of microplastics in the influent was approximately 16.5 particles per liter. Furthermore, the concentration of microplastics size ranging from 106–300 μm was found to be higher than that of microplastics size more than 300 μm in the studied processes. Consequently, smaller microplastics enter the system at a higher rate compared to larger microplastics. Hence, employing a mesh that has reduced dimension from 106 μm utilized in this case would be advantageous. The amount of MPs present in the wastewater effluent of SBR was approximately 0.14 particles/liter. The usefulness of the treatment involving microplastics was evaluated by analyzing the amount of MPs in both the influent and effluent. The treatment efficiency for microplastics wider than 106µm was reported to be approximately 98–99% by Lee and Kim in 2018, while Kazour et al. noted 98% efficiency for this process in 2019. SBR activated sludge, cyclic AS, and anaerobic/anoxic technologies had respective elimination rates 73.0%, 75.6%, and 83.9%, is more efficient than traditional ASMs (Sethulekshmi&Shriwastav, 2024; Azeez et al., 2023; Tang et al., 2022; Ding et al., 2020; Zhang & Chen, 2020).

PE and acrylic resin made up the majority of the MPs detected in SBR treatment outflow and in fine screen inlet. The uninterrupted application of grit chamber, fine screen, and enhanced processes of SBR achieved a removal rate of 72% for pellets and 43% for MPs overall, although a significant amount remained after the secondary treatment.

13.2.4 Anaerobic Upflow Sludge Blanket (UASB)

The development of anaerobic granular sludge is regarded as an essential requirement for the effective functioning of a UASB reactor. The performance of the reactor is primarily determined by the concentration of active biomass and the rate at which influent flows into the reactor. Once the reactor reaches a stable condition, a compact layer consisting of fragmented sludge forms at the bottom. Directly above the sludge bed, a layer composed of finely dispersed particles known as the sludge blanket is present. The area above the sludge blanket that is free from any particles or turbidity is referred to as the settling zone. In order to maintain the sludge granules in a state of suspension within the reactor, it is necessary for the inflow flow rate to remain below a certain threshold limit, often below 1 m/h.

MPs were detected and examined the nature and allocation of microplastics in the sludge of large-scale municipal wastewater treatment plants. Moreover, another study showed that a more contained anaerobic system known as the UASB captured dynamic processes involving the fate of MPs (Wang et al., 2023). The analysis focused on specific sites that indicated that polyethylene (PE) and polypropylene (PP) were the predominant microplastics. The effects of PP-MPs on anaerobic treatment systems have yet to be understood, whereas the inhibitory effect of PE-MPs on methanogenesis has been established (Wei et al., 2019). As a result, the tests were considered to maintain continuous operation of the pilot-scale UASB reactor to examine the influence of PP-MPs upon its performance. Hu et al. (2019) demonstrated that the effectiveness of WWTPs in eliminating MPs can be improved by implementing modern treatment methods. In this investigation, the pilot-scale UASB –AnMBR configuration demonstrated a higher overall elimination of MPs compared to the full-scale CAS method. The primary factor in the complete reduction of MPPs from the PRE-TREAT IN was the ultrafiltration unit. However, there was no alteration in the ratio of MPPs to MPFs in comparison to the final effluent of the CAS system (Pittura et al., 2021).

It is important to acknowledge that AnMBR exhibits distinct fouling and permeability properties (Foglia et al., 2020) in comparison to aerobic MBR. Although the presence of only five MPFs in the PERMEATE was surprising given the membrane's small pore size, it is important to highlight that previous research has similarly identified fibres and MPPs in the permeate (Lares et al., 2018; Michielssen et al., 2016;

TABLE 13.1
Various types of the bioreactors used its removal of microplastic from the wastewater

S.No.	Bioreactor type	Removal efficiency	Microorganisms	Functions	References
1	Aerobic/Anoxic/Aerobic (A2O) Bioreactor	72–98%	Mixed microbial communities (aerobic and denitrifying bacteria such as *Nitrosomonas*, Denitrifying bacteria)	Alternating aerobic and anoxic phases help in breaking down microplastics, nitrification-denitrification process.	Yang et al., 2019; Masiá et al., 2020
2	Membrane Bioreactor (MBR)	99.69%	*Pseudomonas aeruginosa, Bacillus subtilis, Sphingomonas*	Combines biological degradation with filtration, capturing microplastics while microbes degrade particles.	Mintenig et al., 2017; Talvitie et al. 2017; Bayo et al. 2020
3	Sequenctial Batch Reactor (SBR)	92.74%	*Pseudomonas putida, Rhodococcusruber*	Periodic aeration promotes microbial degradation, and physical settling of microplastics.	Wakode& Sayyad 2014; Li et al., 2020
4	Anaerobic Upflow Sludge Blanket (UASB)	94%	Methanogenicarchaea, *Clostridium, Methanosarcina*	Anaerobic digestion of organic compounds and plastic polymers by methanogenic archaea.	Mainardis et al., 2020; Collivignarelli et al., 2021; Mariraj Mohan et al., 2022
5	Anaerobic Fluidized Bed Bioreactor (AFBR)	81.6%	*Methanobacterium, Methanosarcina, Lactococcus, Ethanoligenens*	Enhanced contact between microbes and wastewater allows for greater breakdown of microplastics.	Sun et al., 2021; Ren et al., 2022

6	Circulating Fluidized Bed Biofilm Reactor (CFBBR)	90%	Biofilms with bacteria like *Shewanella*, *Pseudomonas*	Biofilms capture microplastics, and microbial enzymes break down the polymers.	Nelson et al., 2017
7	Expanded Granular Sludge Bed Reactor (EGSBR)	89%	Anaerobic bacteria, Methanogenic archaea	High-rate system, promoting removal of microplastics with simultaneous organic matter degradation.	Zhang et al., 2020; Xu et al., 2018
8	Fluidized Bed Reactor	95.2%	*Nitrosomonas, Nitrosospira, Nitrobacter,Nitrospira*	High contact between wastewater and microbes enhances removal of small microplastic particles.	He et al., 2021; He et al., 2019
9	Conventional Activated Sludge Reactor	95–99%	Mixed microbial communities (bacteria such as *Pseudomonas, Sphingomonas,* and fungi)	Physical adsorption and biodegradation of microplastics by microbial enzymes.	Lares et al., 2018; Sol et al., 2021; Jiang et al., 2021

Talvitie et al., 2017; Ziajahromi et al., 2017). This phenomenon may be attributed to intermittent breaches of filters caused by minor leaks or to the presence of airborne contaminants in uncovered tanks where the permeate is gathered (Talvitieet al., 2017). Anaerobic membrane bioreactors (AnMBRs) have emerged as the central component of water resource recovery facilities. One notable advantage of AnMBRs is their ability to ensure a high-quality effluent (permeate). Our findings further validate the effectiveness of AnMBRs in removing microplastics (MPs), thereby confirming the potential reuse of the permeate for irrigation or fertigation purposes (Akyol et al., 2020).

13.2.5 Anaerobic Fluidized Bed Bioreactor (AFBR)

Fixed-bed bioreactors have also been adopted as replacements for the anaerobic method to treat industrial and domestic waste waters. The inert materials (such as ceramic, low density polyethylene, and polyurethane foam) supports the biomass in a biofilm which is formed in a fixed bed reactor. This type of reactor possesses a number of advantages: a reduced concentration of suspended solids in the effluent along with high performance and stable operation as well as the ability to keep a high solid retention time, that is a high concentration of bio-mass and so on, even at low hydraulic retention times (Cuel et al., 2011; Mockaitis et al., 2014; Manetti& Tomei, 2023; Arabi et al., 2020; He et al., 2019). However, for an efficient performance of anaerobic microorganisms, association system temperature and pH need to be control at 370°C and 6.8–7.4 respectively (Metcalf and Eddy, 2003).

A new composite particle of modified-zeolite and microplastic (RZ) was developed using thermal alteration and it proved effective in reducing fecal matter. Based on the results of a 171-day study on biological nutrient removal conducted in a single integrated fluidized bed bioreactor, it has been established that the nitrogen removal efficiency of the bioreactor with the modified-zeolite/microplastic particle (MZ) bioreactor was 10.0% greater than that of the RZ particle bioreactor. The specific changes that occurred in MZ biofilm composition were promoted by the reduction in free ammonia concentration experienced during chronic exposure (He et al., 2021).

Specialized AFRB was created by Chen et al., in 2019 and is patented as the Spiral Symmetry Stream Anaerobic Bioreactor (SSSAB). In the reaction zone of SSSAB three elliptic plates were symmetrically and spirally stacked. These elliptic plates helped the reaction zone to divide into three separate zones. To make it easier to collect biogas, each zone had a gas collection pipe. The classification was based on the three-stage anaerobic theory. Reactors are improved by the concentration of the dominant bacteria at various stages of microbial anaerobic fermentation in different zones of reactor processing efficiency (Song et al., 2021).

13.2.6 Circulating Fluidized Bed Biofilm Reactor (CFBBR)

The CFBBR, a dual-column system designed by the University of Western Ontario, is capable of supporting two different environments and is particularly effective in eliminating organic pollutants, nitrogen and phosphorus from wastewaters. It is a sophisticated water treatment that includes fundamentally a coffee cup from the fluidized bed reactors and a postcard from the biofilm systems. It consists of an anoxic column with low oxygen and high nitrates and an aerobic column with medium to high oxygen. This means that both technologies can be performed at the same time inside a single process.

Effort to foster transfer between the two columns may be used to improve phosphorus removal due to the fact that cycling particles between aerobic and anaerobic conditions may also facilitate transfer. Laboratory and pilot-scale reactors employing the CFBBR have treated municipal wastewater and leachate (Nelson et al., 2017). The major carriers for the biofilm of the bacteria and other microorganisms which colonies the carrier particles are formed during said process. Waste that may originate from municipal or industrial sources such as food, beverage, pharmaceutical and textile manufacturing, can be treated very effectively by this technology as it has inherent properties of fluidization which include better mass transfer and mixing characteristics. It is able to achieve such efficiencies while creating less than a third of the biological sludge that the activated sludge process would typically generate, and around 90% of the original organic load and up to 80% of nitrogen can be removed from it.

The CFBBR can treat wastewaters with increased organic solid concentration. Conventional methods struggled to address these issues due to their long residence times. But there could be important challenges such as the likelihood of clogging, and the need for regular maintenance and cleaning of the carrier media. It might also be true that the cost for setting up a CFBBR system is higher than the conventional systems but these cost can be offset by lower sludge handling costs and operating efficiency in the long run (Chowdhury et al., 2009; Andalib et al., 2010).

13.2.7 Expanded Granular Sludge Bed Reactor (EGSBR)

Granular sludge is used in an Expanded Granular Sludge Bed (EGSB) reactor, an anaerobic wastewater treatment system, to transform organic pollutants into biogas, mostly CO2 and methane. This technology is widely used to treat industrial effluent from the dairy, brewery, distillery, paper, and chemical sectors, including phenols, amines, and organic acids with a high organic content. The granular sludge it contains has a higher settling velocity than conventional anaerobic sludge, allowing for better biomass retention and separation. EGSBs are suitable for treating wastewater with a high concentration of organic pollutants because of their small size, which allows them to handle high organic loading rates.

The wastewater rises through the bed of granular sludge after entering the reactor at the bottom. This configuration allows for adequate mixing and interaction between the wastewater and biomass (Xu et al., 2018). According to Xu et al. (2018), the acetotrophic methanogen (Methanosaeta) and hydrogenotrophic methanogens (Methanobaterium) are responsible for the sludge granulation process because, following the formation of granular sludge, the amount of tightly bound EPS (TB-EPS) has been steadily increasing while the amount of loosely bound EPS (LB-EPS) has remained relatively constant (Suarez et al., 2018; Torres et al., 2018).

13.3 SUMMARY AND CONCLUSION

In conclusion, bioreactors offer a sustainable method of lowering wastewater microplastic pollution and laying the groundwork for healthier, cleaner aquatic environments. Technical advancements and continued research will be crucial to the development and implementation of bioreactor technologies for effective microplastic removal. When MPs are present in wastewater, conventional wastewater treatment technologies face serious challenges. The persistence of these tiny pollutants highlights the pressing need for innovative and efficient remediation methods. Using the enzymatic capabilities of microorganisms, bioreactors offer a practical means of decomposing and getting rid of MPs. Numerous varieties of bioreactors, including anaerobic/anoxic/aerobic (A2O) bioreactors, membrane bioreactors (MBR), sequencing batch reactors (SBR), anaerobic upflow sludge blankets (UASB), anaerobic fluidized bed bioreactors (AFBR), circulating expanded granular sludge bed reactors (EGSBR) and fluidized bed biofilm reactors (CFBBR). Specialized techniques are required to maximize the effectiveness of bioreactors for MP removal since each type has unique advantages and challenges. The efficiency of MP removal in wastewater treatment plants (WWTPs) has been shown to be increased by integrating membrane systems, filtering technologies, and advanced oxidation processes.

However, additional research and development is required to address the limitations of current approaches and ensure the full removal of MPs, especially in light of the growing global water demand and industrial wastewater creation. By creating bioreactor technologies and enhancing treatment processes, we can better protect human health and aquatic ecosystems from the harmful consequences of microplastic contamination.

REFERENCES

Acarer, S. (2023). Microplastics in wastewater treatment plants: sources, properties, removal efficiency, removal mechanisms, and interactions with pollutants. *Water Science & Technology*, 87(3), 685–710.

Akyol, Ç., Foglia, A., Ozbayram, E. G., Frison, N., Katsou, E., Eusebi, A. L., & Fatone, F. (2020). Validated innovative approaches for energy-efficient resource recovery and re-use from municipal wastewater: From anaerobic treatment systems to a biorefinery concept. *Critical Reviews in Environmental Science and Technology*, 50(9), 869–902.

Andalib, M., Nakhla, G., & Zhu, J. (2010). Biological nutrient removal using a novel laboratory-scale twin fluidized-bed bioreactor. *Chemical Engineering & Technology*, 33(7), 1125–1136.

Arabi, S., Pellegrin, M. L., Aguinaldo, J., Sadler, M. E., McCandless, R., Sadreddini, S., ...& Saldanha, V. (2020). Membrane processes. *Water Environment Research*, 92(10), 1447–1498.

Azeez, G. K., Al-Jaberi, F., Ahmed, S. A., & Hussain, A. A. (2023). Sequencing batch reactor (SBR) technology in wastewater treatment: A mini-review. In AIP Conference Proceedings (Vol. 2806, No. 1). AIP Publishing.

Ballent, A., Corcoran, P. L., Madden, O., Helm, P. A., & Longstaffe, F. J. (2016). Sources and sinks of microplastics in Canadian Lake Ontario nearshore, tributary and beach sediments. *Marine Pollution Bulletin*, 110(1), 383–395.

Baresel, C., Harding, M., & Fang, J. (2019). Ultrafiltration/granulated active carbon-biofilter: efficient removal of a broad range of micropollutants. *Applied Sciences*, 9(4), 710.

Bayo, J., López-Castellanos, J., & Olmos, S. (2020). Membrane bioreactor and rapid sand filtration for the removal of microplastics in an urban wastewater treatment plant. *Marine Pollution Bulletin*, 156, 111211–111219.

Blair, R. M., Waldron, S., & Gauchotte-Lindsay, C. (2019). Average daily flow of microplastics through a tertiary wastewater treatment plant over a ten-month period. *Water Research*, 163, 114909.

Cai, Y., Wu, J., Lu, J., Wang, J. & Zhang, C. (2022). Fate of microplastics in a coastal wastewater treatment plant: microfibers could partially break through the integrated membrane system. *Frontiers of Environmental Science and Engineering*, 16(7), 96. https://doi.org/10.1007/ s11783-021-1517-0.

Cairone, S., Hasan, S. W., Choo, K. H., Li, C. W., Zarra, T., Belgiorno, V., & Naddeo, V. (2024). Integrating artificial intelligence modeling and membrane technologies for advanced wastewater treatment: Research progress and future perspectives. *Science of The Total Environment*, 173999.

Corpuz, M. V. A., Cairone, S., Natale, M., Giannattasio, A., Iuliano, V., Grassi, A., ...& Naddeo, V. (2024). Sustainable control of microplastics in wastewater using the electrochemically enhanced living membrane bioreactor. *Journal of Environmental Management*, 370, 122649.

Carpenter, E.J., & Smith, K.L. (1972). Plastics on the Sargasso Sea surface. *Science* 175, 1240–1241.

Chen, X., Wang, Y., Wang, Z., & Liu, S. (2019). Efficient treatment of traditional Chinese pharmaceutical wastewater using a pilot-scale spiral symmetry stream anaerobic bioreactor compared with internal circulation reactor. *Chemosphere*, 228, 437–443.

Chowdhury, N., Zhu, J., Nakhla, G., Patel, A., & Islam, M. J. C. E. (2009). A Novel Liquid-Solid Circulating Fluidized-Bed Bioreactor for Biological Nutrient Removal from Municipal Wastewater. *Chemical Engineering & Technology: Industrial Chemistry-Plant Equipment-Process Engineering-Biotechnology*, 32(3), 364–372.

Colella, M., Ripa, M., Cocozza, A., Panfilo, C., & Ulgiati, S. (2021). Challenges and opportunities for more efficient water use and circular wastewater management. The case of Campania Region, Italy. *Journal of Environmental Management*, 297, 11317.

Collivignarelli, M. C., Abba, A., Caccamo, F. M., Calatroni, S., Torretta, V., Katsoyiannis, I. A., ...& Rada, E. C. (2021). Applications of up-flow anaerobic sludge blanket (UASB) and characteristics of its microbial community: a review of bibliometric trend and recent findings. *International Journal of Environmental Research and Public Health*, 18(19), 10326.

Cuel, M. F., Kwong, W. H., Zaiat, M., & Foresti, E. (2011). Solution of a heterogeneous modeling of a horizontal-flow anaerobic immobilized biomass (HAIB) reactor by the sequencing method. *Canadian Journal of Civil Engineering*, 38(4), 415–421.

Ding, N., An, D., Yin, X., & Sun, Y. (2020). Detection and evaluation of microbeads and other microplastics in wastewater treatment plant samples. *Environmental Science and Pollution Research*, 27, 15878–15887.

Dvořák, L., Svojitka, J., Wanner, J., & Wintgens, T. (2013). Nitrification performance in a membrane bioreactor treating industrial wastewater. *Water Research*, 47(13), 4412–4421.

Egea-Corbacho, A., Martín-García, A. P., Franco, A. A., Quiroga, J. M., Andreasen, R. R., Jørgensen, M. K., & Christensen, M. L. (2023). Occurrence, identification and removal of microplastics in a Wastewater Treatment Plant compared to an advanced MBR technology: Full-scale pilot plant. *Journal of Environmental Chemical Engineering*, 11(3), 109644.

Foglia, A., Akyol, Ç., Frison, N., Katsou, E., Eusebi, A. L., & Fatone, F. (2020). Long-term operation of a pilot-scale anaerobic membrane bioreactor (AnMBR) treating high salinity low loaded municipal wastewater in real environment. *Separation and Purification Technology*, 236, 116279.

Gatidou, G., Arvaniti, O. S., & Stasinakis, A. S. (2019). Review on the occurrence and fate of microplastics in Sewage Treatment Plants. *Journal of Hazardous Materials*, 367, 504–512.

Ghumra, D. P., Agarkoti, C., & Gogate, P. R. (2021). Improvements in effluent treatment technologies in Common Effluent Treatment Plants (CETPs): Review and recent advances. *Process Safety and Environmental Protection*, 147, 1018–1051.

Haq, I., Kumar, S., Kumar, V., Singh, S.K., & Raj, A. (2016a). Evaluation of bioremediation potentiality of ligninolytic Serratia liquefaciens for detoxification of pulp and paper mill effluent. *Journal of Hazardous Materials* 305, 190–199.

Haq, I., Kumari, V., Kumar, S., Raj, A., Lohani, M., & Bhargava, R.N. (2016b). Evaluation of the phytotoxic and genotoxic potential of pulp and paper mill effluent using Vigna radiata and Allium cepa. *Advanced Biology,* 8065736.

Haq, I., Kumar, S., Raj, A., Lohani, M., & Satyanarayana, G.N.V. (2017). Genotoxicity assessment of pulp and paper mill effluent before and after bacterial degradation using Allium cepa test. *Chemosphere* 169: 642–650.

Haq, I., Raj, A., & Markandeya (2018). Biodegradation of Azure-B dye by Serratia liquefaciens and its validation by phytotoxicity, genotoxicity and cytotoxicity studies. *Chemosphere* 196, 58–68.

Haq, I., Mazumder, P., & Kalamdhad, A.S. (2020). Recent advances in removal of lignin from paper industry wastewater and its industrial applications–A review. *Bioresources Technology* 312, 123636.

Haq, I, & Kalamdhad, A.S. (2021). Phytotoxicity and cyto-genotoxicity evaluation of organic and inorganic pollutants containing petroleum refinery wastewater using plant bioassay. *Environmental Technology and Innovations* 23, 101651.

Haq, I., Kalamdhad, A.S., & Pandey, A. (2022a). Genotoxicity evaluation of paper industry wastewater prior and post-treatment with laccase producing Pseudomonas putida MTCC 7525. *Journal of Cleaner Production* 342, 130981.

Haq, I., Singh, A., & Kalamdhad, A.S. (2022b). Integrated technologies for the remediation of paper industry waste in a circular bioeconomy In Editor(s): Sunita Varjani, Ashok Pandey, Mohammad J. Taherzadeh, Huu Hao Ngo, R.D. Tyagi, *Biomass, Biofuels, Biochemicals*, Elsevier, pp. 351–363, ISBN 9780323885119.

Haq, I., Roy, S., & Kalamdhad, A.S. (2022c). Characterization of Pulp and Paper Mill Wastewater and Its Toxicity Analysis Using Vigna Radiata. In *Environmental Degradation: Monitoring, Assessment and Treatment Technologies*; Springer International Publishing: Cham, Switzerland, pp. 261–269. ISBN 9783030941475.

Haq, I., & Kalamdhad, A.S. (2022). Advanced and Ecofriendly Technologies for the Treatment of Industrial Wastewater to Constrain Environmental Pollution. In: Arora, S., Kumar, A., Ogita, S., & Yau, Y.Y. (eds) *Innovations in Environmental Biotechnology.* Springer, Singapore. https://doi.org/10.1007/978-981-16-4445-0_22.

Haq, I., & Kalamdhad, A.S. (2023). Enhanced biodegradation of toxic pollutants from paper industry wastewater using Pseudomonas sp. immobilized in composite biocarriers and its toxicity evaluation. *Bioresources Technology and Reports,* 24, 101674.

Haq, I., Kalamdhad, A.S., & Malik, A. (2024). Bioremediation of petroleum refinery wastewater using Bacillus subtilis IH-1 and assessment of its toxicity. *Archives of Microbiology*, 206, 296.

He, D., Luo, Y., Lu, S., Liu, M., Song, Y., & Lei, L. (2018). Microplastics in soils: Analytical methods, pollution characteristics and ecological risks. *TrAC Trends in Analytical Chemistry*, 109, 163–172.

He, X., Li, H., & Zhu, J. (2019). A value-added insight of reusing microplastic waste: Carrier particle in fluidized bed bioreactor for simultaneous carbon and nitrogen removal from septic wastewater. *Biochemical Engineering Journal*, 151, 107300.

He, X., Li, H., & Zhu, J. (2021). A value-added step towards promoting the serviceability of fluidized bed bioreactor in treating wastewater with low carbon to nitrogen ratio. *Science of the Total Environment*, 750, 141665.

Hu, Y., Gong, M., Wang, J., & Bassi, A. (2019). Current research trends on microplastic pollution from wastewater systems: a critical review. *Reviews in Environmental Science and Bio/Technology*, 18, 207–230.

Jiang, C., Yin, L., Li, Z., Wen, X., Luo, X., Hu, S., ...& Liu, Y. (2019). Microplastic pollution in the rivers of the Tibet Plateau. *Environmental Pollution*, 249, 91–98.

Kazour, M., Terki, S., Rabhi, K., Jemaa, S., Khalaf, G., & Amara, R. (2019). Sources of microplastics pollution in the marine environment: Importance of wastewater treatment plant and coastal landfill. *Marine Pollution Bulletin*, 146, 608–618.

Kim, J., Bae, E., Park, H., Park, H. J., Shah, S. S. A., Lee, K., ...& Choo, K. H. (2024). Membrane reciprocation and quorum quenching: An innovative combination for fouling control and energy saving in membrane bioreactors. *Water Research*, 250, 121035.

Kishor, R., Purchase, D., Saratale, G. D., Saratale, R. G., Ferreira, L. F. R., Bilal, M., ...& Bharagava, R. N. (2021). Ecotoxicological and health concerns of persistent coloring pollutants of textile industry wastewater and treatment approaches for environmental safety. *Journal of Environmental Chemical Engineering*, 9(2), 105012.

Lai, T. M., Dang, H. V., Nguyen, D. D., Yim, S., & Hur, J. (2011). Wastewater treatment using a modified A2O process based on fiber polypropylene media. *Journal of Environmental Science and Health*, Part A, 46(10), 1068–1074.

Lares, M., Ncibi, M. C., Sillanpää, M., & Sillanpää, M. (2018). Occurrence, identification and removal of microplastic particles and fibers in conventional activated sludge process and advanced MBR technology. *Water Research*, 133, 236–246.

Lee, H., & Kim, Y. (2018). Treatment characteristics of microplastics at biological sewage treatment facilities in Korea. *Marine Pollution Bulletin*, 137, 1–8.

Leslie, H. A., Brandsma, S. H., Van Velzen, M. J. M., & Vethaak, A. D. (2017). Microplastics en route: Field measurements in the Dutch river delta and Amsterdam canals, wastewater treatment plants, North Sea sediments and biota. *Environment International*, 101, 133–142.

Li, L., Song, K., Yeerken, S., Geng, S., Liu, D., Dai, Z., ...& Wang, Q. (2020). Effect evaluation of microplastics on activated sludge nitrification and denitrification. *Science of the Total Environment*, 707, 135953.

Magni, S., Binelli, A., Pittura, L., Avio, C. G., Della Torre, C., Parenti, C. C., ...& Regoli, F. (2019). The fate of microplastics in an ItalianWastewater Treatment Plant. *Science of the Total Environment*, 652, 602–610.

Mainardis, M., Buttazzoni, M., & Goi, D. (2020). Up-flow anaerobic sludge blanket (UASB) technology for energy recovery: a review on state-of-the-art and recent technological advances. *Bioengineering*, 7(2), 43.

Manetti, M., & Tomei, M. C. (2023). Anaerobic removal of contaminants of emerging concern in municipal wastewater: Eco-toxicological risk evaluation and strategic selection of optimal treatment. *Science of The Total Environment*, 168895.

Mariraj Mohan, S., & Swathi, T. (2022). A review on upflow anaerobic sludge blanket reactor: Factors affecting performance, modification of configuration and its derivatives. *Water Environment Research*, 94(1), e1665.

Masiá, P., Sol, D., Ardura, A., Laca, A., Borrell, Y. J., Dopico, E., ...& Garcia-Vazquez, E. (2020). Bioremediation as a promising strategy for microplastics removal in wastewater treatment plants. *Marine Pollution Bulletin*, 156, 111252–111258.

McQuarrie, J. P., & Boltz, J. P. (2011). Moving bed biofilm reactor technology: process applications, design, and performance. *Water Environment Research*, 83(6), 560–575.

Metcalf, L. (2003). Wastewater engineering: treatment and reuse. Metcalf & Eddy Inc. McGraw-Hill Inc., New York. Mohammed, AN and ElBably, MA (2016). Technologies of domestic wastewater treatment and reuse: options of application in developing countries. *JSM Environ Sci Ecol*, 4(3), 1033.

Michielssen, M. R., Michielssen, E. R., Ni, J., & Duhaime, M. B. (2016). Fate of microplastics and other small anthropogenic litter (SAL) in wastewater treatment plants depends on unit processes employed. *Environmental Science: Water Research & Technology*, 2(6), 1064–1073.

Mintenig, S. M., Int-Veen, I., Löder, M. G., Primpke, S., & Gerdts, G. (2017). Identification of microplastic in effluents of waste water treatment plants using focal plane array-based micro-Fourier-transform infrared imaging. *Water Research*, 108, 365–372.

Mishra, S., Singh, R. P., Rout, P. K., & Das, A. P. (2022). Membrane bioreactor (MBR) as an advanced wastewater treatment technology for removal of synthetic microplastics. *Development in Wastewater Treatment Research and Processes*, 45–60.

Mockaitis, G., Pantoja, J. L. R., Rodrigues, J. A. D., Foresti, E., & Zaiat, M. (2014). Continuous anaerobic bioreactor with a fixed-structure bed (ABFSB) for wastewater treatment with low solids and low applied organic loading content. *Bioprocess and Biosystems Engineering*, 37(7), 1361–1368.

Mulkerrins, D., Jordan, C., McMahon, S., & Colleran, E. (2000). Evaluation of the parameters affecting nitrogen and phosphorus removal in anaerobic/anoxic/oxic (A/A/O) biological nutrient removal systems. *Journal of Chemical Technology & Biotechnology*, 75(4), 261–268.

Nasir, M. S., Tahir, I., Ali, A., Ayub, I., Nasir, A., Abbas, N., ...& Hamid, K. (2024). Innovative technologies for removal of micro plastic: A review of recent advances. *Heliyon.* https://doi.org/10.1016/j.heliyon.2024.e25883

Nelson, M. J., Nakhla, G., & Zhu, J. (2017). Fluidized-bed bioreactor applications for biological wastewater treatment: a review of research and developments. *Engineering*, 3(3), 330–342.

Ogonowski, M., Gerdes, Z., & Gorokhova, E. (2018). What we know and what we think we know about microplastic effects–A critical perspective. *Current Opinion in Environmental Science & Health*, 1, 41.

Pittura, L., Foglia, A., Akyol, Ç., Cipolletta, G., Benedetti, M., Regoli, F., ...& Fatone, F. (2021). Microplastics in real wastewater treatment schemes: Comparative assessment and relevant inhibition effects on anaerobic processes. *Chemosphere*, 262, 128415.

Rashid, R., Shafiq, I., Akhter, P., Iqbal, M. J., & Hussain, M. (2021). A state-of-the-art review on wastewater treatment techniques: The effectiveness of adsorption method. Environmental Science and Pollution Research, 28(8), 1–17.

Ren, H. Y., Wang, X. W., Kong, F., Zhao, L., Xing, D., Ren, N. Q., & Liu, B. F. (2022). Enhanced semi-continuous hydrogen production by addition of microplastics under mesophilic and thermophilic fermentation. *Fuel*, 315, 123247.

Rubiños, S. L., Grados, J. H., Medina, J. T., Chávez, E. N., Huarcaya, E., Rocha, V. E., ...& Neyra, M. R. (2023). Microplastics in the ecosystem: A systematic review of the methods for their detection and removal. *International Journal of Ecology*, 2023(1), 8826716.https://doi.org/10.1155/2023/8826716

Saidulu, D., Majumder, A., & Gupta, A. K. (2021). A systematic review of moving bed biofilm reactor, membrane bioreactor, and moving bed membrane bioreactor for wastewater treatment: Comparison of research trends, removal mechanisms, and performance. *Journal of Environmental Chemical Engineering*, 9(5), 106112.

Sethulekshmi, S., & Shriwastav, A. (2024). Long-term presence of microplastics in aerobic and anaerobic sequential batch reactors: Effect on treatment, microbial diversity, and microplastics morphology. *Water Research*, 250, 121029.

Singh, M., & Srivastava, R. K. (2011). Sequencing batch reactor technology for biological wastewater treatment: a review. *Asia-Pacific Journal of Chemical Engineering*, 6(1), 3–13.

Singh, A., Srivastava, A., Saidulu, D., & Gupta, A. K. (2022). Advancements of sequencing batch reactor for industrial wastewater treatment: Major focus on modifications, critical operational parameters, and future perspectives. *Journal of Environmental Management*, 317, 115305.

Sol, D., Laca, A., Laca, A., & Díaz, M. (2021). Microplastics in wastewater and drinking water treatment plants: occurrence and removal of microfibres. *Applied Sciences*, 11(21), 10109–10121.

Song, Q., Sun, Z., Chang, Y., Zhang, W., Lv, Y., Wang, J., ...& Chen, X. (2021). Efficient degradation of polyacrylate containing wastewater by combined anaerobic–aerobic fluidized bed bioreactors. *Bioresource Technology*, 332, 125108.

Suárez, W. A. B., da Silva Vantini, J., Duda, R. M., Giachetto, P. F., Cintra, L. C., Ferro, M. I. T., & de Oliveira, R. A. (2018). Predominance of syntrophic bacteria, Methanosaeta and Methanoculleus in a two-stage up-flow anaerobic sludge blanket reactor treating coffee processing wastewater at high organic loading rate. *Bioresource Technology*, 268, 158–168.

Sun, J., Dai, X., Wang, Q., Van Loosdrecht, M. C., & Ni, B. J. (2019). Microplastics in wastewater treatment plants: Detection, occurrence and removal. *Water Research*, 152, 21–37.

Sun, J., Zhu, Z. R., Li, W. H., Yan, X., Wang, L. K., Zhang, L., ...& Ni, B. J. (2021). Revisiting microplastics in landfill leachate: unnoticed tiny microplastics and their fate in treatment works. Water Research, 190, 116784.

Talvitie, J., Mikola, A., Koistinen, A., & Setälä, O. (2017). Solutions to microplastic pollution–Removal of microplastics from wastewater effluent with advanced wastewater treatment technologies. *Water Research*, 123, 401–407.

Talvitie, J., Mikola, A., Setälä, O., Heinonen, M., & Koistinen, A. (2017). How well is microlitter purified from wastewater?–A detailed study on the stepwise removal of microlitter in a tertiary level wastewater treatment plant. *Water Research*, 109, 164–172.

Tang, Q., Wu, M., Zhang, Y., Li, J., Liang, J., Zhou, H., ...& Zhang, X. (2022). Performance and bacterial community profiles of sequencing batch reactors during long-term exposure to polyethylene terephthalate and polyethylene microplastics. *Bioresource Technology*, 347, 126393.

Thompson, R.C., Olsen, Y., Mitchell, R.P., Davis, A., Rowland, S.J., John, A.W.G., McGonigle, D., & Russell, A.E. (2004). Lost at sea: where is all the plastic? *Science* 304 (5672), 838.

Torres, K., Álvarez-Hornos, F. J., San-Valero, P., Gabaldón, C., & Marzal, P. (2018). Granulation and microbial community dynamics in the chitosan-supplemented anaerobic treatment of wastewater polluted with organic solvents. *Water Research*, 130, 376–387

Wakode, P. N., & Sayyad, S. U. (2014). Performance evaluation of 25MLD sewage treatment plant (STP) at Kalyan. *American Journal of Engineering Research*, 3(03), 310–316.

Wang, Q., Li, Y., Liu, Y., Zhou, Z., Hu, W., Lin, L., & Wu, Z. (2022). Effects of microplastics accumulation on performance of membrane bioreactor for wastewater treatment. *Chemosphere*, 287, 131968.

Wang, C., Wei, W., Zhang, Y. T., Chen, X., & Ni, B. J. (2023). Hydrochar alleviated the inhibitory effects of polyvinyl chloride microplastics and nanoplastics on anaerobic granular sludge for wastewater treatment. *Chemical Engineering Journal*, 452, 139302.

Wei, W., Huang, Q. S., Sun, J., Dai, X., & Ni, B. J. (2019). Revealing the mechanisms of polyethylene microplastics affecting anaerobic digestion of waste activated sludge. *Environmental Science & Technology*, 53(16), 9604–9613.

Wilderer, P. A., Irvine, R. L., & Goronszy, M. C. (Eds.). (2001). *Sequencing batch reactor technology*. IWA Publishing.

Xu, S., Bernards, M., & Hu, Z. (2014). Evaluation of anaerobic/anoxic/oxic (A2/O) and reverse A2/O processes in biological nutrient removal. *Water Environment Research*, 86(11), 2186–2193.

Xu, H., Liu, Y., Gao, Y., Li, F., Yang, B., Wang, M., ...& Sand, W. (2018). Granulation process in an expanded granular sludge blanket (EGSB) reactor for domestic sewage treatment: impact of extracellular polymeric substances compositions and evolution of microbial population. *Bioresource Technology*, 269, 153–161.

Xu, X., Zhang, L., Jian, Y., Xue, Y., Gao, Y., Peng, M., ...& Zhang, Q. (2021). Influence of wastewater treatment process on pollution characteristics and fate of microplastics. *Marine Pollution Bulletin*, 169, 112448.

Xu, Q., Gao, Y., Xu, L., Shi, W., Wang, F., LeBlanc, G. A., ...& Lei, K. (2020). Investigation of the microplastics profile in sludge from China's largest Water reclamation plant using a feasible isolation device. *Journal of Hazardous Materials*, 388, 122067.

Yang, L., Li, K., Cui, S., Kang, Y., An, L., & Lei, K. (2019). Removal of microplastics in municipal sewage from China's largest water reclamation plant. *Water Research*, 155, 175–181.

Yuan, F., Zhao, H., Sun, H., Zhao, J., & Sun, Y. (2021). Abundance, morphology, and removal efficiency of microplastics in two wastewater treatment plants in Nanjing, China. *Environmental Science and Pollution Research*, 28, 9327–9337.

Zhang, Z., & Chen, Y. (2020). Effects of microplastics on wastewater and sewage sludge treatment and their removal: a review. *Chemical Engineering Journal*, 382, 122955

Zhang, X., Chen, J., & Li, J. (2020). The removal of microplastics in the wastewater treatment process and their potential impact on anaerobic digestion due to pollutants association. *Chemosphere*, 251, 126360.

Ziajahromi, S., Neale, P. A., Rintoul, L. & Leusch, F. D. L. (2017). Wastewater treatment plants as a pathway for microplastics: development of a new approach to sample wastewater-based microplastics. *Water Research*, 112, 93–99. https://doi.org/10.1016/j.watres.2017.01.042.

Index

For Product Safety Concerns and Information please contact our EU representative GPSR@taylorandfrancis.com
Taylor & Francis Verlag GmbH, Kaufingerstraße 24, 80331 München, Germany

www.ingramcontent.com/pod-product-compliance
Ingram Content Group UK Ltd.
Pitfield, Milton Keynes, MK11 3LW, UK
UKHW022218190926
475702UK00003B/423
9781032459479